THE
AMERICAN EXPERIENCE WITH ALCOHOL

Contrasting Cultural Perspectives

THE
AMERICAN
EXPERIENCE
WITH
ALCOHOL

Contrasting Cultural Perspectives

Edited by

Linda A. Bennett

George Washington University Medical Center
Washington, D.C.

and

Genevieve M. Ames

Prevention Research Center
Pacific Institute for Research and Evaluation
Berkeley, California

Plenum Press • New York and London

Library of Congress Cataloging in Publication Data

Main entry under title:

The American experience with alcohol.

   Includes bibliographies and index.
   1. Alcoholism—United States—Case studies. 2. Ethnic groups—Case studies. 3. Alcoholism—United States—Cross-cultural studies. I. Bennett, Linda A. II. Ames, Genevieve M. [DNLM: 1. Alcohol Drinking. 2. Cross-Cultural Comparison. 3. Ethnic Groups—psychology. Wm 274 A509]
HV5292.A45   1985   362.2′92′0973   85-9302
ISBN 0-306-41945-9

First Printing—July 1985
Second Printing—August 1986

© 1985 Plenum Press, New York
A Division of Plenum Publishing Corporation
233 Spring Street, New York, N.Y. 10013

Printed in the United States of America

To Sir Raymond Firth
for his inspiration

Contributors

JOAN ABLON, Medical Anthropology Program, University of California, San Francisco, California

GENEVIEVE M. AMES, Prevention Research Center, Pacific Institute for Research and Evaluation, Berkeley, California

LINDA A. BENNETT, Center for Family Research, George Washington University Medical Center, Washington, D.C.

BRUCE BERG, School of Criminology, Florida State University, Tallahassee, Florida

NOEL J. CHRISMAN, School of Nursing, University of Washington, Seattle, Washington

GRACE TONEY EDWARDS, Appalachian Studies Program, Radford University, Radford, Virginia

PAUL J. FREUND, Institute for African Studies, Community Health Research Unit, University of Zambia, Lusaka, Zambia

ATWOOD D. GAINES, Department of Anthropology, Case Western Reserve University, Cleveland, Ohio

M. JEAN GILBERT, Spanish Speaking Mental Health Center, University of California, Los Angeles, California

BARRY GLASSNER, Department of Sociology, Syracuse University, Syracuse, New York

ANDREW J. GORDON, Division of Sociomedical Sciences, School of Public Health, Columbia University, New York, New York

HERB HATANAKA, Special Service for Groups, Los Angeles, California

DWIGHT B. HEATH, Department of Anthropology, Brown University, Providence, Rhode Island

DENISE HERD, Alcohol Research Group, Institute of Epidemiology and Behavioral Medicine, Berkeley, California

HARRY H. L. KITANO, School of Social Welfare, University of California, Los Angeles, California

CAROLYN MCKAY, Department of Psychiatry, University of Miami Medical School, Miami, Florida

J. BRYAN PAGE, Department of Psychiatry, University of Miami Medical School, Miami, Florida

LUCY RIO, Department of Psychiatry, University of Miami Medical School, Miami, Florida

MIRIAM B. RODIN, School of Public Health, University of Illinois Health Sciences Center, Chicago, Illinois

ROBIN ROOM, Alcohol Research Group, Institute of Epidemiology and Behavioral Medicine, Berkeley, California

BEN JAMES SIMBOLI, Department of Health Services, State of California Health and Welfare Agency, Berkeley, California

RICHARD STIVERS, Department of Sociology, Anthropology, and Social Work, Illinois State University, Normal, Illinois

STANLEY SUE, Department of Psychology, University of California, Los Angeles, California

JACQUELINE SWEENEY, Department of Psychiatry, University of Miami Medical School, Miami, Florida

MARTIN D. TOPPER, Winslow, Arizona

ROBERT T. TROTTER II, Department of Psychology and Anthropology, Pan American University, Edinburg, Texas

WAI-TSANG YEUNG, School of Social Work, Chinese University of Hong Kong, Shatin, Hong Kong

JOAN WEIBEL-ORLANDO, Neuropsychiatric Institute, University of California, Los Angeles, California

JOSEPH WESTERMEYER, Department of Psychiatry, University of Minnesota, Minneapolis, Minnesota

Foreword

This volume is an important contribution to our understanding of culture and alcohol in the United States. Its appearance is also a milestone in the history of alcohol studies in American anthropology. Over the last six years, the volume's editors, initially along with Miriam Rodin, have served as the coorganizers of the Alcohol and Drug Study Group of the American Anthropological Association (AAA). In this capacity, they have organized sessions at the AAA and other meetings, greatly strengthened the research network with a regular and informative newsletter, and painstakingly promoted the publication of anthropological work on alcohol and drugs. Appearing just as the responsibility for the Study Group is passed on to others, this book is a fitting emblem of the care and energy with which its editors have built an institutional nexus for alcohol and drug anthropology in North America.

The contents of this volume offer a uniquely wide sampling of the diversity of cultural patterns that make up the American experience with alcohol. The collective portrait the editors have assembled extends in several dimensions: through time and history, across such social differentiations as gender, age-grade, and social class, and through such major social institutions as the church and the family. Clearly the dominant dimension of variation in the material that follows, however, is ethnicity. The book offers us a sampler of unprecedented richness of the different experiences with alcohol of American ethnoreligious groups. Beyond this, we can discern the outline of a new understanding of American ethnicities and of their experiences with alcohol, an understanding that emphasizes change as well as stasis in cultural patterns, creation as well as inheritance, and interchange as well as immanence. My comments thus focus on this new understanding of ethnicity.

The variations in drinking among American ethnicities is not in itself a new phenomenon. The social sciences have considered ethnic differences in the United States at least as far back as the Progressive era, when the descendants of earlier immigrants became concerned about the living conditions and cultural patterns of the new waves of immigrants crowding into American cities. Not least among the concerns, both of "know-nothing" nativists and of well-wishing social reformers were the drinking patterns and problems of the new immigrants; thus, several studies in the early years of the century focused on the interplay of ethnicity and alcoholism in the urban environment.

In social science generally, studies of ethnic variation, and particularly of ethnic variation within the white population of the United States went out of style in the 1940s and 1950s. It was not only that such a focus went against the prevailing melting-pot ideology of American society, but also that any emphasis on ethnic differentiation was uncomfortably reminiscent of the Nazis' racist ideology and its genocidal consequences. Alcohol studies, however, were the exception. Perhaps the most strongly developed strand of social science research at the Yale Center for Alcohol Studies in the 1940s and 1950s was a series of studies of drinking in American ethnicities, chosen, as Snyder put it, as "strategically significant for furthering the understanding of normal as well as abnormal drinking" (1978, p. 2). The tradition began with the 1944 dissertation of Robert Bales (1980), in which Irish and Jewish drinking habits, and interviews of Irish-Americans, were presented. It continued with studies by Snyder (1978; originally published 1958) of alcohol and the Jews, by Williams and Straus (1950) and by Lolli and associates (1952) of Italian-Americans, by Barnett (1955) of Chinese-Americans, and by Glad (1947) and by Skolnick (1957) of the ethnoreligious variation in drinking patterns in American youth. By the 1960s, this social research tradition had led to a theory of drinking problems and their prevention, the "cultural integration" theory, which became unusually influential in the United States and elsewhere (see Chafetz & Demone, 1962, pp. 76–99; Plaut, 1967; Mäkelä, 1975; Room, 1976; Frankel & Whitehead, 1981).

One reason this tradition could flourish in an era that discouraged studies of ethnic variation within the white population was that it cut against the grain of racist and nativist assumptions of cultural superiority. In terms of their drinking practices, ethnicities that had borne the brunt of racist and nativist attacks—such as Italians, Jews, and the Chinese—could be presented as paragons others might aspire to copy. The contrast was not only with ethnicities represented in the older migrations—notably, the Irish-Americans—but also particularly and pointedly with the "old stock" Anglo-American, later known as the WASP, the White Anglo-Saxon Protestant.

In the light of this half-hidden agenda of enlightened antitheses, it is remarkable how little empirical work on the WASP "reference culture" is found in the literature of the time. In codifying the cross-ethnic contrasts of this literature into an explicitly cultural-integrationist theory, Ullman (1958) was forced back on Bacon's ideal-type characterization (1957), admittedly "clearly over-simplified and often without sufficient data," to fill in the position in his typology reserved for "the United States American of the northeast quarter of the nation—Protestant, middle-class, urban, white, from Anglo-Saxon background of three or more generations in this country." Not the least of the virtues of the present compilation is that, on the ethnic map of American drinking, the long-standing mythological beast is finally replaced by some empirical material on alcohol and the non-fundamentalist WASP.

By the 1960s, the general field of studies of alcohol in American culture had come to include a relatively new tradition of survey research of adult drinking practices. With the importance of ethnic variation already established in the literature, those of us working in this tradition struggled to include the ethnic dimension in our analyses. But from a survey researcher's point of view, the problem with American ethnicities is that there are so many of them: it would take an impossibly large general population survey sample, for instance, to be able to say anything about the Hmong immigrant group described by Westermeyer in these pages. Even "large" ethnicities such as black Americans, can be only cursorily analyzed on the basis of conventionally sized random national samples. Survey researchers have surmounted this problem in several ways: by abandoning a strict probability design; by limiting their study to communities with a restricted range of ethnicities; or by aggregating samples from studies conducted for other purposes (see Knupfer & Room, 1967; Greeley, McCready, & Theisen, 1980; Caetano, 1984; and the chapters by Simboli, Kitano *et al.* and Sue *et al.* in this volume for various permutations of these strategies).

Dealing with the population as a whole also brought alcohol survey researchers up against several facts that tended at the time to be viewed primarily as methodological embarrassments: that ethnicity as a self-identification was sought by some, but rejected by others; that many Americans could choose between several ethnic identifications; and that the few large "ethnoreligious groups," which conventional probability surveys could deal with, included categories not encountered in the literature. For instance, the second-largest "ethnoreligious" category in Cahalan, Cisin, and Crossley's nationwide survey (1969, p. 54)—even excluding those who identified their main national origin as "Scotch-Irish"—was "Irish Conservative Protestants." As we noted concerning the latter category, "their 'Irishness' is in most cases filtered through many generations of residence in the rural South" (Cahalan & Room, 1974, p. 100).

Viewed in hindsight and in the perspective of this volume, these awkward facts were significant in that they told us something substantive, something about the mutability, reactivity, and constructibility of ethnicity and ethnic cultures. Along with the general literature of 15 years ago, we had tended to regard ethnicities as unchanging. Even if we recognized that there was some passage by individuals across ethnic boundaries (Cahalan & Room, 1974, p. 72), an ethnicity was, by assumption, something that did not change either in its self-definition or in its relation to alcohol. In the general literature on American ethnicity, the only mechanism for change was "acculturation," viewed as a process by which ethnic particularities were leached out generation by generation in the melting pot. With respect to alcohol as with respect to identity, then, an ethnicity was seen as having only two options: to maintain its patrimony, or to disappear into a residually defined general "American" pattern.

These assumptions about American ethnicity began to disappear with "the rise of the unmeltable ethnics" (Novak, 1973)—the revival of ethnic consciousness in parts of the white population in the wake of the black civil rights movement. Recent years have taught us that ethnic identity, particularly in a multiethnic society, is often a definitely self-conscious construction or retention. The advent of constructionist paradigms in social science and the new emphasis on historical studies of the formation of consciousness—national, class, or ethnic—have helped us to transcend the old assumptions about and understandings of ethnicities and their meaning and place in the world. An anthropological study of the rise of nationalist ideology in Quebec insists "on the ongoing construction of cultural 'content.' . . . Cultural action is semiotic: it is perpetually reinvented in the present" (Handler, 1984, p. 63). In a far-reaching analysis of the historical processes of the mobilization and incorporation of tribal and village peoples into the industrialized work forces of a global market, Eric Wolf (1984) emphasizes the collective forging of new ethnic identities and consciousnesses by those thus uprooted. In the light of such phenomena, we must "rethink the concept of culture." He proposes:

> Once we locate the reality of society in historically changing, imperfectly bounded, multiple and branching social alignments, . . . the concept of a fixed, unitary and bounded culture must give way to a sense of fluidity and permeability of cultural sets. In the rough-and-tumble of social interaction, groups are known to exploit the ambiguities of inherited forms, to impart

new evaluations or valences to them, to borrow forms more expressive of
their interests, or to create wholly new forms in answer to changed circumstances. . . . "A culture" is thus better seen as a series of processes that
construct, reconstruct, and dismantle cultural materials, in response to identifiable determinants. (Wolf, 1982, p. 387)

The new, more complicated ideas of the relationship of ethnicity
and drinking began to come into the alcohol literature around 1970. In
independent analyses, several anthropologists reinterpreted the formation of American Indian drinking patterns as in part a self-conscious
acting out of the stereotypes held by another culture, whether as a
statement of identity or for material purposes, in frontier situations of
culture contact and conflict (MacAndrew & Edgerton, 1969; Honigmann
& Honigmann, 1970; Lurie 1971). Such analyses took the crucial step of
seeing the possibility of an ethnicity's alcohol culture being formed in
part as a reaction or in reference to other ethnicities. But these analyses
all involved groups with no precontact history of drinking, and which,
unlike other American ethnicities, were not immigrants who had left an
ancestral culture on another continent. The perceptions of these analyses
were not, therefore, clearly applicable to American ethnicities in general.
But in a 1971 dissertation, and in later work represented in this volume,
Richard Stivers applied the same style of analysis to an American ethnic
group—Irish-Americans—with a clearly defined premigration alcohol
culture. Stivers' demonstration that Irish-American drinking clearly differed from Irish drinking in Ireland, and his argument that Irish-American drinking could be seen in part as playing up to the Anglo-Saxon
cultural stereotype of the "stage Irishman" (Stivers 1976), embodied
several substantially new ideas about an ethnicity's drinking culture:
that it could change substantially under new circumstances; that the
changes could reflect conscious and purposive collective decisions and
actions; and that, in a multiethnic society, the changes and redefinitions
might well be made with reference to other ethnicities.

Reading over the studies contained in this book, I was struck by
their frequent and fruitful use of these new perspectives, which see the
relation of alcohol and culture in terms of flux and fluidity as well as
inertia and stasis, in terms of intercultural conflict, accommodation, and
antithesis, as well as of the immanent unfolding of a culture, and in
terms of self-conscious symbolization and ideology, as well as of an
unconscious cultural reproduction or reactions to material circumstances.

We are also reminded time and again of alcohol's pervasiveness not
only as a material artifact, but also as an object and occasion of symbolism. In ways as various as the ethnicities represented here, the use

of alcohol—or abstention—becomes an icon or a prop or an excuse in intimate dramas within the family, in daily performances within an ethnicity, and on the wider stage of interethnic contacts and relations. Picking up and extending this emphasis on the interplay of ethnicity and the symbolic dimensions of drinking is a worthy agenda for future research for anthropologists and other social scientists. If we wish to understand and reduce the occurrence of problems related to alcohol, we must know more about the cultural meanings attached to its use.

ROBIN ROOM

REFERENCES

Bacon, S. D. Social settings conducive to alcoholism: A sociological approach to a medical problem. *Journal of the American Medical Association, 1957, 164,* 177–181.

Bales, R. F. *The "fixation factor" in alcohol addiction: An hypothesis derived from a comparative study of Irish and Jewish social norms.* New York: Arno Press, 1980. Originally a Harvard doctoral dissertation, 1944.

Barnett, M. L. Alcoholism in the Cantonese of New York City: An anthropological study. In O. Diethelm (Ed.), *Etiology of chronic alcoholism,* Springfield, IL: Thomas, 1955, pp. 179–227.

Caetano, R. Ethnicity and drinking in northern California: A comparison among whites, blacks and Hispanics. *Alcohol and Alcoholism, 1984, 19,* 31–44.

Cahalan, D., Cisin, I. H., & Crossley, H. M. *American drinking practices: A national study of drinking behavior and attitudes,* Monograph No. 6. New Brunswick, NJ: Rutgers Center of Alcohol Studies, 1969.

Cahalan, D., & Room, R. *Problem drinking among American Men,* Monograph No. 7. New Brunswick, NJ: Rutgers Center of Alcohol Studies, 1974.

Chafetz, M. E., & Demone, H. W. *Alcoholism and society.* New York: Oxford University Press, 1962.

Frankel, B., & Whitehead, P. C. *Drinking and damage: Theoretical advances and implications for prevention,* Monograph No. 14. New Brunswick, NJ: Rutgers Center of Alcohol Studies, 1981.

Glad, D. D. Attitudes and experiences of American-Jewish and American-Irish male youths as related to differences in adult rates of inebriety. *Quarterly Journal of Studies on Alcohol,* 1947, *8,* 406–472.

Greeley, A. M., McCready, W. C., & Theisen, G. *Ethnic drinking subcultures.* New York: Praeger Special Studies, 1980.

Handler, R. On sociocultural discontinuity: Nationalism and cultural objectification in Quebec, with comments and a reply. *Current Anthropology, 1984, 25,* 55–71.

Honigmann, J. J., & Honigmann, I. *Arctic townsmen.* Ottawa: St. Paul University Press, 1970.

Knupfer, G., & Room, R. Drinking patterns and attitudes of Irish, Jewish and white Protestant American men. *Quarterly Journal of Studies on Alcohol, 1967, 28,* 676–699.

Lolli, G., Serianni, E., Banissoni, F., Golder, G., Mariani, A., McCarthy, R. G., & Toner, M. The use of wine and other alcoholic beverages by a group of Italians and Americans of Italian extraction. *Quarterly Journal of Studies on Alcohol, 1952, 12,* 27–48.

Lurie, N. O. The world's oldest on-going protest demonstration: North American Indian drinking patterns. *Pacific Historical Review,* 1971, *40,* 311–332.

MacAndrew, C., & Edgerton, R. *Drunken comportment.* Chicago: Aldine, 1961.

Mäkelä, K. Consumption level and cultural drinking patterns as determinants of alcohol problems. *Journal of Drug Issues,* 1970, *5,* 348–357.

Novak, M. *The rise of the unmeltable ethnics: Politics and culture in the seventies.* New York: Macmillan, 1973.

Plaut, T. F. A. *Alcohol problems: A report to the nation.* New York: Oxford University Press, 1967.

Room, R. Ambivalence as a sociological explanation: The case of cultural explanations of alcohol problems. *American Sociological Review,* 1976, *41,* 1047–1065.

Skolnick, J. H. *The stumbling block: A sociological study of the relationship between selected religious norms and drinking behavior.* Unpublished doctoral dissertation, Yale University, New Haven, CT, 1957.

Snyder, C. R. *Alcohol and the Jews.* Carbondale and Edwardsville: Southern Illinois University Press, 1978. Arcturus paperback originally published in 1958.

Stivers, R. *A Hair of the Dog: Irish drinking and American stereotype.* University Park and London: Pennsylvania State University Press, 1976.

Ullman, A. D. Sociocultural backgrounds of alcoholism. *Annals of the American Academy of Political and Social Science,* 1958, *315,* 48–54.

Williams, P. H., & Straus, R. Drinking patterns of Italians in New Haven. *Quarterly Journal of Studies on Alcohol, 11,* 51–91; 250–308; 452–483; 586–629.

Wolf, E. *Europe and the people without history.* Berkeley, Los Angeles, and London: University of California Press, 1982.

Wolf, E. *Incorporation and identity in the making of the modern world.* 2nd Annual Edward Westermarck Memorial Lecture, Helsinki, Finland, May 1984.

Preface

This volume was inspired by discussions held during the 1981 American Anthropological Association meetings. Following the annual gathering of the Alcohol and Drug Study Group (of the AAA), several individuals suggested that we—as coorganizers of the Group—sponsor symposia on the general theme of alcohol and cultural variability. In the process of developing a symposium for the 1982 meetings, we conceptualized the broader notion of an edited volume of originally written chapters on the topic of American society and alcohol use and alcoholism. Eight of the authors included in the volume took part in that 1982 AAA session. In a sense, the symposium served as a "trial run" for developing a conceptual format for writing the case studies and the overall prospectus for the book.

The 29 contributors to the book encompass several disciplines, with anthropology predominating. Other fields represented include sociology, psychiatry, folklore, public health, and social work. Although some of the material in this volume draws from previously published studies by the authors, all chapters have been written with the goals of this book in mind. Its overall objective is to clarify how culture—religion and ethnicity, for example—is a significant force in the way people think about and respond to the use of alcohol and alcoholism.

Our decision to compile this volume was based, in part, upon the conspicuous absence of published material on alcohol and contemporary American culture. The project was planned, more specifically, for use by professionals in alcohol research and the treatment and prevention of alcoholism. Furthermore, we know of no textbook on the topic of American society and alcohol suitable for the classroom. The publication of this volume is particularly timely, considering the rapidly developing interest in the issue of alcohol and culture in colleges, universities, and professional schools. Because of the currency of the topic, lay audiences and special alcohol-related interest groups should also find the volume valuable.

Subdivided into eight sections, the book has been organized according to three basic types of material. It begins with a section on theory and history that, in essence, sets the stage for the rest of the book. Nineteen case studies follow in the next six sections, categorized loosely around different cultural groupings—Americans of European heritage; Black Americans; American Indians; Spanish-speaking populations; Asian groups—and religion and family. In the final section, Dwight B. Heath synthesizes the case study material and reviews the diverse patterns of belief and action presented.

GENEVIEVE M. AMES
LINDA A. BENNETT

Contents

III   CASE STUDIES: BLACK AMERICANS

IV   CASE STUDIES: AMERICAN INDIANS

V CASE STUDIES: SPANISH-SPEAKING POPULATIONS

VI CASE STUDIES: ASIAN GROUPS

I

Theoretical and Historical Framework

1

Alcohol Belief Systems in a Culturally Pluralistic Society

LINDA A. BENNETT and GENEVIEVE M. AMES

"The American experience with alcohol" is a multi-cultural mosaic of pattern and variablilty. Children in the United States are socialized to distinctive outlooks toward alcohol that, in turn, play a major role in their drinking practices in adulthood. Such outlooks are based upon ethno-religious heritage, family culture, and personal experience. In this anthology, the primary emphasis is on the ethnic and religious pluralism of alcohol beliefs in American society, the historical roots of contemporary alcohol patterns, and changes in the beliefs and practices due to acculturation over generations in the United States.

It is commonplace to note that Americans live in a pluralistic society and that cultural variability is part and parcel of the way of life. However, it is no small challenge to account for the role that culture plays in determining alcohol-related beliefs and practices. A central aim of this book is to advance our understanding of alcohol and culture in regard to a subset of ethnic and religious groups. We focus on perspectives

LINDA A. BENNETT • Center for Family Research, Department of Psychiatry and Behavioral Sciences, George Washington University Medical Center, Washington, D.C. 20037. GENEVIEVE M. AMES • Prevention Research Center, Pacific Institute for Research and Evaluation, Berkeley, California 94704.

toward alcohol use generally, alcoholism more specifically, and to some extent, alcoholism treatment.

To address these concerns, contributors have structured their chapters around three major areas of inquiry, each having to do with alcohol-related belief systems. Beliefs are defined as learned understandings or assumptions about cause and effect relationships within the natural, social, and symbolic worlds. We endeavor, in conducting this project, to better comprehend (1) the range of alcohol-related beliefs as found in certain cultural groups; (2) the experience of problem drinking and chronic alcoholism as influenced by cultural and familial heritage; and (3) the significance of cultural patterning as it affects attempts to seek treatment for alcohol problems.

In respect to the first area of inquiry, the following types of questions are posed: Is drinking alcoholic beverages considered to be normal and natural behavior to the cultural group under discussion? Is it thought to be deviant? Under what circumstances is drinking appropriate and acceptable? What kinds of alcoholic beverages are ordinarily consumed, with whom, where, and in what amounts? To what extent does alcohol play a role in the ritual life of these subcultures? And how are the specific beliefs regarding alcohol use integrated into the wider cultural values of the group?

We then attempt to clarify how different cultural groups define and interpret alcoholism or problem drinking. Although from an "outsider" point of view, we can theoretically specify criteria for determining when alcohol consumption enters the "problematic realm" on the basis of medical and social consequences, in reality our perspective is likely to be at odds with the individuals being studied, who have contrasting cultural traditions. In other words, "alcoholism" is in large measure a cultural category and not simply a concretely defined, specifically diagnosed illness. Deep-seated values and cultural pressures to conform constitute a major force in determining if and when problem drinking or alcoholism is an appropriate label. In American society, especially, cultural differences are critical factors in the recognition by family, friends, and coworkers that one's drinking has reached problematic proportions. Tremendous variability persists—even in the wave of widespread publicity regarding the nature of alcoholism as a "disease"—in people's inclination to deny, ignore, adjust to, or confront the drinking patterns of those around them.

This reality of American society, in turn, leads to our third consideration, and that is the impact of cultural variability on the capacity of individuals and their families to benefit from clinical resources for treatment of alcoholism. Two aspects are involved. The first is the seeking out of services, once the alcoholic or family member has come to recognize that an alcohol problem exists. The second is the integration of the cultural perspective of the treatment modality with those of the alcoholic regarding acceptable treatment goals and processes.

It is no wonder that so much confusion surrounds our ability as a society to identify alcoholism as alcoholism in specific cases. This is not simply the result of the ambiguous nature of the illness and the variable courses it takes, although these features are of great consequence. The confusion is also the result of conflicting viewpoints held by different cultural groups regarding acceptable drinking norms. In health-related matters generally, the notions we hold about the etiology of illness are based upon a combination of Western "scientific" rationale and popular or folk beliefs. Considering that this is very much the case in respect to the more readily diagnosable "medical" illnesses, such as heart disease and cancer, it is an even greater challenge to establish the line between normal and problematic drinking patterns.

For some ethnic and religious groups, any drinking whatsoever is deviant; for others, drinking may have a positive religious and social significance. In some cultural contexts, alcohol is considered to be an instrument of evil, leading to other highly undesirable behavior. Yet in others, it enhances the ritual life of the group. The case material encompassed here documents the complexity of the belief systems of American society having to do with alcohol. It also points to the futility of superimposing a dogmatic model for diagnosis and treatment of alcohol problems without paying serious attention to the cultural context in which they occur.

This anthology details the ethnographic information on the alcohol beliefs and practices of approximately 20 American ethnic and religious groups. The question naturally arises as to the implications of this rich material for action. In Chapter 24, Dwight Heath addresses this issue in his comprehensive synthesis of the commonalities and contrasts among the case materials. He states that it would be presumptuous of him to propose practical implications from the data on a case-by-case basis, since these populations "are far better known and understood by the contributing authors." We concur with his position, and let the individual contributors speak for themselves regarding the applicability of their data to alcohol treatment and policy.

However, it would be extremely valuable to have a theoretical model for culture-specific alcohol treatment and prevention, which is at the same time workable in specific cultural contexts while universally applicable. The sad fact is that no such model exists. In planning this book, we have undertaken a more modest venture. We view the presentation of the extensive ethnographic data on alcohol and American culture and Heath's thorough review of the case studies to be an initial, though substantial step in that direction.

2

Alcoholism
Illness or Disease?

NOEL J. CHRISMAN

Alcoholism is a major health problem in the United States that affects most families in some way and vexes most health practitioners. In addition, it is a moral and spiritual problem, stimulating concern and outcry from the nation's pulpits. And finally, it is the source of numerous intra- and interpersonal troubles for people, people who wonder what the problem is. As illustrated in Chapter 1, what alcoholism "is" has been the subject of heated debate. My contribution to this book is designed to explicate some of the cultural issues involved in this debate, particularly those having to do with the status of alcoholism as a disease. My approach will be to explore the relevance of the illness–disease distinction to an understanding of alcoholism.

ILLNESS AND DISEASE

The illness–disease distinction is significant in anthropology as a way of comparing native conceptions of sickness with medical conceptions. Over the last decade, it has been of great value in analyses of indigenous medical systems. In fact it is one of the simple, yet central,

NOEL J. CHRISMAN • Department of Community Health Care Systems, School of Nursing, University of Washington, Seattle, Washington 98195.

ideas that has facilitated the movement of medical anthropology from an anthropology of religion with Western biomedical or public health conceptions to a cross-cultural science for examining different health beliefs and practices and systems of health care. In essence the illness–disease distinction has released us from looking at health beliefs and rituals as cultural patterns or as cultural features that contrast with biomedical beliefs. This distinction allows anthropologists the flexibility to see all conceptions of sickness as culturally and socially constructed. Thus, we can compare conceptions from the professional or folk sectors of the health care system with those of the popular sector more evenhandedly.

An explicit comparison of illness and disease is of particular importance in anthropological research carried out in complex societies. In contrast to small scale, non-Western cultures, industrialized and urbanized societies are characterized by a great deal of intra-cultural variation. The three sectors of the health care system—popular, professional, and folk—each hold competing and sometimes conflicting notions of what sickness is and how to provide proper care (Kleinman, 1978, 1980). In addition, the variation within sectors is considerable, particularly in such populous contries as China, India, and the United States (Kleinman, Kunstadter, Alexander, & Gale, 1975; Leslie, 1976; Chrisman & Kleinman, 1983). Thus, we have come to expect multiple interpretations of sickness and care across and within sectors.

The anthropological definitions of illness and disease also vary. In his review of medical anthropology, Fabrega (1972) defined disease as "altered body states or processes that deviate from norms as established by Western biomedical science" (p. 213); an *etic* view based on biomedicine.[1] Illness was seen from an *emic* standpoint as involving a change in the state of being, labeled as discontinuous with everyday life, believed to be caused by "socioculturally defined agents or circumstances, for which culturally appropriate treatments are available" (p. 185). Similar, but simpler is Eisenberg's statement (1977) that "patients suffer 'illnesses'; physicians diagnose and treat 'diseases'" (p. 11). Although it is clear that the illness–disease distinction can be an emic–etic contrast, Western observers can create conceptual problems by assuming that the etic category must be based on Western biomedicine.

Foster and Anderson (1978) provide another view of the illness–disease distinction. For them disease is a pathological concept and

---

[1] *Etic* and *emic* are terms used by anthropologists to contrast the cognitive categories used by members of a particular society to describe their reality (emic) with categories used by the scientific observer to describe that same reality (etic).

illness is a cultural one. "We speak, for example, of plant and animal diseases, quite divorced from culture. But man's diseases become socially significant only when they are identified as illness, a physiological malfunctioning that is seen to threaten the individual and his society" (p. 40). On the individual level, illness "is the social recognition that a person is unable to fulfill his normal roles adequately, and that something must be done about the situation" (1978, p. 40). In this approach Foster and Anderson join with others in thinking of disease as biological (the same thing that affects plants and animals), adding that the human, and therefore cultural, experience of this "physiological malfunctioning" is illness. Similarly Kleinman (1980) defined disease as referring "to a malfunctioning of biological and/or psychosocial processes, while the term *illness* refers to the psychosocial experience and meaning of perceived disease" (p. 72), a definition of one in terms of the other.

The Foster and Anderson assertion that illness is defined in terms of its threat to individual and society through the individual's inability to carry out social roles deserves additional attention. This is similar to the argument used by Parsons (1951) to substantiate his view that sickness is a societal deviance requiring a sick role and subsequent social action to rectify the situation. Other social scientists, such as Fabrega (1972, 1973) and Mechanic (1978), tend to refer to the social reactions to illness, such as impaired role performance, as *illness behavior* rather than illness. Chrisman (1977) also differentiates symptom definition (illness classification) from role shift (the role-related consequences of illness) in his concept of the health-seeking process. Maintaining a conceptual distinction between cognitive aspects of sickness (illness) and behavioral aspects of sickness (illness behavior) is useful in anthropological theory so that materialist and idealist perspectives are not confused (Hahn & Kleinman, 1983).

Medical theory has also produced versions of the illness–disease distinction. One of the most important, that of Alvan Feinstein, focuses on disease, host, and illness levels of clinical data. Feinstein (1967) describes disease data as morphological, physiological, or biochemical, for example. Host features include both personal qualities and aspects of the external environment. Finally, the illness level of data refers to such clinical phenomena as the person's subjective sense of symptoms and the objective findings of a clinical examination (pp. 24–25). The cultural features of sickness would be considered host characteristics in this conception of illness and disease. Illness is the result of host–disease interaction and is cultural only insofar as host characteristics affect the interaction.

The Feinstein (1967), Kleinman (1980), and Foster and Anderson (1978) perspectives are roughly similar in that all three assume a biological (or psychosocial) basis of sickness. However, the latter two (anthropological) approaches focus more on the person's experience and the meaning of sickness for the self and society. In contrast a significant feature of Feinstein's view is his attention to the *interaction* of disease with the host to produce illness. Although this position is epidemiological in its dynamics, Feinstein does not distinguish between host and environment—a key element in the epidemiological triad of host, agent, and environment; this is an issue to which we will return below.

With the assistance of Hahn (1984), Kleinman (1982) has recently suggested another approach in which illness and disease are seen as alternative constructions of clinical reality, each embedded within a different cultural subsystem. "When we become sick, we first experience *illness:* i.e., we perceive, label, communicate, interpret, and cope with symptoms" usually in the company of family members (p. 169). Then "during the clinical transaction the practitioner begins to construe the patient's problem as *disease:* i.e., he perceives, labels, interprets and treats it as a specific abnormality in his profession's nosological system" (p. 170). This illness–disease perspective in which there can be variations in disease categories in the professional sector and variations in illness categories in the popular sector shows both as cultural constructions of clinical reality. "They are explanatory models anchored in the different explanatory systems and social structural arrangements comprising the separate sectors (and subsectors) of local health care systems" (Kleinman, 1980, p. 73).

In societies in which multiple professional nosologies exist in the professional sector, usually because of the introduction of Western biomedicine to a traditional medical system, there are alternative disease conceptions. Thus, traditional Chinese medicine and Western biomedicine are options for care in both Taiwan (Kleinman, 1980) and the People's Republic of China (Kleinman, 1982). Ayurvedic (and to a lesser extent, Yunani) medicine coexists with biomedicine in India (Leslie, 1975). In the United States, where naturopathy and particularly chiropractic are professionalizing with more demanding and lengthier educational programs (perhaps in an attempt to achieve greater legitimacy), allopathic medicine may not provide the only professional sector nosology for the categorization of sickness as disease.

A fundamental aspect of this view of disease is the recognition that it is embedded within a sociocultural system: the professional sector of the health care system. For social, political, and economic reasons, the professional sector, and its knowledge base on which disease constructions are based, carries great weight in society. In addition, this sector possesses societal legitimacy and the associated power to prevail when competing views of sickness obtain. In part professional sector legitimacy is also based on its connection with a society's Great Tradition of science, religion, and/or philosophy. The knowledge is found in esoteric, and frequently sacred, texts that are closely related to the society's fundamental assumptions about the nature of reality.

In our society we seem to make the assumption that the fundamental order of things is biophysical. We also believe that science is the mechanism for discovering and explicating this order. Thus, it is no surprise that, in our culture, the most powerful explanations of disease are based on scientific biophysical knowledge. In contrast to India, for example, American explanations of sickness in which social, psychological, or spiritual considerations take precedence are less credible. This is easily seen in psychiatry in which there is a resurgence of interest in the biochemical underpinnings of mental illness or in medicine in which the biopsychosocial model (Engle, 1977) is waging an uphill battle against a purely biophysical disease construction.

Illness is also embedded within a sociocultural system, the popular sector. In contrast to the narrowness of purpose found in the professional sector, in which the focus is on health care delivery, popular sector purposes are more diverse and diffuse, much more a part of the conduct of everyday life. Beliefs about the way things are, social relationships, feelings, and the like are all influential in the construction of both personal and group meanings. Because it is experiential and so much goes into everyday experience, a broad knowledge base is brought to bear on the explanation of illness. In addition, this knowledge base does not have the prestige of the formal explications of knowledge found in the Great Tradition linked to the professional sector.

Like the professional sector, however, popular health culture conceptions of illness are also strongly biological, adhering to the same general beliefs about a biophysical reality. In addition, the demonstrated curative power of biomedicine, believed to be based on science, promotes the diffusion of biologically based conceptions of sickness to the popular sector. As is also true in other cultures when spiritual, humoral, or magical causes of illness are part of the explanatory models in the popular sector, they are frequently combined with biophysical conceptions of pathophysiology (Snow, 1974). Illness categories included in the popular health culture are not as well organized as the professional nosologies because there is no educational or bureaucratic structure to sustain an organized body of knowledge. Nor is there the need for such organization; lay people are patients or clients, not healers. However, 70% to 90% of all illness episodes are taken care of in family and community contexts (Demers, Altamore, Mustin, Kleinman, & Leonardi, 1980; Zola, 1972). Such popular sector care requires a body of knowledge—a system of health beliefs and practices that help categorize what the health problem is and what to do about it.

Illness and disease then are different constructions of clinical reality. Both are based on a commonsense phenomenon of sickness, ranging from general malaise to such health crises as heart attacks. As implied in the phrase "clinical reality," the purpose of constructing sickness into a culturally appropriate cognitive pattern is to explicate the nature of the sickness, to relate it to other similar phenomena, and to take care of it—either through treatment or by ignoring it, and hoping that it will go away (Chrisman & Kleinman, 1983). In addition, illness and disease constructions of sickness are embedded in the social realities of their respective health care sectors: the popular sector with its powerful association with everyday life and the professional sector with its association with the Great Tradition and its social legitimacy. This approach frees anthropologists from Western biomedical sickness categories as *the* etic standard for discussing cultural variation in conceptions of sickness. Yet in this freedom, there is also the loss of an automatic etic standard. This is as it should be. Like other aspects of methodology, standards of measurement should be chosen, not assumed. Foster and Anderson (1978) imply that biological measurements are relevant, but as Ritenbaugh explains (1982), the line between abnormal and normal in the spectrum of biological variation is still culturally defined. At least for the present, I imagine that most anthropologists will retain Western biomedical categories of disease as an appropriate etic standard.

In contrast to the system level of discussion of illness and disease and their relationships to the health care sectors, sickness at the individual level begins as a perception of deviation from some personal level of normal health (Freidson, 1970). Sickness categorizations provide people with ways of assigning meaning to the experience, meanings that form a semantic network with socioculturally important features of life (Good, 1977). "Shared cultural beliefs (about the body, specific symptoms, and illness generally), constraints in our concrete social situation, and aspects of the self (personality, coping style, prior experience, etc.) together with our explanatory model for the particular episode of illness will orient us how to act when ill, how to diagnose and treat, and how to regard and manage the life problems illness creates" (Kleinman, 1982, p. 169).

It is important to recognize and distinguish analytically three aspects of an illness at the individual level: the label or term applied to the illness; the description of the illness, its explanatory model; and its consequences for everyday life, including its semantic links with other cultural features and its social effects, such as the sick role (Segall, 1976) or secondary gain (Alexander, 1982). The meaning and usage of the explanatory model concept have been the subject of debate in medical anthropology (e.g., Comments, 1981). In this chapter, an explanatory model is a person's "notions about an episode of sickness and its treatment" (Kleinman, 1980, p. 105), composed of statements about onset, etiology, pathophysiology, course, and treatment (Kleinman, Eisenberg, & Good, 1978).

An illness explanatory model of greater or lesser complexity and completeness is constructed for symptoms that have been personally marked for clinical attention, that is, pathologized (Chrisman & Kleinman, 1983). This happens during a cognitive process when symptoms are linked with beliefs about sickness in a person's repertoire of health beliefs and practices, based on the plethora of health knowledge available in the popular health culture.

The process of constructing an explanatory model is not simply the transaction of knowledge and symptoms, however. Stigmatizing or frightening sicknesses and their labels may be avoided as options for illness construction. For various reasons, such illnesses as leprosy, cancer, mental illness, and alcoholism may be "denied" as viable labels and descriptions of sickness for long periods of time. Moreover, symptoms that might otherwise be normalized and ignored may become the subject of cultural attention because of personal distress (Katon, Kleinman, & Rosen, 1982; Nichter, 1981; Zola, 1972) or social problems (Zola, 1973).

At some point in 10% to 30% of illness episodes, the person or family finds that the illness—the symptoms, the label, the beliefs about pathological processes and/or the effects on everyday life—has become unmanageable within existing family and social network resources. Greater healing power is necessary and usually this power is sought within the professional sector (e.g., Young, 1980). When this occurs, the popular and professional sectors interact. During the encounter with the practitioner, "illness as a lay experience becomes transformed into disease as a biomedical explanation and a clinical entity is created. That is to say, the practitioner constructs a new social reality, the disease, which will come to influence how the patient thinks and lives his illness" (Kleinman, 1982, p. 170). The new information only *influences* the patient's thought and action, however. For example, although there was a statistically significant shift toward greater agreement in patient and doctor explanatory models after the patient visit in my study of 101 primary care patients, the shift was small. Moreover, about one half of the non-compliance with physicians' recommendations made in those visits was the result of an incongruence of patient and practitioner explanatory models. Thus, patient perspectives continue to be important following the doctor–patient encounter (Chrisman, 1983).

Illness perspectives are also persistent elements during at least the initial stages of the clinical encounter. Patients call only limited and culturally formed aspects of their illnesses to the attention of physicians (Zola, 1966). What the practitioner hears from the patient and responds to is an *illness statement*, "the biological 'tokens' and a brief history that aid in gaining entry to the examination room" (Chrisman, 1983). These patient statements are embedded in a culturally shared set of meanings within the popular culture. (It is easy for anthropologists to forget that physicians were raised in the popular health culture and thus are likely to respond knowledgeably to popular culture statements.) Once complaints have been elicited from the patient and the physical examination has begun, the practitioner engages in the interpretive process of transforming illness into disease (Good & Good, 1981). In the symbol-laden interaction of the clinical encounter lies the experiential and analytic problem with alcoholism as a sickness category.

ALCOHOLISM

Neither the popular health culture nor the biomedical health culture have a well-defined or limited cluster of meanings for alcoholism and both are strongly influenced by prevailing attitudes and emotions linked to this health problem. Neither alcoholism as illness nor as disease is well defined in these health cultures. Moreover, when sociocultural variations across ethnic and social class groups and disciplinary orientations across professional specialties are highlighted, definitional problems are exacerbated.

Within the professional sector, the principal difficulty seems to be an inability to settle upon a biological core definition. Kane (1981) cites two related problems: (1) problems in comprehending the wide variability in the ways alcoholism is presented to practitioners as a health problem and (2) difficulties in sorting out host versus environment influences in the etiology and pathophysiology of alcoholism. For example, health scientists continue to debate such issues as the role of genetic and psychosocial factors in host susceptibility and the relative importance of culturally patterned drinking behavior as an environmental influence. (Anthropologists tend to focus their research on drinking patterns, Heath, 1978.) I see this professional definitional uncertainty as a cultural problem. The significance of the epidemiological model of host, agent, and environment notwithstanding, biomedicine tends to locate disease within the individual—such as a lesion, a pathophysiological process, or an anatomical or genetic malformation (e.g., Hahn, 1982). Moreover, the canons of biomedicine require an accurate, or scientific, description of reality as a prerequisite for wide acceptance of a diagnostic category and for full confidence in treatment regimens. The biological and psychosocial complexity of alcoholism has defied such description.

Of course, in our pluralistic society, lack of consensus has not prevented health practitioners and other professionals from conceptualizing—and thus diagnosing and treating—alcoholism in a variety of ways. These include biological, social, moral, and spiritual approaches alone or in combination. But without a bureaucratically sanctioned standard—that is, disease—these conceptualizations exist as competing alternatives. From a lay perspective, the range of alternatives provides for elaborate hierarchies of resort (Schwartz, 1969), hope, and occasionally confusion and anger.

Competing professional conceptualizations of disease and their accompanying therapies do not always cause problems for the professional sector. For example, Gaines (1979) noted very different etiologies of mental illness among a group of psychiatrists. However, their explanatory models included acceptable terminologies for explaining disease and recognizable treatment and did not cause noticeable problems. Cancer therapies also vary significantly and include almost the same range of variation as alcohol abuse therapies. In addition, cancer resembles alcoholism in that there are many types of cancer and different organ systems can be involved. However, cancer is dissimilar in that there is a common cluster of biologically described pathophysiologies. Variations occur in etiological conceptions and in treatment.

The absence of a well-accepted disease conception for alcoholism underlies the uncertainty in the development of illness models in the popular sector. It is a well-advertised health problem, much like diabetes, cancer, and heart disease. But in the absence of a consistent biomedical explanatory model, the broader cultural meanings of everyday life continue to play a strong role in popular conceptions of the illness. In particular, social and moral explanations related to a lack of personal control or moral weakness persist as important components in popular culture explanatory models. In addition, alcoholism resembles the diseases listed above in its chronicity. We Americans, with our high technology and cure-oriented value system, have not yet developed a clear cultural pattern for understanding chronic illness, nor have we developed consistent expectations for a chronic sick role (Alexander,

1982). Cultural influences on the formation of explanatory models and the difficulties involving illness behaviors and sick role behaviors seem to contribute to both patient and practitioner reluctance to consider a diagnosis of alcoholism except in the most flagrant cases.

Currently in the United States, the terms "alcoholic" and "alcoholism" are shared among lay and professional people. This is in part the result of a diffusion from the professional sector, but is mostly the result of historical developments in the popular health culture itself, as outlined by Ames in Chapter 3. This situation of shared popular and professional terminology with variations in meaning is similar to that found by Blumhagen (1980) for hypertension. Although the name of the sickness—hypertension—was used by both doctors and patients, explanatory models differed. A large proportion of Blumhagen's sample of patients ascribed their hypertension to too much stress (hyper-tension), whereas practitioners focused on morphological changes in the circulatory system.

To some extent, this meaning problem results from shared *illness* perspectives among doctors and patients. Both believe that stress plays a role in high blood pressure, even though specific mechanisms are not well spelled out biologically. Yet the transformation to disease is facilitated with the existence of a physical finding: consistently elevated blood pressure as measured by a sphygmomanometer. This contrasts sharply with the alcohol-related, patient–practitioner encounter. Both participants begin the curing interaction with illness perspectives that include all the sociocultural baggage common in the popular culture. But the expected transformation to disease becomes problematic. Without a well-developed disease perspective, including accepted mechanisms for gathering "objective" data to confirm a diagnosis, the practitioner must use a popular culture knowledge base involving numerous lay stereotypes, of which moral stigma is probably the most important. Thus, the transformation to disease is more strongly influenced by patient report and a commonsense knowledge base than is usual for other health problems, including psychiatric problems.

Alcoholism is one of the very few health problems for which the constellation of beliefs embedded in popular health culture so strongly influences both patients and practitioners. This is related to its recency as a disease and its centuries-long recognition by the popular culture. Alcoholism has "become" a disease only in the last few decades, a process that is part of the general trend of "medicalizing" social problems (Zola, 1977). In turn, this trend is a part of a longer history of rational and scientific thought replacing "outmoded" ways of thinking.

Although medicalization is viewed negatively by most social scientists as an encroachment on the cognitive and behavioral territory of lay people (Eisenberg & Kleinman, 1981), categorizing alcoholism as a disease rather than a social or moral problem is promoted by such elements of society as Alcoholics Anonymous (AA). For much of its history, this organization has used the disease status of alcoholism to reduce the feelings of moral culpability experienced by their client population, that is, labeling alcoholism as a disease is part of the successful treatment pattern of AA.

A disease label is not without its problems, however. Patients who desire help for a drinking problem, but who do not want to boldly state that fact become upset when their physicians overlook or ignore seemingly obvious signs of the problem, such as reported blackouts or recurring cuts and abrasions (from falling down or running into things while drunk). An alcoholism diagnosis does not usually emerge from the standard detective work of history and physical examination as do ulcers, gallstones, or tumors in the case of stomachaches. An alcoholism diagnosis usually depends on the very thing the patient wants to hide (and avoid moral culpability for): the admission of heavy drinking and its consequent effects on everyday life. In addition, the treatment for alcoholism does not fit one of the major lay stereotypes of what a disease treatment from a doctor should be: medication or surgery. Alcoholism treatment instead frequently includes an admission of alcoholism, psychotherapy (or its lay equivalent), and behavior modification—all "psychological," perhaps indicative of a problem in the mind, not the body conceived in a biophysical fashion. Only hospitalization during the drying-out period approximates treatment for a "disease."

CONCLUSION

Two major issues emerge from this discussion of alcoholism as illness or disease. One concerns the ambiguity and complexity of popular sector explanatory models of the illness. The degree to which alcoholism is seen as a sickness or a social or moral failing, or some mixture of the two, varies. And when it is seen as a sickness, the degree to which biological, humoral, or spiritual components are included in individual explanatory models also varies. Much more anthropological research is required on variations in popular culture explanatory models of alcoholism. For example, anthropologists doing ethnographic research on drinking patterns in various settings should also elicit data on their informants' alcoholism explanatory models, their social and moral understandings of the illness and their beliefs about the nature and the efficacy of treatment. Along with information drawn from such community samples, we need comparative data from clinic samples—from patients in both hospital and ambulatory care facilities. Patients, because they are more involved with the illness, tend to have more complete and elaborate explanatory models that can be compared with the explanatory models of community members. In addition, their personal histories of seeking and receiving care for alcoholism (help-seeking trajectories) can illuminate our understanding of popular–professional sector interactions, including evaluations of diagnostic and treatment events, variations in cultural responses to treatment modalities, and the role of stigma. Similarly, practitioner explanatory models and experiences with patient outcomes can be explored.

Such descriptive research would add a great deal to our understanding of how beliefs in the popular health culture are variously combined across different ethnic and other subcultural groups (including religious groups) in American society, particularly because of the importance of the social and moral underpinnings of alcoholism. This knowledge could have the practical outcome of clarifying for health professionals the degree to which sociocultural factors influence compliance with therapeutic regimens. Of equal importance would be a clarification of the popular health culture beliefs of health practitioners and the influence of these beliefs on the ways practitioners interact with alcoholic or suspected alcoholic patients.

The second issue is related to ambiguity in professional sector conceptions of alcoholism as a disease and the consequent difficulties in professional diagnosis and treatment. A wide variety of symptoms may be associated with alcohol abuse, but need not be identified by patient or family as the consequence of alcohol consumption. That is a person may experience blackouts, family dysfunction, or gastrointestinal upsets and generate an explanatory model unrelated to alcohol. The person seeking professional help for such complaints is relatively unlikely to receive an alcoholism diagnosis if the complaint is not phrased in terms of alcohol use because of the confusion and lack of knowledge in biomedicine and the persistent unwillingness to diagnose alcoholism without the patient's acquiescence. In addition family and other lay consultants may also not encourage an alcohol-related explanatory model either for protective (and ultimately moral) reasons or because they do not include alcoholism in their conceptions of common illnesses.

Research to explore these problems requires a more complex and less cost-effective design than the more typical ethnographic research described above. The central element in such investigations involves audio or video recordings of practitioner–patient interactions in primary care settings, coupled with intensive interviews with patients about the presenting health problem and its possible relationship with alcohol consumption. When alcohol abuse is discovered from ethnographic interviews, tape recordings can be analyzed to discover the conditions under which alcoholism diagnoses are or are not made.

An anthropological discussion of alcoholism using the illness–disease distinction can raise issues in ways that make cultural analyses less dependent upon Western biomedicine as an implicit standard. This is valuable for anthropology, in that the sociocultural meanings of this health problem, including the disease meanings of alcoholism, may be conceived as cultural and social, without tacit biomedical underpinnings. The results of such analyses are valuable to our clinical colleagues because anthropologists can help them sort out which aspects of their practice are common sense, based on popular health culture knowledge, and which are professional sense, based on biomedical knowledge. Clinically oriented anthropologists can use sociocultural understandings of alcoholism in a number of ways to help clinicians: didactic instruction in health science schools, consultations involving non-compliance and recidivism, instruction in patient interviewing skills, and seminars or rounds in which practitioner frustration emerges. Moreover, anthropologists can use these data to help formulate health education programs designed to clarify contemporary biomedical understandings of and approaches to alcohol abuse and to reduce societal stigma.

In demarcating what is cultural about alcoholism, the illness–disease distinction raises a series of important clinical and research problems, three of which stand out for me: (1) the need for descriptive data on popular culture explanatory models of problem drinking; (2) the importance of helping health practitioners examine their own and their patients' conceptions of alcoholism; and (3) the opportunity to document and analyze culture change in both illness and disease conceptions in the years ahead.

ACKNOWLEDGMENTS

I would like to acknowledge the valuable help of Katherine Carlson in writing this chapter. I have also benefited from the editorial advice of Arthur Kleinman, Robert Hahn, Sandra Counts, and Genevieve Ames.

REFERENCES

Alexander, L. Illness maintenance and the new American sick role. In N. Chrisman &
    T. Maretzki (Eds.), *Clinically applied anthropology.* Dordrecht, Holland: D. Reidel, 1982,
    351–369.
Blumhagen, D. Hyper-tension: A folk illness with a medical name. *Culture, Medicine, and
    Psychiatry,* 1980, *4,* 197–229.
Chrisman, N. J. The health seeking process: An approach to the natural history of illness.
    Culture, Medicine, and Psychiatry, 1977, *1,* 351–377.
Chrisman, N. J. Popular culture explanatory models in primary care. Paper presented at
    the XIth International Congress of Anthropological and Ethnological Sciences, Van-
    couver, British Columbia, 1983.
Chrisman, N. J., & Kleinman, A. Popular health care, social networks, and cultural mean-
    ings: The orientation of medical anthropology. In D. Mechanic (Ed.), *Handbook of
    health, health care, and the health professions.* New York: The Free Press, 1983, 569–590.
Comments: Culture, Medicine, and Psychiatry. *Culture, Medicine, and Psychiatry,* 1981, *5,*
    337–379.
Demers, R., Altamore, R., Mustin, H., Kleinman, A., & Leonardi, D. An exploration of
    the depth and dimensions of illness behavior. *Journal of Family Practice,* 1980, *11,*
    1085–1092.
Eisenberg, L. Disease and illness: Distinctions between professional and popular ideas of
    sickness. *Culture, Medicine, and Psychiatry,* 1977, *1,* 9–23.
Eisenberg, L., & Kleinman, A. Clinical social science. In L. Eisenberg & A. Kleinman
    (Eds.), *The relevance of social science for medicine.* Dordrecht, Holland: D. Reidel, 1981,
    1–23.
Engle, G. The need for a new medical model: A challenge for biomedicine. *Science,* 1977,
    196, 129–136.
Fabrega, H. Medical anthropology. In B. J. Siegel (Ed.), *Biennial review of anthropology,
    1971.* Stanford: Stanford University Press, 1972, 167–229.
Fabrega, H. Toward a model of illness behavior. *Medical Care,* 1973, *11,* 470–484.
Feinstein, A. R. *Clinical judgment.* Huntington, New York: Robert E. Krieger, 1967.
Foster, G. M., & Anderson, B. *Medical anthropolopgy.* New York: Wiley, 1978.
Freidson, E. *Profession of medicine.* New York: Dodd, Mead, 1970.
Gaines, A. Definitions and diagnoses: Cultural implications of psychiatric help-seeking
    and psychiatrists' definitions of the situation in psychiatric emergencies. *Culture,
    Medicine, and Psychiatry,* 1979, *3,* 381–419.
Good, B. The heart of what's the matter. The semantics of illness in Iran. *Culture, Medicine,
    and Psychiatry,* 1977, *1,* 25–58.
Good, B., & Good, M. D. The meaning of symptoms: A cultural hermeneutic model for
    clinical practice. In L. Eisenberg & A. Kleinman (Eds.), *The relevance of social science
    for medicine.* Dordrecht, Holland: D. Reidel, 1981, 165–196.
Hahn, R. "Treat the patient, not the lab": Internal medicine and the concept of "person."
    Culture, Medicine, and Psychiatry, 1982, *6,* 219–237.
Hahn, R. Rethinking "illness" and "disease." In E. V. Daniel & J. Pugh (Eds.), South
    Asian systems of healing. *Contributions to Asian Studies,* 1984, *18,* 1–23.
Hahn, R., & Kleinman, A. Biomedical practice and anthropological theory: Frameworks
    and directions. *Annual Reviews of Anthropology,* 1983, *12,* 305–333.
Heath, D. Forward, Ethnicity and alcohol use. *Medical Anthropology,* 1978, *2,* 3–9.

Kane, G. *Inner-city alcoholism: An ecological analysis and cross cultural study.* New York: Human Sciences Press, 1981.

Katon, W., Kleinman, A., & Rosen, G. Depression and somatization: A review, Parts I and II. *American Journal of Medicine,* 1982, *72,* 127–135, 241–247.

Kleinman, A. International health care planning from an ethnomedical perspective: Critique and recommendations for change. *Medical Anthropology,* 1978, *2,* 71–94.

Kleinman, A. *Patients and healers in the context of culture.* Berkeley: University of California Press, 1980.

Kleinman, A. Neurasthenia and depression: A study of somatization and culture in China. *Culture, Medicine, and Psychiatry,* 1982, *6,* 117–190.

Kleinman, A., Kunstadter, P., Alexander, E. R., & Gale, J. L. (Eds.), *Medicine in Chinese cultures: Comparative studies of health care in Chinese and other societies.* DHEW Pub. No. (NIH) 1975, 75–653.

Kleinman, A., Eisenberg, L., & Good, B. Culture, illness, and care. *Annals of Internal Medicine,* 1978, *88,* 251–258.

Leslie, C. Pluralism and integration in the Indian and Chinese cultures. In A. Kleinman, P. Kunstadter, E. R. Alexander, & J. L. Gale (Eds.), *Medicine in Chinese cultures.* DHEW Pub. No. (NIH) 75-653, 1975, 401–419.

Leslie, C. (Ed.) *Asian medical systems.* Berkeley: University of California Press, 1976.

Mechanic, D. *Medical sociology,* 2nd Ed. New York: The Free Press, 1978.

Nichter, M. Idioms of distress. Alternatives in the expression of psychosocial distress. *Culture, Medicine, and Psychiatry,* 1981, *5,* 379–409.

Parsons, T. *The social system.* Glencoe, Ill.: The Free Press of Glencoe, 1951.

Ritenbaugh, C. Obesity as a culture-bound syndrome. *Culture, Medicine, and Psychiatry,* 1982, *6,* 347–363.

Schwartz, L. R. The hierarchy of resort in curative practice. *Journal of Health and Social Behavior,* 1969, *10,* 201–208.

Segall, A. The sick role concept: Understanding illness behavior. *Journal of Health and Social Behavior,* 1976, *17,* 163–170.

Snow, L. Folk medical beliefs and their implications for care of patients. *Annals of Internal Medicine,* 1974, *81,* 82–96.

Young, J. A model of illness treatment decisions in a Tarascan town. *American Ethnologist,* 1980, *7,* 106–131.

Zola, I. K. Culture and symptoms—an analysis of patients' presenting complaints. *American Sociological Review,* 1966, *31,* 615–630.

Zola, I. K. The concept of trouble and sources of medical assistance. *Social Science and Medicine,* 1972, *6,* 216–236.

Zola, I. K. Pathways to the doctor—from person to patient. *Social Science and Medicine,* 1973, *7,* 677–689.

Zola, I. K. Healthism and disabling medicalization. In I. Ilich (Ed.), *Disabling professions.* London: Marion Boyars, 1977.

3

American Beliefs about Alcoholism
Historical Perspectives on the Medical-Moral Controversy

GENEVIEVE M. AMES

A PROBLEM OF CONFLICTING BELIEFS

One of the most significant difficulties in dealing with alcoholism on any level is that so few people agree on what it is. In American society, there are differing and controversial perspectives current in the medical professions, the academic and scientific communities, and the population at large as to whether habitual and excessive consumers of alcohol are suffering from a disease, are victims of conflicting cultural norms, are weak-willed and irresponsible persons, or a combination of these possibilities. The stigmatizing aspects of alcoholism and the general paradigmatic differences among professionals have had far-reaching implications not only for prevention and treatment programs, but also for the affective behavior of the afflicted drinker and his or her family.

GENEVIEVE M. AMES • Prevention Research Center, Pacific Institute for Research and Evaluation, Berkeley, California 94704. Preparation of this chapter was in part supported by a fellowship from the National Institute on Alcohol Abuse and Alcoholism (5 T32 AA07240) through the Alcohol Research Group, School of Public Health, University of California, Berkeley and by a National Alcohol Research Center Grant (AA06282-02) from the NIAAA to the Prevention Research Center, Pacific Institute for Research and Evaluation, Berkeley.

The voluminous scientific literature on the subject leaves little argument as to the concomitant importance of the biomedical, psychological, and sociocultural aspects of alcoholism or its status as a serious health problem in its own right or as a precursor of other health problems. Nevertheless, two basic definitional questions remain, and these frame the general direction of all research and treatment: Is alcoholism a "disease" with biological predispositions, and therefore a pathological concept, or is it an "illness" described in terms of learned behavior, and therefore a cultural or psychosocial concept? But when we look beyond the entrenched positions of most alcohol researchers, clinicians, and policymakers to the ways in which traditional cultural beliefs and emotional responses to drinking behavior affect the alcoholic experience, the debate can be seen as a medical-moral controversy arising from the fact that the dominant medical and moral belief systems in this country do not always agree as to the cause, effects, and treatment of chronic alcoholism.

THE MEDICAL-MORAL DILEMMA

From a medical perspective, the majority of treatment modalities for alcoholism utilize in one form or another what is generally referred to as the medical or disease model of alcoholism.[1] Today, almost 40 years after its first restatement in the modern scientific literature (Jellinek, 1946), the precise content of the disease concept is still vague and uncertain (see Room, 1983). However, this view in general recognizes alcoholism as a progressive disease that can be divided into symptomatic phases, and it differentiates between the addictive and the non-addictive drinker (Jellinek, 1952). The main differentiating criterion between the two categories is that the addictive drinker experiences loss of control

---

[1] There are two other working models of alcoholism, neither of which will be discussed here because they have not had as significant an impact on the specific belief systems of the general public as the medical and moral models have. These are generally referred to as the psychological and sociocultural models of alcoholism. The psychological model (as derived from the field of psychoanalysis, particularly Freudian theory) assumes that certain personality traits established in eary childhood predispose the individual to alcoholism (Zwerling, 1959; Barry, 1974). As does the disease model, this approach views the alcoholic as victimized. Drinking is a circumstance of a preexisting psychopathic personality and thereby precludes early recognition and treatment of the drinking problem. This approach is still utilized by many psychiatrists. Another psychological approach is the behavioral learning model from the field of experimental learning psychology. This model assumes that alcohol consumption is both caused and maintained by the association of alcohol intake with positive rewarding experiences; thus individuals who are

and the non-addictive drinker does not. The predisposition to loss of control for the addictive alcoholic, according to Jellinek (1960), is a chain reaction stemming from (1) psychological conflicts or social situations that initiate the drinking, eventually leading to (2) a physical demand for alcohol. In this process, the body tissues acquire a tolerance to alcohol and cell metabolism adapts to it. Withdrawal symptoms, "craving," and loss of control over the amount consumed result.

In 1952, the World Health Organization's Committee on Mental Health (World Health Organization, 1952) and in 1957 the American Medical Association (1961) accepted this diagnostic framework when they officially recognized alcoholism as a disease and as a treatable illness. Alcoholics Anonymous (AA), a highly successful self-help organization, also uses the disease concept as one component of its working model.

In keeping with this "established" medical model for a widely prevalent health problem, it would seem that alcoholism should no longer be viewed as a sign of weakness, moral dissipation, or basic mental instability. In twentieth-century America, where the medical system is based primarily on biophysical explanations for disease and illness and where the people generally accept scientific explanations for their health problems, it should follow that people, willingly and with some relief, accept "scientific" biomedical explanations for the complex condition of alcoholism. Yet clinicians, and the alcoholic patients they treat, often operate with conflicting or at best overlapping beliefs about and explanations of alcoholism. Millions of Americans who have been medically diagnosed as alcoholics still perceive themselves as "deviant" rather than "diseased" and take measures to conceal their drinking and to protect themselves and their families from the widespread social opprobrium associated with alcoholism. Similarly to people suffering from mental illness, cancer, or more recently AIDS (acquired immune deficiency syndrome), the alcoholic must deal with negative attitudinal responses to

---

subjected to stressful situations may obtain relief from stress through alcohol use because of its pharmaceutical effects. The drinking habit is progressively strengthened by repetitive use of alcohol to combat anxiety and alleviate stress (see Armor et al., 1978 for a review of psychological models).

The sociocultural models cover broad areas of external factors in relation to alcohol use and alcoholism, ranging from consumption rates of whole societies to individual responses to a particular environmental situation; psychological and physiological factors are not always excluded. In its emphasis on the effect of an individual's environment, learned attitudes and beliefs about alcohol consumption rates, drinking behavior, and problem drinking, this approach empirically challenges (either implicitly or explicitly) the inevitability of the disease progression theory (see Heath, 1978 and Bacon, 1957 for reviews of the sociocultural approach).

the problem, in addition to the physically debilitating symptoms of the illness.

In many respects, even the terminology for drinking problems, whether used in relation to disease, illness, or affective behavior, is controversial and contributes to moralizing conceptions of alcoholic behavior. "Alcoholic" and "alcoholism," concepts that are widely used in relation to alcohol problems and the subsequent physical and social effects of drinking, conjure up differing emotional responses, not only among the general population, but also among researchers and clinicians (Clark, 1975; Room, 1970). The prevailing negative response to these terms is usually related to unresolved issues underlying the primary identifying symptom of chronic alcoholism: loss of control over one's own behavior (Cahalan, 1970).

However, "loss of control" may be variably perceived and experienced among various cultures worldwide; in American society it is a culturally unacceptable condition. A frequently proposed explanation for negative reactions to the loss of control is that such behavior is incompatible with the dominant Protestant ethic. Heavy drinking and frequent drunkenness, recognized symptoms of chronic alcoholism, are not acceptable in a culture that values a disciplined will as the key to "morality, success and respectability" (Lemert, 1951).

Although this explanation may be true to some degree, we know that the old Calvinistic theology is not the only guiding force for the persistence of the moral model nor can it account for the general ambivalence about the validity of the medical model. Differing and oftentimes conflicting beliefs about alcohol use and drinking behavior, emotional and moralistic responses to problem drinkers, ongoing scientific arguments about the exact nature of the disease concept, and the significance of these issues for all persons affected by alcoholism are problems that have deep historical roots. In the following pages, I will review these historical roots by tracing the emergence of belief systems that conceptualize alcoholism, on the one hand, as a health problem and therefore treatable by the established medical professions, and on the other hand, as a moral problem and therefore subject to social ostracism and attempts at religious salvation. With the recognition that these are complex issues and that there are many and differing approaches to them, this review is limited exclusively to a history of the so-called medical and moral models of alcoholism.[2]

___

[2] For a comprehensive review of the medical model as it has been conceptualized in the sociological tradition, see Robin Room (1983).

THE EVOLUTION OF THE MORAL MODEL

In colonial America, heavy and habitual drinking was perceived not as a morally disallowed compulsion or a medically diagnosed addiction, as it is variably defined today, but rather as a choice and therefore as natural and normal behavior.[3] As a beverage, alcohol was viewed not as an intoxicating substance, but rather as a healthy "gift" of God, with many curative and life-sustaining properties. Drinking was pervasive among the colonists, and alcohol served practical purposes in everyday life.

Alcohol was readily available. Until the late seventeenth century, most people made their own cider, ale, and brandy from locally grown produce. An abundant stock of alcoholic beverages, laid in along with other consumable foodstuffs, was considered both a necessary staple of subsistence and a symbol of civility. Alcohol was often used as a substitute for drinking water for the reason that well water was considered impure and too "thin" for healthful human consumption. In addition to quenching thirst, alcohol served as a household medicinal substance—useful to fight fatigue, soothe indigestion, ward off fever, and relieve aches and pains (Aaron & Musto, 1981, p. 131). Alcohol also served important social purposes and enhanced conviviality in both formal and informal gatherings, whether such activities were related to church, work, home, or government, or the tavern, the focal point of every town's social and political life (Levine, 1978).

The consumption of alcohol was an integral part of normal, everyday living in colonial America and was not particularly problematic. There were those who drank habitually to the point of drunkenness, but a large percentage of the population drank moderate amounts of alcohol. Gusfield (1962) noted that the general population's lack of concern about alcohol-related problems can be attributed to the fact that colonial drinking took place within a social system in which it was limited and controlled, a system wherein "drunkenness occurred and was punished but was seldom frequent or widespread" (p. 104). Since no empirical surveys of the incidence or prevalence of drinking problems exist for this period of American history, it is difficult to judge the accuracy of such statements. However, early evidence of the widespread use of alcohol can be found in Puritan ministers' sermons, which spoke out

---

[3] My assessment of the evolution of the moral model relies primarily on existing reviews and perspectives of Keller (1976), Clark (1976), Levine (1978), Gusfield (1962, 1963, 1967), Rorabaugh (1979), Tyrrell (1979), Glaser (1976), and Aaron and Musto (1981).

against the misuse of alcohol, and in colonial laws which made public drunkenness a punishable offense.

Unlike the general population, Puritan ministers were seriously concerned about habitual drunkenness, but this was only because they considered such behavior to be an abuse of the "good creature of God"; in their belief system, the "good creature of God" was the drink, not the drinker (Keller, 1976, p. 16). To them, drinking in excess was a personal indiscretion, and alcohol itself was no more to blame for drunkenness than food was to blame for gluttony (Aaron & Musto, 1981, p. 132). In colonial law, drunkards became deviants (or problematic) only when they became dependent on society, and at that point they were not distinguished as a class from criminals, the poverty-stricken, or the insane. A problem drinker (or alcoholic) who was self-supporting or who owned property was treated with deference, like anyone else of his class, and was exempt from punishment for public drunkenness.

In his historical analysis of the pre-temperance movement, Levine (1978) concluded that the most radical distinction between colonial and contemporary ideas about alcohol misuse is that in colonial times "addicted meant habituated and one was habituated to drunkenness, not to liquor" (p. 147). In other words, whereas contemporary thought considers alcoholics to be devoid of willpower or hopelessly addicted to alcohol, the traditional colonial view was that drinkers loved to drink and to get drunk because it was a self-indulgent pleasant experience (albeit an abuse of a "good" thing).

By the end of the eighteenth century, rum was arriving in abundance from the West Indies and various alcoholic substances were being manufactured locally.[4] With the increasing availability of both imported hard liquor and cheap domestic alcohol (whiskey, rum, and rye), people drank more and in a context that was less strictly controlled than when taverns and drinking patterns "had been under the aegis of a proprietary civic elite" (Aaron & Musto, 1981, p. 134). For example between 1800 and 1830, the annual per capita consumption increased until it exceeded five gallons—a rate nearly three times that of today (Rorabaugh, 1979, p. 8). Other estimates place the consumption of distilled spirits even higher, at 9.5 gallons per capita in 1830, as compared to 2.6 in 1978 (Lender & Martin, 1982, p. 196).

During the time of this increase in alcohol production and consumption, America was undergoing rapid demographic and economic changes. Although the precise reasons for the spread of heavy drinking

---

[4] William Penn was one of the first local manufacturers of alcoholic beverages (Glaser, 1976).

can only be speculated upon, it is surely not a coincidence that the nineteenth-century drinking "epidemic" occurred at a time when America was shedding its agrarian, hierarchial, and colonial past, moving from the old world form to a new and radically different destiny. Aaron and Musto (1981) noted that during this period of rapid disjuncture and discontinuity, at a time when the network of deference and respect for the civic elite was breaking down, limits and controls on drinking also changed. As hard drinking became increasingly prevalent and problematic on both the American frontier and in urban slums, traditional thinking about alcohol use and misuse began to change, and out of this transformation of beliefs and values the 100-year temperance movement began to emerge. Levine (1984) has traced that transformation to a 50-year period between 1785 and 1835:

> At the beginning of that period, almost no one believed alcohol to be a dangerous or destructive substance, and the idea of life-long abstinence from liquor, and regular consumption of water, would have seemed ludicrous, bizarre. At the end of that period, a "temperance movement" advocating exactly that programme had grown into a massive reform movement. The temperance movement demonized alcohol, literally referring to it as a "demonic" substance, and became the largest enduring middle-class mass movement of the 19th century. (p. 110)

The initial impetus of the Temperance Era has been collectively defined relative to various social factors. As described earlier, increased availability of alcohol was a major factor underlying the increased per capita consumption rate to the highest ever in American history. In the context of the post-Revolution political, economic, and social upheaval, hard drinking and habitual drunkenness came to be seen as a genuine social problem and there was legitimate cause for concern. People from various social classes and from both urban and rural societies were spurred by genuine humanitarian concerns to "liberate society from enslavement to drink" (Dannenbaum, 1984, p. 11).

Another factor in the early development of the Temperance Movement was the encouragement of anti-alcohol sentiment by religious men. The Quakers, as an organized religious group, were among the first to speak out against alcohol misuse. It was largely due to the Quaker influence on colonial culture that the Puritan definition of alcohol changed from "God's good creature" to "demon rum" (Keller, 1976, p. 16). In fact the first published objections to the widespread trend toward heavy drinking, and urgings toward total abstinence, were introduced to public thought in religious pamphlets written by the Quaker reformer Anthony Benezet.

In addition to the Quakers, frontier religious revivalists of the eighteenth century, and later ministers of the more conservative Protestant churches, spoke out against alcohol. The propensity for heavy drinking and drunkenness in that period led traveling preachers and urban ministers to preach sermons that portrayed the "defiant and unrepentant drinker" as a person of depravity and wickedness, in contrast to the pious and non-drinking churchgoer (Rorabaugh, 1979, p. 206). Abstinence became the symbol for conversion to God's way, salvation from sin, and the keystone to the virtuous life. Following a long period of religious quiescence and at a time when Americans were searching for a revitalization of purpose, many ministers and preachers, and the laity, embraced the anti-alcohol movement as a purposeful cause to follow (Rorabaugh, 1979). The stage was set for a moral reformation dedicated to abstinence, and religious groups, in rising to that cause, provided organizational techniques that eventually were incorporated into the nationwide temperance movement (Tyrrell, 1979).

The Temperance Movement was also influenced by the publications of Benjamin Rush, a physician and a highly regarded leader. A professor of medicine, a member of the Continental Congress, and one of the signers of the Declaration of Independence, Rush published a pamphlet in 1784 with the title "An Inquiry into the Effects of Ardent Spirits on the Mind and Body" (Glaser, 1976). In this and other writings, he conceptualized excessive drinking and drunkenness in terms of disease and alcohol addiction. However, the disease model of Rush and other physicians of that era (similarly to current concepts on alcoholism as a disease) was defined not in terms of disease pathology as we know it today; it was conceptualized as a "disease of the will" and an addiction brought on by a gradual breaking down of moral willpower. Treatment and total recovery could be accomplished by compassion, understanding, and total abstinence (Rush, 1785, cited in Glaser, 1976). Often recognized as the intellectual and political founder of the temperance movement, Rush voiced fundamental themes that were elaborated upon by temperance advocates throughout the nineteenth century. Levine (1984) summarized these themes as follows:

> Rush argued that distilled liquors were physically toxic, morally destructive, and addictive. Regular drinkers, he said, ran the serious risk of many diseases (he listed jaundice, epilepsy, madness, and diabetes, among others); they also tended to engage in many forms of antisocial, immoral, and criminal behavior. Further, he argued, regular drinkers become "addicted" to alcohol, and he described alcohol addiction in the full contemporary sense of the term: those addicted experience uncontrollable, overwhelming, and irresistible desires for drink—in other words, they experience what today is referred to as "loss of control." Finally, Rush called this condition a "disease" and recommended total abstinence as the only remedy for the addicted individual. (p. 110)

The writings of Rush, combined with the exhortations by men of religion, changed the perspective on heavy drinking from normal and natural behavior to the context of addiction, disease, and sin. It also set off the movement that eventually denounced not only intemperance, but also any use of alcoholic beverages. By 1830, more than one half a million people had signed pledges for temperance or total abstinence. The assumption that the desire for alcohol or the habit of intoxication was a disease that could be controlled only through moral willpower was gaining acceptance in the belief system of the established culture. From this point on, as the temperance movement gained momentum it was influenced not only by the complexities of an emerging, expanding, and proliferating nation, but also by concerns about health and morality.

Finally, there is historical evidence that the organization and promotion of the Temperance Era stemmed mainly from the established middle-class population. According to Gusfield (1963), the American middle class was alarmed by the growing number of immigrants and fearful of the increased political power of these "ignorant masses." They were searching for ways to inhibit cultural integration of the new arrivals from Europe, and the alcohol issue became a convenient means to that end (p. 5). By 1840, nativist Americans had promoted abstinence and a religion-oriented life-style to the degree that it became a status symbol; temperance became the touchstone of respectability, distinguishing the middle class from the lower working class, most of whom were Catholic Irish and German immigrants. Drinking was presented as a threat to the character, as well as the values of hard work, obedience, and of course sobriety. The ethnocentric attitudes on alcohol use, not to mention the indifference to the health-related aspects of heavy drinking, were implicit in such temperance slogans as "Keep the temperate people temperate; the drunkards will soon die, and the land will be free" (Maxwell, 1950, cited in Glaser, 1976).

In a historical analysis of this period, Gusfield (1963) noted that by the late nineteenth century, the temperance movement had evolved into a self-serving political mechanism by which American Protestantism and the mainstream middle-class culture sought to deprive the growing immigrant population of power with conflicting notions of values and ideology underlying the surface conflict over drinking behavior. The high point of this struggle was the passing of the Eighteenth Amendment to prohibit the sale of alcohol. From a social perspective, it could be said that the old Protestant, Victorian, middle-class culture was fighting for survival, and the alcohol issue evolved as a political tool to attain that end (Clark, 1976).

In the wake of socially charged causes and responses to patterns of increased drinking and to drinking behavior, the emphasis on health-related concerns of heavy alcohol consumption greatly diminished. Although the association of alcohol with disease and addiction, as originally outlined by Benjamin Rush and his colleagues, never disappeared altogether from American thought, by the beginning of the twentieth century the temperance movement entrepreneurs had lost interest in advancing the idea. Levine (1978) has noted:

> One aspect of the shift away from a focus on the addicting qualities of alcohol was the weakening, and in many cases, the loss of the movement's long-standing sympathetic attitude toward the habitual drunkard. The drunkard came to be viewed less and less as a victim, and more and more as simply a pest and menace. (p. 161)

From a health perspective, the ethnocentric thinking of the late-nineteenth and early-twentieth centuries profoundly influenced that era's eventual treatment of alcoholics as criminal, immoral, depraved, or insane. One serious, health-related consequence of Prohibition was that the existing private medical institutions that specialized in the treatment of alcoholism went out of business; medical professionals themselves no longer recognized alcoholism as a health problem; and physicians and hospitals stopped treating "identified" alcoholic patients (Keller, 1976, p. 20). In terms of help-seeking, alcoholism was viewed wholly as a problem for the judiciary and religious institutions.

Relative to beliefs and attitudes, the most important issue in this pre-1940 historical account is the transition of frequent and heavy drinking—or "chronic alcoholism" as it is known today—from a habituated to an addictive status (Levine, 1978), and from a natural, if unfortunate condition of one's life to a condition that suggests immoral or deviant behavior. The latter had the distinction of political, legislative sanctions in the passing of the Prohibition Amendment outlawing the sale and public use of alcoholic beverages.

THE BIRTH OF THE MEDICAL MODEL

Following the repeal of Prohibition, popular sentiment about alcohol use gradually changed from "dry" to "wet" values. In their study of alcohol images in American films, Herd and Room (1982) illustrate how movies and, to some degree, novels and magazines served as major carriers of wet values during this period. However, in the minds and hearts of the average person, and in terms of overall cultural conscious-ness, alcohol problems as recently as 1940 (and even later) were still viewed as private rather than public concerns. In the context of the lingering temperance ideology, neither the general populace nor the scientific community was greatly concerned with the alcoholic as an individual or with alcoholism as a health problem. Alcoholism as a con-cept—or research on alcohol-related problems and on whether alcohol-ism was or was not a disease—were simply not issues of scientific con-cern. There were a few studies in the 1930s and early 1940s, but these were limited to captive populations of alcoholic men in jails, in mental hospitals, and on Skid Row; alcohol problems were explained in terms of the characteristics of these groups. The general public found comfort in the thought that the core alcoholics were "out there" among the mental and social misfits. The average person who drank excessively denied the problem by disassociation from this established stereotype, and professional efforts to help alcoholics as late as the mid-1940s were al-most nonexistent (Straus, 1976).

The stage was set for yet another reconceptualization of problem drinking in American society. It was during this transitional period, again in response to an established need for change, that a small group of medical and behavioral scientists started the movement toward an updated medical model of alcoholism. This turn to a more definitive medical model of alcoholism, and what is known today as the "alco-holism movement," has been viewed as part of a calculated strategy to combat stigma and prejudice, to encourage alcoholics to seek help, and to change the negative attitudes toward alcoholics that prevailed among physicians and other helping professions (Straus, 1976; Cahalan, 1970).

The first major center for these new alcohol studies was founded at Yale (it was later moved to Rutgers). Under the leadership of E. M. Jellinek, a new concept of alcoholism as a progressive disease, with recognizable, symptomatic phases, emerged.[5] As described earlier, Jel-linek, the principal author of this model, differentiated between the addictive and the non-addictive drinker, the criterion being the predis-position to loss of control. As a paradigm for medical treatment, the disease concept is vague and it is difficult to pin down exactly what it means and implies; "Jellinek himself used 'alcoholism' in at least three

---

[5] Other significant leaders in the movement to explain alcohol-related problems in terms of other than deviant behavior (but not necessarily in terms of disease) were H. W. Haggard, Selden Bacon, and Mark Keller. Along with Jellinek, they founded the first major Center for Alcohol Studies at Yale University.

quite different meanings in different periods of his work" (Room, 1983, p. 54).[6]

Among alcohol researchers (and to some degree, the medical professions), there are differing opinions and ongoing debates about the assumptions and validity of the disease model (Albrecht, 1973; Cahalan, 1970; Room, 1970; Straus, 1976); it is by no means universally accepted.[7] However, in terms of treatment, and from a medical perspective, the implications of Jellinek's model were far reaching. The World Health Organization (1952), the American Medical Association (1961), and Alcoholics Anonymous use the disease concept as a working model. With respect to the alcohol movement's overall goal to effect changes in current notions on alcoholism, it could be said that the declaration of alcoholism as a disease accomplished its purpose (see Gitlow, 1973; Keller, 1977). Millions of dollars in federal and local funds are allocated each year for research on, treatment of, and education about alcoholism as an illness. Public drunkenness—in the absence of "public nuisance" or illegal behavior (such as drunk driving)—is in many states no longer a

---

[6] The following is an abbreviated version of Jellinek's 1960 model for alcoholism; it is the revised version of his original 1946 model. (1) *Alpha alcoholism* represents a purely psychological and continual dependence on the effect of alcohol to relieve bodily or emotional pain. The drinking, albeit "undisciplined," does not lead to loss of control, withdrawal symptoms, or interference wih personal life (dependence is *not* physical). This is a developmental stage that can remain static for years. There are no signs of progression. (2) *Beta alcoholism* constitutes complications, which may occur as "polyneuropathy, gastritis, and cirrhosis of the liver." The incentive to heavy drinking may be social custom, and health complications may come from poor nutritional habits. Transition to gamma or delta alcoholism is less likely in this phase than in the instance of alpha alcoholism. (3) *Gamma alcoholism* is a phase in which progression from psychological to physical dependence has occurred. This includes (a) acquired tissue tolerance to alcoholism; (b) adaptive cell metabolism; (c) withdrawal symptoms and "craving"; and (d) loss of control of intake. (This type of alcoholism predominates in the United States.) (4) *Delta alcoholism* has the first three characteristics of gamma alcoholism. In place of the fourth (loss of control), there is inability to abstain. Although the drinker can still control the amount of intake on any given occasion, he or she cannot go more than two days without withdrawal symptoms. (This kind of alcoholism predominates in France. It is sometimes referred to as an "endemic alcoholic condition." It differs form the gamma phase seen in American alcoholics in its social and psychological experiences and behavioral changes.) (5) *Epsilon alcoholism* refers to periodic drinking bouts. It is known in Europe and Latin America as "dipsomania" (Jellinek, 1960, pp. 33–41).

[7] The widely differing approaches to alcoholism (other than the disease model) are evidenced in the voluminous literature on etiological theories. There is evidence that a predisposition to alcoholism may be genetic, neurophysiological, developmental, physiological, psychological, sociocultural, or the consequence of long-term drinking. For an excellent review of the range of possible cause and effect factors, see Kissin (1974).

punishable offense, and most insurance companies now pay alcohol-related medical expenses.

But with respect to the prevailing beliefs and attitudes of the general public and successful treatment outcomes, the movement was a dismal failure. In both the alcohol literature and population surveys, the themes of stigma, denial, and reluctance to seek treatment persist. In his analysis of three general population surveys in various sections of the United States, taken over a 25-year span (Cumming & Cumming, 1957; Mulford & Miller, 1964, and Orcutt, 1976), Room (1983) noted that, although a growing number of Americans are accepting the disease concept or medical model of alcoholism, there is not a corresponding decline in belief in the "moral weakness" concept. In the abovementioned 1976 nationwide survey, roughly two thirds of the respondents (including those who were aware of the disease concept of alcoholism) did not "absolve the alcoholic from moral responsibility" for the problem (Room, 1983).

OVERLAPPING BELIEF SYSTEMS AND TREATMENT

The paradox of these two opposing belief systems—or models for explaining alcoholism—is that in practice they are not mutually exclusive; rather, they overlap. The "birth" of the medical model in the 1940s was not accompanied by an abrupt change in medical beliefs about alcoholism. The beliefs that prevailed prior to 1940, and the connotations of immorality formed in the temperance era may in fact be subsumed within the medical model (Straus, 1976). As early as 1945, the redefinition of alcoholism as a medical problem created a dilemma for physicians, as more alcoholics and families turned to them for help with a condition that in the past had received little serious medical regard (Haggard, 1945, p. 213). Surveys show that many health professionals and social workers not only lack adequate training for treatment of alcoholism—a situation that has persisted from the 1940s until recent years—but also hold a negative attitude toward alcoholics in general. Their complaints focus on such factors as the complexity of alcohol-related problems, too great a demand on intensive-care facilities, and the futility of treating addicts and "degenerates"—or those who have the potential to be such (Riley and Marden, 1945 and Middleton, 1971, as reviewed by Straus, 1976). Although professional associations have recommended changes in medical school curricula (American Medical Association, 1972), their practicing constituents seem to lag behind in actual efforts toward change, and many who are working within established health fields are disinclined to work with alcoholics. The pervasive fear is that those who become clearly identified as helpers of persons who are chronic alcoholics will be stigmatized because of their patients and their status among their colleagues will be diminished—thus the persistence of the moral model. Straus (1976) presented clear evidence that this "shadow of derived stigma" has also delayed much-needed interdisciplinary research efforts, deterred many young scientists from becoming interested in alcohol research, and prevented certain prestigious academic journals from publishing or even encouraging scholarly articles on alcohol-related subjects.

With respect to treatment, the problem is one of cultural incompatibility; the medical and moral models cannot be integrated, but as mentioned earlier, they do overlap, and the traditional moral view seems to have an edge over the newer medical view.

In the moral model or belief system about alcoholism, drunkenness is perceived as primarily immoral or irresponsible or derelict behavior (or all of those) and only secondarily as a health problem. A natural side effect of such "moral flaws" is loss of self-esteem and status, not only for the problem drinker, but in most cases for the drinker's family as well. Inasmuch as immoral behavior belongs in the realm of spiritual degeneracy, it logically follows that those affected by such beliefs assume that attempts to arrest the drinking problem—or the treatment process—should be of a spiritual (as promoted in AA) or religious nature. Those who are of the disposition to seek help from a spiritual source most often go to their priests, ministers, or church-related counselors. Either directly or indirectly, spiritually oriented healers utilize religious rather than medical models. Although in recent years some American religions have made efforts to educate their ministers in various established approaches to alcohol treatment, for most churches the treatment procedure would logically focus on abstinence from the instrument of evil (alcohol), possibly prayer or some form of penitence, and encouragement to resume expected roles in relation to family, church, job, and community. As the responsibility for the excessive drinking rests solely on the drinker, so does the responsibility for seeking and completing the healing–cleansing process (although encouragement to do so may come from significant others).

Alcoholics Anonymous, a highly successful self-help organization that utilizes the disease theory, is also in a sense a religious or spiritually oriented healing resource. Their "treatment" includes the acknowledgment of a higher power for purposes of gaining self-acceptance, strength, and serenity. In the course of group meetings, AA also encourages members to give testimonials on past drinking experiences—a practice that can be viewed as a form of cleansing oneself of past sins or irresponsible alcohol-related behavior.

The established medical model depicts alcoholism as a progressive disease process that, if not arrested, leads to death. In contrast to the moral view, the afflicted person is perceived as victimized rather than weak willed, and therefore to some degree relieved of the responsibility for excessive drinking and for becoming an alcoholic. In this view responsibility for treatment rests with the established medical system, and treatment can be provided by a variety of healers trained in the various physiological and psychological disciplines. Theoretically, if the treatment modality utilizes the medical model, diagnosis of the problem is in terms of the extent of the disease process or phases outlined in Jellinek's model (for example, does the patient have delirium tremens, dependence on alcohol, craving, or withdrawal symptoms). If followed to its logical end, treatment would focus on the only means of arresting the disease—total abstinence—and, when necessary, on the physiological and psychological effects of long-term drinking.

Historically, then, the establishment of a medical model of alcoholism has not overcome the nearly 200-year molding force of societal and traditional cultural beliefs among either the general public or health professionals. Both groups are still strongly influenced by the moral model, although the health professionals seem to fall behind as the public view of alcohol problems changes to one that is more realistic and accepting of humanistic and environmental considerations.

What is significant is not the voluminous medical and sociological debates on the subject, but rather the existential reality of people afflicted with a serious, but ambiguous health problem. In terms of alcohol treatment and prevention, perhaps we should consider how the knowledge of historical determinants of both the moral and the medical model can be of theoretical and practical use not only to research and treatment professionals, but to problem drinkers and their families as well.

ACKNOWLEDGMENTS

Robin Room, Jim Roberts, and Ron Roizen provided helpful comments.

REFERENCES

Aaron, P., & Musto, D. Temperance and prohibition in America: A historical overview. In M. Moore & D. Gerstein (Eds.), *Alcohol and public policy: Beyond the shadow of prohibition*. Washington, D. C.: National Academy Press, 1981.

Albrecht, G. L. The alcoholism process: A social learning viewpoint. In P. Bourne & R. Fox (Eds.), *Alcoholism: Progress, research, and treatment.* New York: Academic Press, 1973.

American Medical Association. In E. T. Thompson & A. C. Hayden (Eds.), *Standard nomenclature of diseases and operations,* 5th Ed. New York: Blakiston, 1961.

American Medical Association. *Manual on alcoholism.* Chicago, IL: AMA, 1968.

American Medical Association. Council on Mental Health and Committee on Alcoholism and Drug Dependence. Medical school education on abuse of alcohol and other psychoactive drugs. *Journal of the American Medical Association,* 1972, *219,* 1746–1749.

Armor, D. J., Polich, M., & Stambul, H. *Alcoholism and treatment.* New York: Wiley, 1978.

Bacon, S. Social settings conducive to alcoholism: A sociological approach to a medical problem. *Journal of the American Medical Association,* 1957, *164,* 177–81.

Barry, H., III. Psychological factors in alcoholism. In B. Kissin & H. Begleiter (Eds.), *The biology of alcoholism,* Vol. 3. New York: Plenum Press, 1974.

Cahalan, D. *Problem drinkers.* San Francisco: Jossey-Bass, Inc., 1970.

Clark, N. H. *Deliver us from evil: An interpretation of American prohibition.* New York: Norton, 1976.

Clark, W. Conceptions of alcoholism: Consequences for research. *Addictive Diseases: An International Journal,* 1975, *1,* 395–430.

Cumming, E., & Cumming, J. *Closed ranks: An experiment in mental health education.* Cambridge, MA: Harvard University Press, 1957.

Dannenbaum, J. The social history of alcohol. *The Drinking and Drug Practices Surveyor,* 1984, *19,* 7–11.

Gitlow, S. E. Alcoholism: A disease. In P. Bourne & R. Fox (Eds.), *Alcoholism: Progress, research, and treatment.* New York: Academic Press, 1973.

Glaser, F. B. Alcoholism in Pennsylvania—a bicentennial perspective. Reprinted from *Pennsylvania Medicine* (July), 1976.

Gusfield, J. R. Status conflicts and the changing ideologies of the American temperance movement. In D. J. Pittman & C. R. Snyder (Eds.), *Society, culture, and drinking patterns.* New York: Wiley, 1962.

Gusfield, J. R. *Symbolic crusade: Status politics and the American temperance movement.* Champaign, IL: University of Illinois Press, 1963.

Gusfield, J. R. Moral passage: The symbolic process in public designations of deviance. *Social Problems,* 1967, *15,* 175–188.

Haggard, H. W. The physician and the alcoholic. *Quarterly Journal of Studies on Alcohol,* 1945, *6,* 213–21.

Heath, D. B. The sociocultural model of alcohol use: Problems and prospects. *Journal of Operation Psychiatry,* 1978, *9,* 56–66.

Herd, D., & Room, R. Alcohol images in American film 1909–1960. *The Drinking and Drug Practices Surveyor,* 1982, *18,* 24–34.

Jellinek, E. M. Phases in the drinking history of alcoholics. *Quarterly Journal of Studies on Alcohol,* 1946, *7,* 1–88.

Jellinek, E. M. Phases of alcohol addiction. *Quarterly Journal of Studies on Alcohol,* 1952, *13,* 673–684.

Jellinek, E. M. *The disease concept of alcoholism.* New Haven, CT: College and University Press, 1960.

Keller, M. Problems with alcohol: An historical perspective. In W. J. Filsted, J. J. Rossi, & M. Keller (Eds.), *Alcohol and alcohol problems: New thinking and new directions.* Cambridge, MA: Ballinger Publishing, 1976.

Keller, M. A lexicon of disablements related to alcohol consumption. In G. Edwards,

M. M. Gross, M. Keller, J. Moser, & R. Room (Eds.), *Alcohol-related disabilities.* WHO Offset Publication No. 32, Geneva: World Health Organization, 1977.

Kissin, B. The pharmacodynamics and natural history of alcoholism. In B. Kissin & H. Begleiter (Eds.), *The biology of alcoholism, Vol. 3: Clinical Pathology.* New York: Plenum Press, 1974.

Lemert, E. M. *Social pathology: A systematic approach to the theory of sociopathic behavior.* New York: McGraw-Hill, 1951.

Lender, M. E., & Martin, J. K. *Drinking in America, a history.* New York: The Free Press, 1982.

Levine, H. The discovery of addiction: Changing conceptions of habitual drunkenness in America. *Journal of Studies on Alcohol,* 1978, *39,* 143–174.

Levine, H. The alcohol problem in America: From temperance to alcoholism. *British Journal of Addiction,* 1984, *79,* 109–119.

Mulford, H., & Miller, D. Measuring public acceptance of the alcoholic as a sick person. *Quarterly Journal of Studies on Alcohol,* 1964, *25,* 314–323.

Orcutt, J. D. Ideological variations in the structure of deviant types: A multivariate comparison of alcoholism and heroin addiction. *Social Forces,* 1976, *55,* 419–437.

Room, R. *Assumptions and implications of disease concepts of alcoholism.* Paper presented at the 29th International Congress on Alcoholism and Drug Dependence, Sydney, Australia, February, 1970.

Room, R. Sociological aspects of the disease concept of alcoholism. In R. Smart, F. Glaser, Y. Israel, H. Kalant, R. Popham, & W. Schmidt (Eds.), *Research advances in alcohol and drug problems,* Vol. 7. New York & London: Plenum Press, 1983.

Rorabaugh, W. J. *The alcoholic republic: An American tradition.* New York: Oxford University Press, 1979.

Straus, R. Problem drinking in the perspective of social change, 1940–1973. In W. J. Filstead, J. J. Rossi, & M. Keller (Eds.), *Alcohol and alcohol problems: New thinking and new directions.* Cambridge, MA: Ballinger Publishing, 1976.

Tyrrell, I. *Sobering up: From temperance to prohibition in antebellum America, 1800–1860.* Westport, CT: Greenwood Press, 1979.

World Health Organization. Expert Committee on Mental Health, Alcoholism Subcommittee. Second Report. World Health Organization Technical Service Report No. 18. In H. Milt (Ed.), *Basic handbook on alcoholism.* Fairhaven, NJ: Scientific Aids Publications, 1952.

Zwerling, I. Psychiatric findings in an interdisciplinary study of 46 alcoholic patients. *Quarterly Journal of Studies on Alcohol,* 1959, *20,* 543–554.

4

Getting on the Program
A Biocultural Analysis of
Alcoholics Anonymous

MIRIAM B. RODIN

Alcoholics Anonymous (AA) is a large, well-known self-help organization. A coherent set of beliefs forms the core of what AA members call "The Program." These beliefs offer alcoholics a path to recovery. My purpose here is to examine the convergence of two aspects of recovery from alcoholism. The first aspect is the process by which alcoholics begin to restructure their identities and reinterpret their experiences in terms of the AA ideology. The second aspect is the physiological event occurring during the early stages of sobriety, when most AA members first come to the organization. My approach will be to show how the AA meeting constitutes a dramatic performance that is tailored to the biological state of alcoholics in early recovery.

ALCOHOLICS ANONYMOUS AND THE RECONSTRUCTION OF SOCIAL IDENTITY

In seeking to understand the means by which many alcoholics have found help in AA, previous reseachers have been drawn to the similarity

MIRIAM B. RODIN • School of Public Health, University of Illinois Health Sciences Center, Chicago, Illinois 60680. This research was supported in part by an extramural research grant from the State of Illinois Department of Mental Health and Developmental Disabilities, Drs. D.B. Shimkin, M.B. Rodin, and J. Lowe co-principal investigators.

between AA and certain religious sects. Despite public disclaimers in AA publications concerning the role of religious belief, the overriding tone of the organization is religious and unmistakably fundamentalist and Protestant (Forrest, 1982). Several researchers have compared the social structural characteristics of AA and religious sects and cults (Adler & Hammet, 1973; Galanter, 1982; Madsen, 1979; Sadler, 1977). In the framework of viewing AA as an essentially religious organization, other researchers have compared the mechanisms of healing by conversion as they occur in AA with those in religious cults (Galanter, 1980; Favazza, 1982; Singer, 1982).

In a related vein, the roots of AA ideology are plainly set forth in the writings of Bill W., the founder (Alcoholics Anonymous, 1955, 1957). Even at the outset, Bill W. took care to separate AA ideology from conventional religion by defining the beliefs as a "spiritual program for recovery." The difference, however, is connotative. One of the earliest students of AA healing was a psychiatrist, Harry Tiebout (1943), who was intrigued by the convergence of Jungian insights into religious experience with those he observed in his alcoholic patients. An extensive analysis of AA ideology by Kurtz (1982) demonstrates the conscious incorporation of the thinking of Jung and William James on religious experience by Bill W. in his formulation of the spiritual program. Comparison of the structure of AA meetings with traditional rural American Methodist meetings further establishes AA's place in the religious cultural domain (Whitley, 1977). Yet none of these studies has attempted to explain the fit between the cultural forms of AA and its mechanism for evoking change in alcoholics specifically. We do not know why AA, for all its apparent similarities to other forms of American religious experience, seems uniquely fitted to the task of preventing drinkers from drinking.

METHODS

A variety of research methods have been applied to the task of discovering whether the AA program does in fact promote recovery, or the related task of discovering the characteristics of subgroups who might benefit most from the AA program. In regard to the latter, survey studies have yielded reasonably consistent sociological profiles of AA affiliators as compared with AA drop-outs (Leach & Norris, 1977; Norris, 1976; Leach, 1973; Ogborn & Glaser, 1981; Alford, 1980; Bebbington, 1976; Boscarino, 1980). The natural history of AA affiliation has also been shown to follow a reasonably consistent course as determined both by survey methods (Carlson, 1982; Trice, 1957, 1959; Trice & Roman, 1970) and by life history methods (Rudy, 1977). These methods however introduce a consistent recall bias in that recovery is reported retrospectively. Informants' accounts are irrevocably shaped by the context of data gathering. We have accounts only as they are told to non-alcoholics in an artificial setting. The accounts are separated in time and in space from the context in which AA healing occurs.

To avoid this bias, our study is based on the analysis of transcripts of tape-recorded AA meetings. Consent to record speakers was obtained at open AA meetings held in various hospitals and communities in Chicago. Discussion was not recorded at the request of the members. The particular groups were selected after several years of occasional attendance at open and closed meetings of six different groups that represent a spectrum of sociodemographic types including young adults, professionals, veterans, workingmen, and hospitalized patients. The tape recordings were collected over a six-month period in the spring and summer of 1981. All names are fictitious. Unfortunately, no women's talks were obtained in this study.

THE RESEARCH STUDY: THE AA PROGRAM

It is particularly important to analyze AA speakers' texts, since the texts and their public performance constitute the central ritual for AA recovery. This is stated quite explicitly in the founding documents, which include sample stories of how Bill W. and others discovered the AA program of recovery (1955). The texts of "Alcoholics Anonymous" and "The Twelve Steps and The Twelve Traditions" (Alcoholics Anonymous, 1957) constitute core statements of values, beliefs, and symbols that are the shared knowledge of the group. The alcoholics' stories contained in the texts are offered as comfort to other alcoholics and are used as models for speakers and for private counseling between recovering alcoholics and those they sponsor. We have no access to these private conversations, but the speaker meeting, in which one alcoholic tells his story to a larger audience of alcoholics, is the principal means of recruiting new members. The speaker meeting is thus the setting for an alcoholic's first induction into AA (Sorenson & Cutter, 1982). Although other aspects of the program may be explored by a non-alcoholic researcher, they are less accessible both ethically and practically.

Alcoholics Anonymous involves a variety of formal and informal activities designed to support alcoholics through the early stages of sobriety and to encourage their acquisition of new friendships, habits, and philosophical perspectives as recovery proceeds. Thus, AA offers both group and dyadic social relationships. Initially, a newcomer may attend an open speaker meeting. Subsequently, the recruit seeks a personal sponsor and home group. The group sponsors both speaker meetings and step meetings.

Meeting chairmen are selected either for a period of service, or for one meeting. The chairman convenes the meeting by welcoming everyone and introducing himself or herself. The chair then asks a member to read the serenity prayer, the central statement of AA belief from the Big Book and the Twelve Steps. At a speaker meeting, the chair then welcomes and introduces the speaker. Speakers can be recruited in a variety of ways and often are not members of the group they address. Speakers are usually but not always experienced members who have been sober for several months. Other speakers, particularly at large downtown meetings, are members of many years standing and have considerable public speaking ability. Speakers' texts may be from a few minutes to three-quarters of an hour long. Fifteen to twenty minutes is more usual. Following the speaker, the chairman may make a few personal remarks (called a comeback), call a coffee break, or go directly to audience response. Responses to a speaker vary, but are intended to give members a chance to tell how the speaker's story has taught them something new or reminded them of related aspects of their own recovery. The chair's opening statement, the first line a speaker says, and all responses begin with the formula "My name is [first name] and I am an alcoholic." To this, the assembled reply as a group "Hi, [first name]." Meetings are closed with a recitation of the Lord's Prayer, as all hold hands.

Step meetings are generally smaller, closed to non-alcoholics, and follow a different order. Whereas speaker meetings are listed in local AA directories for time and place, step meetings are generally closed. In a step meeting, the opening ritual for the chairman's introduction, reading the AA dogma, and Twelve Steps is similar, but the meeting focuses on one of the Twelve Steps. The chair tells what the particular step means to him or her, illustrating the point with personal experiences. The discussion proceeds around the table as each member introduces himself in the AA formula and discusses his interpretation of the step or of problems in mastering the step. As in a speaker meeting, personal advice and support occurs after meetings or during breaks, never in the context of a meeting in session. Step meetings also close with the Lord's Prayer.

Newcomers are recruited from speaker meetings. They are coached through the program by sponsors. The are encouraged to attend at least one meeting a day of either variety. They are encouraged to discuss the steps with their sponsors, to write essays about the steps, and to read daily from the Big Book. It is the sum of meetings, reading, and intense sharing with a sponsor that constitutes "working the program."

THE SPEAKER ROLE AND RECOVERY

The above short description of the components of the AA program points out that speakers at AA meetings were once the newcomers to whom their talks are addressed. The speaker role is a role assumed by most AA members in the course of their recovery. It is one part of the activity called *twelfth stepping*, encoded in the Twelfth Step of the AA program: "Having had a spiritual awakening as a result of these steps, we tried to carry this message to alcoholics, and to practice these principles in all our affairs." Being a speaker is the public extension of the fifth step: "We admitted to God, to ourselves, and to another human being, the exact nature of our wrongs." Twelfth stepping comes at a point in an alcoholic's AA career after the previous steps have been "worked." That is to say, the speaker has striven to apply the steps, if not actually mastered them.

Taylor (1977) interviewed experienced AA sponsors in order to derive a conceptual model of the stages of AA recovery. She identified six stages:

1. Hitting bottom;
2. Identification with AA (or at least one self-identified alcoholic in AA);
3. Conversion, which is equivalent to affiliation with AA;
4. The AA honeymoon, a period of elation and rising self-confidence following the first few months of sustained sobriety;
5. Life problem re-emergence, a dangerous time for recovering alcoholics when stresses begin to impinge on their newfound confidence; and
6. Perpetual recovery, a lifelong commitment to working the program.

The speaker role figures prominently in several of these stages. A novice's first AA experience typically occurs after having "hit bottom" when he or she attends a first meeting. Hitting bottom, as described by

AA members, is a time of despair, desperation, and loneliness. Generally after a period of intense drinking, a newcomer is physically and psychologically depleted. Sitting in a first meeting, he or she is often still intoxicated or ill with the early phases of withdrawal. In such a depleted state, the newcomer first hears another alcoholic's story, that of the speaker.

The second stage of recovery, "identification," is initiated when the newcomer hears a speaker's own account of hitting bottom and recognizes his or her current situation in that story. The moment of recognition is followed within minutes by the speaker's offer of hope through the AA program.

The third stage of recovery, "conversion," begins when the newcomer begins to actively practice components of the program with the aid of a sponsor who guides and instructs. Sponsors often give directive advice for getting through withdrawal and postponing drinking when the desire arises (Alibrandi, 1977). One component of conversion, according to Taylor's informants, is "getting your story straight." With the sponsor's assistance, the alcoholic practices telling his or her drinking history according to AA guidelines and is taught to reinterpret the past in the light of AA doctrine. This is no less than a rehearsal for assuming the role of the speaker at a later time. Public practice occurs during responses to speakers at speaker meetings and at step meetings.

One speaker I was unable to record was a thin, anxious man who forgot to introduce himself. Clearly a novice, he began by apologizing for his nervousness, saying he had only two months of sobriety, but that his sponsor felt he was ready to speak. He began his story, stopped, apologized again for "not having my story straight," and began again by introducing himself as "Larry, and I guess I'm an alcoholic or I wouldn't be here."

The speaker role is also an important part of conceptual stage six, "Perpetual Recovery." One of the obligations of twelfth stepping is to help others and by so doing to reinforce one's own recovery. Even after years of sobriety, alcoholics safeguard their recovery by reliving the past as sponsors and as speakers.

Especially during the first three stages of recovery, those with which I am principally concerned here, Taylor (1977) identified three strategies for staying sober: simplifying, avoiding, and substituting. In the first strategy, the alcoholic is instructed to consciously terminate introspection (or "stinking thinking" as it is called by AA sponsors). Sponsors teach their recruits cognitive tricks to silence their doubt (Alibrandi, 1977). They are told to replace intrusive thoughts with cliches: "I'm only one drink away from a drunk"; "One day at a time"; "Keep it simple stupid"; "Let go and let God"; "Shut up and listen"; "Look for similarities not differences"; "Easy does it"; "First things first"; "Stay sober for yourself"; "Have faith"; "Don't think." They are told to avoid all stresses and temptations. This includes avoiding non-AA friends and acquaintances, places where alcohol and drugs may be obtained, and any situation where they have to make choices and decisions. They are instructed to avoid hunger, thirst, or fatigue by eating, especially candy and sweets, drinking coffee, juice, soda, and soup. They should use "dime therapy," that is to call an AA member when the urge to drink occurs or to read the Big Book, eat, or sleep instead of drinking. Most importantly, they are urged to substitute daily meetings for daily drinking called "doing the 90 in 90," 90 meetings in 90 days. In meetings they will hear over and over again speakers' accounts of how they stayed sober.

SPEAKERS' TEXTS AS DATA

Thus far, three aspects of AA recovery have been summarized, including an outline of the program; stages of recovery; and strategies to be used by alcoholics during the early stages of their recovery. In this section, selections of speakers' text will be analyzed to demonstrate how, in the setting of a meeting, all these aspects converge in a single performance. Additionally, the texts will be analyzed to explain how the rhetorical style contributes to the purpose of fostering identification between the speaker as a representative of the program and the newcomer.

Speaker texts typically follow a standardized narrative outline: What it was like when I was drinking; When I hit bottom; How I got on the program; What my life is like now in AA. With experienced speakers, each narrative section is reasonably well marked. It is elaborated and brought to a conclusion before the next section is begun. Less experienced or less talented speakers take up the same four topics, but the order may be shifted. Alternatively, the four sections may be discussed in order, but several repetitions of the four-part structure may occur as the speaker shifts from past to present and back again. It would take reproduction of complete transcripts to illustrate this point and several dozen examples are easily available in "Alcoholics Anonymous" (1955). What is perhaps more interesting, and what is not transmitted from these edited versions, is the variation in rhetorical strategies employed by speakers in actual performance. As will be shown, the rhetorical strategy corresponds reasonably well to the structural part of the narrative in which it occurs. Examples are given from transcripts.

The first part of an AA story recounts "What it was like when I was drinking." In this recounting, the speaker provides a living example of working steps five and eight. Step Five: We admitted to God, to ourselves and to another human being, the exact nature of our wrongs. Step Eight: We made a list of all persons we had harmed and became willing to make amends to them all. Often, the speaker simply reports the things that he did while drinking, the extremes to which he went to get drunk. He demonstrates the AA belief that he was totally powerless to refrain from doing harm because of his powerlessness over alcohol. Yet despite the grim content, this portion of a talk is frequently hilarious, as the speaker makes himself the butt of stereotypic jokes about alcoholics. The dissonance between content and affect encourages the newcomer to both recognize himself in the events and to identify with the speaker as a likable, foolish, befuddled victim. I have called the strategy "Curiouser and Curiouser," for as the speaker gradually elaborates his account of how his own alcoholism progressed, he tends to show how he failed to see the inevitable progression to disaster.

The following passage is from a talk by Smilin' Jack, the (pseudonym) AA nickname of a middle-aged, black laborer.

> So I progressed very rapidly into alcoholic drinking. I lost control of what I was doing at a very early stage . . . the beer just never seemed to be enough. So naturally when I would go out I would have a few shots of gin or whatever spirit would hasten the feeling I was after.

Smilin' Jack's story of his alcoholic progression demonstrated 12 such transitions. Each transition, as with the one illustrated above, was marked by his comment on the "naturalness" of the change and the reasonableness of his thinking at the time. To Smilin' Jack, it was reasonable that he would drink himself unconscious evenings and weekends until his wife left him. It was natural that he would change to a night shift so his drinking could go on undisturbed by work. It was completely understandable that beer gave way to gin. For Smilin' Jack, hitting bottom came when he could no longer drink without getting sick before he got drunk. In his account, entering a treatment program seemed a "natural" way to convalesce sufficiently to resume his drinking.

In the second part of an AA talk, the speaker describes how he hit bottom. As contrasted with the reasonable life of drinking, and the good time he had before, the speaker describes his moments of terror and degradation. The rhetorical strategy invokes fear and loathing. If he had been a foolish and likable guy, despite the harm he had done to others, once he hits bottom he feels guilt, shame, fear, and despair.

Sean was a college student in his mid-twenties. As his alcoholism progressed, he lost friends due to drinking, he failed in school and he was too ashamed to face his family. Finally, with nowhere to go one night, he asked his brother if he could stay with him. Alone in the apartment, fear and loathing took hold.

> It happens fast. The last six months happened very very fast. I mean that's what blows me away. It happens fast. I really hated myself. It happened really very fast. I mean the self-loathing. . . . My way of crying out for help on my last drunk. I decided that what I'd do was I'd put one bullet in it (his brother's gun). Spin the chamber and put everything up to faith. But before I did that I called somebody.

Sean's account demonstrates severe disorientation. He had called his sister, but was unable to tell her where he was. She found him nonetheless and brought an AA friend along. The AA friend told Sean that by spinning the chamber he had taken his first step in turning his life over to God, but that by making a phone call instead of pulling the trigger he had chosen to live. That morning, the AA member took Sean to a meeting. Hitting bottom and starting on the road to recovery are closely juxtaposed in this narrative.

In the third part of an AA talk, the speaker describes how he first got to AA and what he thought about it at the time. As contrasted with the fear and loathing of hitting bottom, the rhetorical strategy again shifts to the strategy used to describe his drinking days. If the speaker has given a humorous account of his drinking days, he shifts to blatant self-satire of himself as an AA novice. The strategy of "dissonance" reports his bleak thoughts at the time, but the affective tone shifts to a new and upbeat mood.

Tom was a middle-aged factory worker. In his account, the AA slogans and cliches, the simplifications in other words, first appeared in his talk as he described his frame of mind as he waited for his first AA meeting to start.

> When I came to AA I was forty-six years old and I was running scared for forty-six years. I was about two and one-fourth years old emotionally, about twelve years old mentally, and age sixty physically. My brother had died of alcoholism ten months before. My father died of booze when he was thirty-seven years old in Cook County hospital. My brother died in the V.A. I drank with my brother toward the end. I know the four horsemen: Terror, Bewilderment, Frustration and Despair. I found the poor little 'ole me we all have. I was sick and tired of being sick and tired.

Smilin' Jack also shows the rhetorical strategy of dissonance in the account of his affiliation experience.

> I had been in the detox a couple of days. It was a locked ward. On my way back to the ward one of the AA men said, "Jack, what did you think about the AA meeting?" I said, "Boy, I listen to those people down there. They

really had me cold. Those people are really in bad shape." He said, "Now wait a minute. All of those AA's are going home. They just left. They just went out the door. You're going back to the ward to be locked up and you say you're all right and they've got troubles". . . . What I got out of that was the education that you are getting today. I found out what alcoholism was. I found out that there was something I could do about it.

In the last part of an AA talk, speakers tell about their third stage of recovery, conversion. Then they explain how the program works for them. In the last part, many speakers repetitively employ AA sayings, AA cliches, and AA wisdom. They introduce and emphasize the propositions that form the basis of AA conversion, once identification is achieved:

1. It works because it worked for me;
2. It worked for me because I wanted it;
3. If you want it, it will work for you.

The reasoning is simple by any standard. However, speakers have established their experiential credentials by this point. The speakers bolster the argument by admonishing listeners to simplify. They offer tools, cliches, and concrete behavioral recommendations to silence doubt. They offer examples of AAs who questioned and slipped. "I've known a lot of guys too smart for AA, but I've never met anyone too dumb." Interspersed with the admonishments are offers of hope. In general, the narrative style shifts, becoming telegraphic repetitions of "tools," to keep sober. For example, Tom's talk was quite short, under five minutes in length. In the concluding part, which ran just under two minutes, Tom offered "one day at a time" three times, and the related "every day is a new day." He paraphrases the Big Book, "I couldn't change anyone but myself." At that meeting, Al (pseudonym) the second speaker, presented a cogent guide to not drinking.

> You know it's that first drink, that first pill, that first snort. It is that first shot. It's the first one. Get it through your head. It's the first one. Not the second, the third, the fourth. It's the first one. That was a tough thing for me to understand. It took me a couple of years of sobriety to understand that simple slogan, One day at a time. One day at a time. I could make my life one day at a time.

Al's text contains four stanzas, each based on a particular AA tool. He develops stanzas on the themes of humility and gratitude, as well as, "the first drink" and "one day at a time."

We can compare these older, more traditional AA's with Sean. Sean began his talk by saying "I'm not big on drunkalogues and mine is not a particularly pretty one so I'll try to keep it short." Indeed, most of his

30-minute talk was devoted to describing his current feelings about AA and how it works for him. Throughout he used only two AA slogans, carefully bracketed as quotations from his sponsor. He offers them as tools, as gifts to his listeners. However, despite his more introspective personality, he also shifts to a telegraphic, repetitious style. In the first part of his text, the pressure of his words led to wandering and convoluted sentences, much like his life at the time. In the conclusions, his delivery becomes more declarative, falling easily into free verse.

> To forget about this business of living
> and start living
> and my living is dependent on my not drinking
> and on not getting high.
> And that is where it has to start from
> Because that's what always kept me from doing anything
> Kept me from feeling good about Sean
> carping at me to change.
>
> Today I have the choice not to,
> And along with choice comes the opportunity to live.
> To actually live today
> To know something about what it means to be alive.
> To know what it is to be up
> To be down.

At the same group a week later, Brian gave only a brief recounting of his drinking, moving quickly into the last part of the AA talk, choosing the theme of "going to meetings" as his topic. He introduced each part of his talk with a recent event in his life, concluding with a discussion of meetings as the essence of his perpetual recovery.

> I was just in Canada in my home town with all my friends that I came into AA with fourteen and half years ago. There's six of us. The dirty half dozen they called us. We actually don't talk when we see each other now. We just stand there and we kind of giggle. We kind of laugh a little bit and then we go "God can you believe this." You know it's still that magical for me. What I'm suggesting is that it is possible, probable for everyone here in this room tonight. There is no one that can avoid having what we have. That is if they're willing to go to any length. You see, actually the meeting is over when Gino read what the meeting is about (opening selection from the Big Book). This is sort of a little dessert, a little added entree. Gino has read what the meeting is about so actually you can all go home. I mean this is it. That is the program . . . . Now I know that there are intellectually sufficient people here tonight that are going to tell me that if you go to seventy-three meetings and you carry the message to three point five people and you read the Book one and a half hours a year that you're going to stay sober, and you're going to have a successful life. Well, I wish to tell you that is not the truth as I understand it. The truth as I understand it is that we continue and as far as I can tell, for the rest of our natural life to go to meetings.

Each of the excerpts illustrates the rhetorical strategies, the syntax, and the style selected by AA speakers. The texts represent evocative performances. The rigid structure of the interaction context directs even the novice to attend to the performances, which employ a variety of techniques (humor, dissonance, confession, exhortation, simplification, and repetition) to promote identification with the speaker. These strategies intentionally bypass the listener's critical intellect. Once the newcomer has identified with a speaker in passages about drinking and hitting bottom, the stage is set for affiliation. The speaker begins by saying, "I am just like you. I was confused and frightened. I thought AA was silly once too." But he concludes by asserting, "you can be like me, but only if you do not question."

For the AA newcomer, the first meeting is often a confusing experience. Listening to a speaker, expecting to hear about the evils of alcohol, as often as not, they find themselves reassured by humorous or even nostalgic accounts of drinking as a reasonable way of life. This is followed by poignant, empathic accounts of hitting bottom. Then the pace shifts again to reassuring humor. The rapid transitions may evoke attempts to reconcile the emotional content with current feelings of despair. Yet immediately they are exhorted to stop thinking. In other words, after having emotions elicited they are instructed to deny them—to simplify, to accept, and to surrender. This overall strategy contrasts sharply with the usual psychotherapeutic approach of encouraging active introspection and labeling of feelings. Although talking therapies differ considerably in the theoretical and technical aspects of inducing change, the common feature is expression of and insight into feelings and motivations. Alcoholics Anonymous flatly denies these modalities. Why this should be so is clearer when we consider the biological features of withdrawal as they affect cognitive processes.

BIOMEDICAL FEATURES OF EARLY RECOVERY

Ethanol is a central nervous system depressant, as such it is an anesthetic, albeit a poor one. Its effects are systemic and toxic. The first system to be affected, and the last to recover, is the reticular system of the brain stem. The reticular system is a complex, anatomically diffuse network of nerves that receive sensory input from all body systems and send messages both to higher brain centers and to descending motor and autonomic tracts. Ascending reticular connections serve the general function of controlling states of arousal and alertness. By filtering incoming sensory messages, the reticular system permits focused attention, which is necessary for higher cognitive functions. The reticular system also plays a role in suppressing REM sleep (dreaming) during wakefulness and permitting it in deep sleep states. Reticular nerves are also connected to hypothalamic centers concerned with hormonal and autonomic responses to stimuli. Again, these input messages are largely inhibitory.

After prolonged ethanol intoxication, the reticular system adapts to chemical suppression input by increasing the output of chemical messages. When alcohol is withheld, and its suppressing effects are lost, a two-phase rebound phenomenon occurs. Initially, there is a sudden functional increase in the availability of neurotransmitters, which lasts from 24 to 72 hours. This is followed by up to three weeks of profoundly disturbed attention, memory, perception, and fine motor deficits. Disorientation to time, person, and place are common, although hallucinations are rarer. The sudden release of inhibition of convulsion centers in the temporal lobe may result in seizures. Sympathetic autonomic discharges are also evident in heavy sweating, muscular tremor, racing heart beat, elevated body temperature, and increased blood pressure. The latter condition of somatic "fight or flight" arousal has cognitive analogs in subjective anxiety, restlessness, combativeness, and fear. In the initial phase, sleep is largely impossible. In late withdrawal, REM sleep rebounds, so that restful sleep is difficult, but vivid nightmares occur during both sleep and wakefulness. Recovery of normal sleep patterns and higher cognitive functions may take as long as two years, depending upon the amount of neural damage (Fox, Graham, & Gill, 1972; Tabakoff & Rothstein, 1983).

CONVERGENCE OF BIOLOGICAL AND CULTURAL FACTORS

The convergence between the psychological state of the newcomer and the structure of the speaker's text is both complex and remarkably subtle. During the first week of withdrawal, newcomers are highly suggestible, dependent, frightened, and disoriented. The striking absence of blaming messages and the positive tone of much of the text is reassuring. The empathy of the speaker for the newcomer's confusion fosters positive identification, while the illogic and the dissonance between content and affect of the texts play on confusion. Thus, when the text concludes with exhortations to simplify and surrender, the listeners are eager to comply.

During the second phase of withdrawal, the newcomer's psychological state changes. As autonomic discharge recedes, the newcomer feels more in control. Earlier exhaustion and dependency may give way to involuntary outbursts of anger and resentment ("stinking thinking"). In clinical settings, strict control is used both to reassure and to protect patients from these outbursts. In a community AA setting, the ritualization of meetings prevents the direct interaction of members. Appropriate verbalizations receive instant reward ("Hi Jack!"). Inappropriate behavior is frostily ignored. Persistent cognitive deficits are accommodated by an insistance on simplification and obedience to dogma in a structured and supportive environment. The use of cliches that are easily remembered and often repeated are appropriately scaled to a newcomer's actual cognitive ability.

SPIRITUALITY AND THE NEWCOMER

The religious aspects of AA ideology have received considerable attention. Pursuit of spiritual awakening however is felt to be beyond the capacities of newcomers and is more properly delayed until later stages of recovery when introspection is permitted to re-emerge. In fact it is important for newcomers to be protected from feeling religious pressure they may shy away from. Members are cautious to say "God as we understand him" (Step 3). Newcomers are encouraged to think of their "higher power" as the AA group itself or their sponsor or the Big Book or anything that does not invest authority in the established institutions many alcoholics feel have rejected them.

A major point of Western and particularly Protestant religious belief is that religious commitment is validly made only as a rational choice by a rational mind. In the AA view, which is compatible with the biological evidence, such a choice is clearly beyond the capacities of an alcoholic in withdrawal. Little if any time is given to discussion or acknowledgment of religious yearnings at speaker meetings. To the extent that it occurs it is often confined to remarks made in discussion. Speakers on the whole sidestep the issue entirely or clothe it in a kind of ambivalence.

In Tom's text, the only allusion to spirituality came in his closing statment,

> I can stay off the sauce one day at a time. The people in AA and the program and my higher power allow me to do this for which I am grateful. Thank you.

Brian was more to the point about the ambivalent role of religious conversion in AA recovery, demonstrating that success in AA is not conditional upon achieving a state of grace.

> You know I haven't had a great spiritual awakening where I have woken up at four in the morning and said, oh, you know. I had a friend in Halifax, Nova Scotia and apparently it was quite a trip. But I have never had that happen. Maybe He doesn't want to come to my house. I'm not sure He never visits but I would've liked to have more of a deeper understanding of spiritual things and right now I don't. My spiritual life consists of doing what I'm doing here tonight and going to meetings and trying not to be hard on myself.

Smilin' Jack makes no reference at all to the religious side of the AA belief, concluding as follows:

> Everybody I see that's sober helps me stay straight. I know that up here every meeting fucking up your drinking I'm glad. I'm here today because if I can't drink I want to fuck up your drinking. Cause if I ain't drinking there ain't nobody else going to have a party. That's all I got to say [laughter and applause].

Novices can be quite aware of the pervasive religious tone of an AA meeting. The opening and closing prayers, references to spiritual awakening, higher powers, and love are overtly religious. Yet speaker texts are often ambivalent. Naive questions that either reject or embrace the religious aspect of AA are quickly shut off with cliches. The newcomer is told, in essence, he is too "young" to understand such things. Only after several years, when he or she has mastered the idiom, and satisfied a sponsor concerning his or her awakening spirituality, can an AA openly discuss spirituality. This is normally confined to step meetings or private conversation. Interestingly, digressions into psychological interpretation of a novice's experiences are handled in a similar benignly dismissing manner.

CONCLUSIONS

Alcoholics Anonymous speaker texts have been shown to be highly structured cultural performances. Rather than being the extemporaneous monologues they appear to be, speakers' texts are the culmination of intense instruction and rehearsal. The extreme care with which speakers have been prepared is evidence for the importance of the performance for the purpose of evoking identification with and conversion of newcomers. Apart from the stereotypic organization of the narrative, several effective rhetorical strategies are involved that are crafted to the task of dealing with the impaired mental capacities of alcoholics in with-

drawal. The texts rely initially on the strategy of confusion. By juxta-
posing dismal narrative content with positive affect, as seen in the first
and third parts of a talk, the recruit is simultaneously puzzled and
reassured.

Empathy with his situation, most particularly his current hitting
bottom in the second part, establishes the speaker as an authority or at
least a credible source. For the newcomer, then, three-quarters of the
way through an AA talk he is confused and attentive, puzzled, reassured, and suggestible. At the close, specific ritual formulas in the form
of cliches, easy to remember given his impaired memory, are delivered
rapidly. He is promised hope provided he shuts off the confusing thoughts
evoked by earlier passages. He is told the central mystery of AA, which
is to simplify, and he is told how to do it.

It is important that we understand the importance of cognitive impairment in effecting AA convergence. A completely intact observer is
aware, as are the newcomers, that the messages are ambivalent and the
logic shaky. To an intact observer, the appeal is fairly transparent, but
unlike the impaired alcoholic, despondency and disorientation do not
intervene to support the apparent authority of the speaker. What an
intact observer is likely to dismiss as hocus-pocus sloganeering appears
profound and comforting to a newcomer. Sweating and trembling, his
confusion exacerbated by the rhetoric, he grabs for the cliches that stop
his mind from spinning. He is told that this is good, that thinking is
bad. He also knows with his rational mind that this is not beyond him,
for the speaker himself was once a sweating, shaking recruit.

The model of the successful man derives from deep cultural roots.
In mainstream American culture, Protestant forms of meeting are likely
to be familiar to most, regardless of subculture. Opening and closing
prayers, the testimony of the saved, individual choice, and individual
responsibility are a common cultural ground. The particular forms serve
to invest the speaker with considerable cultural credibility. We know
relatively little about how well the forms and the ideology transfer to
other cultural milieus. Among some American Indian groups, for example, whole families attend AA, not the alcoholic alone. The avoidance
of family stresses is encouraged by mainstream AA sponsors. In at least
one Indian setting, group advising for family problem resolution takes
place in the AA meeting (Jilek-Aill, 1981).

We know relatively little about how black and Hispanic cultural
milieus can adapt AA, although occasional references are made in the
AA literature to the difficulty of establishing mixed groups and home
groups in black and Hispanic communities.

This chapter has served the relatively limited purpose of examining how AA speaker texts work to promote sobriety for alcoholics in one cultural milieu. The folk forms of AA are shown to address presumably culture-free aspects of the biology of alcoholism. Before we can generalize these findings, however, much work is needed to clarify both AA and the traditional modes of healing in alcoholism in other cultures.

ACKNOWLEDGMENTS

The author wishes to thank Ms. Carolyne Keatinge for her intelligent, energetic, and cheerful assistance in the research and Dr. J.C. Salloway for inspired editorial suggestions.

REFERENCES

Adler, H. M., & Hammett, V. B. O. Crisis, conversion, and cult formation: An examination of a common psychosocial sequence. *American Journal of Psychiatry*, 1973, *130*, 861–864.

Alcoholics Anonymous. *Alcoholics Anonymous, The story of how many thousands of men and women have recovered from alcoholism*. New, Rev. Ed. New York: Alcoholics Anonymous World Services, 1955.

Alcoholics Anonymous. *Twelve steps and twelve traditions*. New York: Alcoholics Anonymous World Services, 1957.

Alibrandi, L. A. *The recovery process in Alcoholics Anonymous: The sponsor as folk therapist*. Dissertation, University of California, Irvine, 1977.

Alford, G. S. Alcoholics Anonymous: An empirical outcome study. *Addictive Behaviors*, 1980, *5*, 359–370.

Bebbington, P. E. The efficacy of Alcoholics Anonymous: The elusiveness of hard data. *British Journal of Psychiatry*, 1976, *128*, 572–580.

Boscarino, J. Factors related to "stable" and "unstable" affiliation with Alcoholics Anonymous. *International Journal of the Addictions*, 1980, *15*, 839–848.

Carlson, R. *The recovery process in Alcoholics Anonymous: A comparison of two trajectories*. Technical Report No. 10. Project on Community Dynamics, Social Competence and Alcoholism in Illinois. Illinois DMHDD Extramural Research Grant No. 8154-13. Department of Anthropology, University of Illinois, Urbana, 1982.

Favazza, A. R. Modern Christian healing of mental illness. *American Journal of Psychiatry*, 1982,, *139*, 728–735.

Forrest, J. *The ethnography of not drinking*. Paper presented at the American Anthropological Association Annual Meeting, Washington, D. C., December, 1982.

Fox, R. P., Graham, M. B., & Gill, M. J. The therapeutic revolving door. *Archives of General Psychiatry*, 1972, *26*, 179–182.

Galanter, M. Religious conversion: An experimental model for affecting alcoholic denial. In M. Galanter (Ed.), *Currents in alcoholism, Vol. VI. Treatment, Rehabilitation and Epidemiology*. New York: Grune & Stratton, 1980, 69–78.

Galanter, M. Charismatic religious sects and psychiatry: An overview. *American Journal of Psychiatry*, 1982, *139*, 1539–1548.

Jilek-Aill, L. Acculturation, alcoholism, and Indian-style Alcoholics Anonymous. *Journal of Studies on Alcohol*, 1981, (Suppl. 9), 143–158.

Kurtz, E. Why A.A. works. The intellectual significance of Alcoholics Anonymous. *Journal of Studies on Alcohol*, 1982, *43*, 38–80.

Leach, B. Does Alcoholics Anonymous really work? In P. Bourne & R. Fox (Eds.), *Alcoholism: Progress in research and treatment*. New York: Academic Press, 1973, 245–284.

Leach, B., & Norris, J. L. Factors in the development of Alcoholics Anonymous (A.A.). In B. Kissin & H. Begleiter (Eds.), *The Biology of alcoholism, Vol. 5. Treatment and rehabilitation of the chronic alcoholic*. New York: Plenum Press, 1977, 441–519.

Madsen, W. Alcoholics Anonymous as a crisis cult. In M. Marshall (Ed.), *Beliefs, behaviors and alcoholic beverages: A cross-cultural survey*. Ann Arbor: University of Michigan Press, 1979, 382–388.

Norris, J. L. Alcoholics Anonymous and other self-help groups. In R. E. Tarter & A. A. Sugarman (Eds.), *Alcoholism*. Cambridge, MA: Addison-Wesley, 1976, 735–776.

Ogborn, A. C., & Glaser, F. B. Characteristics of affiliates of Alcoholics Anonymous: A review of the literature. *Journal of Studies on Alcohol*, 1981, *42*, 661–675.

Rudy, D. R. *Becoming alcoholic: Accounts of Alcoholics Anonymous members*. Dissertation, Syracuse University, 1977.

Sadler, P. O. The "crisis cult" as a voluntary association: An interactional approach to Alcoholics Anonymous. *Human Organization*, 1977, *36*, 207–210.

Singer, M. Christian Science healing and alcoholism: An anthropological perspective. *Journal of Operational Psychiatry*, 1982, *13*, 2–12.

Sorenson, A. A., & Cutter, H. S. G. Mystical experience, drinking behavior and reasons for drinking. *Journal of Studies on Alcohol*, 1982, *43*, 588–592.

Taylor, M. C. *Alcoholics Anonymous: How it works. Recovery processes in a self-help group*. Dissertation. University of California, San Francisco, 1977.

Tabakoff, B., & Rothstein, J. D. The biology of tolerance and dependence. In B. Tabakoff, P. M. Sutker, & C. L. Randall (Eds.), *Medical and social aspects of alcohol abuse*. New York: Plenum Press, 1983, 187–220.

Tiebout, H. M. Therapeutic mechanisms of Alcoholics Anonymous. *American Journal of Psychiatry*, 1943, *100*, 468–473.

Trice, H. M. A study of the process of affiliation with Alcoholics Anonymous. *Quarterly Journal of Studies on Alcohol*, 1957, *18*, 39–54.

Trice, H. M. The affiliation motive and readiness to join Alcoholics Anonymous. *Quarterly Journal of Studies on Alcohol*, 1959, *20*, 313–320.

Trice, H. M., & Roman, P. M. Sociopsychological predictors of affiliation with Alcoholics Anonymous. A longitudinal study of "treatment success." *Social Psychiatry*, 1970, *5*, 51–59.

Whitley, O. R. Life with Alcoholics Anonymous: The Methodist class meeting as a paradigm. *Journal of Studies on Alcohol*, 1977, *38*, 831–848.

II

Case Studies
Americans of European Heritage

5

Acculturated Italian-American Drinking Behavior

BEN JAMES SIMBOLI

INTRODUCTION

Drinking behavior, like all other behaviors, is culturally learned. This premise has been given greater credence in recent decades as an increasing number of anthropologists and sociologists have focused on drinking customs and their association to problem drinking. Studies that look at the cultural differences in drinking behavior (Bales, 1944; Snyder, 1958; Lolli, Serianni, Golder, & Luzzatto-Fegiz, 1958; Jessor, Young, Young, & Tesi, 1970; Simboli, 1976; Blane, 1977) offer convincing support for the inclusion of cultural factors in a holistic theory of the causes of alcoholism. One of the apparent finds from the "cultural difference" studies is that some societies have been able to enjoy the benefits of alcohol and avoid its severe liabilities.

Italian-Americans are one of the American ethnic groups that have very few abstainers, yet do not have the alcohol-related problems of other ethnic groups. Early studies on Italian-American drinking habits (Lolli *et al.*, 1958; Williams & Straus, 1950) suggest a relationship between acculturation and changes in drinking practices. This connection sug-

BEN JAMES SIMBOLI • State of California Department of Health Services, Berkeley, California 94704.

gests that an increase in problem drinking in each succeeding generation was accompanied by significant changes in the traditional use of wine and beliefs related to wine drinking. Although these early studies on Italian-Americans looked closely at changes in drinking practices, they did not examine the relationships between the acculturation of Italians in the United States and changes in their drinking behavior and associated beliefs in depth, as was done in the study reported in this chapter.

One of the ways to study changes associated with acculturation[1] is to examine differences among ethnic generational groups, with the premise being that each Italian-American generational group shares a common set of values, beliefs, and attitudes regarding the use and abuse of alcohol. Membership in a generational group implies that a certain feeling of "belongingness" may exist, but obviously is not necessary because membership is by birth, parentage, and ancestry. Also, the basic cultural characteristics of the first generation have been carried for the most part unchanged to America, and these cultural traits have been "transplanted" into the new immigrants' life in the United States.

The pressure to change beliefs about drinking can be seen in the context of the general pressure to acculturate. With regard to examining the differences in generational membership, we can assume that the same sort of acculturation pressure was exerted on all generations, but was stronger for the second and strongest for the third. Since wine is the Italian national beverage, we would expect an elaborate belief system associated with its use and abuse; the acculturation of Italian-Americans, first through third generation, would include progressively less wine consumption and gradually less adherence to the associated belief system. Before examining the changes in drinking behavior for three generations of Italian-Americans, it is appropriate to look briefly at the history of Italian immigrants in the United States and the nature of the alcohol belief system transplanted to America.

ITALIAN IMMIGRANTS: "YESTERDAY AND TODAY"

Since 1830, when the government first began keeping immigration records, nearly 50 million immigrants have been admitted to the United States. The first census, in 1790, showed a population that was approximately 75% British in origin. Since then, there have been vast waves of non-British immigrants. Between 1901 and 1910, the nearly 8 million

---

[1] Acculturation is a process that theoretically begins when the first generation (or immigrant group) comes into first-hand continuous contact with the host society and starts changing its traditional behavior patterns to conform to the host society.

arrivals at Ellis Island were mainly from Hungary, Italy, and Russia. According to Foerster (1919) and Schemerhorn (1949), Italians represented the largest of all immigrant groups arriving during this great wave at the turn of the century.

The bulk of Italian immigrants, 80% of whom were males between the ages of 14 and 45, came from southern Italy and Sicily. Principal reasons given for Italian immigration were the ruin of the land, disease, hunger, and poverty. Amfitheatrof (1973) maintains that economic reasons were of course important, but that the main reason was "powerlessness." To the early Italian immigrants, the United States represented above all opportunities for education and work. It meant escape from an intolerable feudal system of social injustice in an age that was bright and promising elsewhere—and that "elsewhere" was America and what it stood for.

Many early immigrants, because of deep family and cultural ties to the "old country," came to the United States with the idea of improving the family's lot by earning enough money to return to Italy and buy land (Vecoli, 1974). After the Italian quota restrictions of the early 1920s, immigration from Italy was cut to a minimum, but through World War II and up to the Immigration Act of 1961 (HR2580), Italians have steadily emigrated to the United States. Italian immigration statistics from the Instituto Central (1973) indicate that the majority (85%) of immigrants to the United States are still coming from Southern Italy.

The past and recent Italian immigrants who represent the first generation come from basically the same cultural environment; however, recent immigrants are better educated, with better skills. In part this is due to the selection factors of the new immigration act requiring sponsorships, which in most cases is based on occupational skills.

ALCOHOL BELIEF SYSTEM TRANSPLANTED IN THE UNITED STATES

We cannot describe the alcohol belief system transplanted by the immigrant generation without first discussing the symbolic meaning and ritualistic use of wine in Italy. Since Greek and Roman times, the Mediterranean area has been a major region for the production and consumption of wine. Wine was then (and for the most part is now) the principal beverage. Through centuries of wine drinking, the beverage has become an integral part of southern European cultures, and as a result many beliefs concerning its qualities and use have evolved.

An important aspect that affects many beliefs associated with wine drinking is the relationship of wine to the holy sacrifice of the Mass. All Catholics, including Italian-American Catholics, believe or have been taught to believe that the wine presented by the priest during the Mass is miraculously transformed into the blood of Christ. This everlasting symbol representing the association between wine and blood permeates many beliefs linked to the use of wine in Italian culture. The belief among Italian-American immigrants that wine makes new blood still exists today.

Italian beliefs concerning the relationships between blood and wine have influenced the close association between the use of wine and an individual's health. The abundant use of wine in Italian folk medicine is one manifestation of this. Silone (1962) describes Cassarola, an Italian folk healer, and her wine prescription. The customer asks for assistance regarding her three-year-old daughter who is suffering from some illness. Cassarola prescribes a glass of wine for the little girl every morning. The girl's mother questions the prescription by saying, "She's only three years old." Cassarola responds confidently, "She already three? Well, then you can give her a glass of wine at night too" (p. 106).

Reinforcement of the belief that wine promotes a healthy body comes from the fact that most of the water in highly populated areas of southern Italy is unpalatable and unsafe because of its high mineral and bacteria content, respectively. Williams (1938) referred to southern Italy's water pollution problem by stating "In some parts of Calabria, the washing of a newborn in wine became another of the adaptations to this everpresent need" (p. 4). Williams goes on to comment that the limited use of water by southern Italians is reinforced by the many proverbs that tell how the human stomach and intestines can be "rusted" by water.

The following excerpts, sampled from Italian survey/interviews[2] done on use of wine as a national beverage, suggest that there is an elaborate health belief system closely associated with food, nourishment, good digestion, strength for hard work, good blood, and of course general well-being and good health.

> A good glass of wine makes good blood, therefore it is better to pay the bartender than the pharmacist (Farmer, from Varese).

> To wine I give the highest importance for living, especially when I think that for my age alone—from it I can have major nutriment (retired administrator from Rome).

Wine has been thought to have medicinal qualities since early Greek

---

[2] The National Surveys were conducted by "DOXA" Instituto per Le Ricerche Statistiche e L'Analisi Dell 'Opinions Publica, Milano, Perpaolo Luzzatto Fegiz, Director (1959).

and Roman times (Lucia, 1954, 1963) and this historical influence may have left an indelible mark on Italian wine drinking.

In terms of beliefs about its usefulness for health purposes, the quality of wines more often relate to clarity, color, and bouquet, with alcohol content being least important. This is suggested by the low alcohol content of many Italian wines, especially those that are made at home. Beliefs concerning the amounts of wine that can be consumed daily, in moderation, without harm are also related to health. Wine is mainly consumed on a daily basis, usually around meal times. Another health aspect is related to a firm belief that the whole family can use wine without any harmful effects. Further evidence of cultural integration of wine use is the fact that children drink wine with their parents in public places.

Home winemaking by many Italian-Americans also indicates wine's importance as a cultural focus. The art of home winemaking involves deep-rooted beliefs that link land and family social organization through viticulture. It takes years of patient care to raise vines so that they can produce a quality grape for the production of wine. It also takes the combined effort of family members and friends to harvest the grapes and make the wine. Great care is taken to ensure that wine does not spoil while it is aging. Proper aging ensures a "healthy" wine, which also means that the body consuming it will be healthy.

PROBLEM DRINKING IN ITALY

Not many studies were done on Italian problem drinking prior to the 1960s. However, American anthropologists and sociologists who were concerned with the social changes undergone by Italian immigrants in America did occasionally comment on Italian problem drinking in Italy. Lolli et al. (1958), in comparing their Rome sample with one from Boston, note: "Although the Italians regularly consumed alcohol in larger quantities, the occurrence of intoxication among this group was relatively uncommon" (p. 100). Other investigators studying Italian culture also noted this phenomenon. Moss and Cappannari (1960) noted that sobriety was more commonplace than drunkenness, even in light of the large amounts of wine being consumed in Cortina d'Aglio (a southern Italian village in the Abruzzi–Molise highlands). Based on their ethnographic work they noted: "Large amounts of poor quality wine are used to wash down the food; though the wine is low in alcohol content (10%–11%), copious amounts are taken. Drunkenness is rare and alcoholism is virtually unknown" (p. 96). Another account of Italian sobriety is found in Banfield (1958). In the findings of his ethnographic work, which was done in a southern Italian village, he comments on the crime statistics of the locale, indicating that out of a total of 553 arrests in one year, only 3% were for drunkenness.

Drunkenness, however, is only one indication of problem drinking and alcoholism. Further down the road, if the problem develops fully, it can lead to alcoholic psychosis and cirrhosis of the liver. In the period 1968–1970, the rate of alcoholics per 100,000 population for Italy compared to the United States was 600 compared to + 4,200, or one-seventh the rate (Efron & Keller, 1972). Of course, a high cirrhosis of the liver rate is another indicator of alcoholism and, according to Whitehead and Harvey (1974), the Italian cirrhosis rate is one of the highest in Europe; however they note that with regard to social problems connected with drinking, Italians have one of the lowest alcoholism rates.

Jellinek (1962) discusses the unique relationship between drunkenness and alcoholism as it pertains to Italian drinking characteristics. In essence, he suggests that because wine is so highly valued, the disassociation between drunkenness and alcoholism is common in the Italian perception of problem drinking; therefore "a drinker is classed as an alcoholic largely when he develops one of the alcoholic mental disorders. In the absence of such a disorder, the excessive drinker is just a drunk."

PROBLEM DRINKING IN THE UNITED STATES

Lolli et al. (1958) performed one of few studies done on Italian-Americans that included measures of problem drinking along with drinking practices. They used drunkenness as the only measure of problem drinking and found that the proportion of subjects who never experienced intoxication was substantially higher among the Italians in Italy than among Italian-Americans. These data suggest that the increased intoxication among the Italian-Americans may be due to a change from ancestral drinking customs (mainly wine drinking with meals) to a more American form of beer and hard liquor drinking (not necessarily with meals). Jessor et al. (1970) also found that Italians in Italy reported a markedly less frequent occurrence of drunkenness than the Italian-Americans they sampled. They state: "Fifty-two percent of the Rome drinkers and 58% of the Palermo drinkers reported no occasions of being 'drunk or pretty high' in the last year; only 24% of the Boston sample could say the same thing" (p. 218). This once again suggests that there is less problem drinking among Italians in Italy than among Italians in the United States. Greeley and McCreedy (1975), looking at the drinking behavior of college students, found that Italian-Americans, compared to Anglo-Americans, drank significantly less hard liquor and also had significantly less drunkenness. This suggests that even though acculturated drinking practices are well underway among Italian-Americans, they still have not reached the level of the host society. Blane's (1977) study of an East Coast Italian-American community found that Italian-Americans still are not fully Americanized with respect to drinking practices. In some ways, this comes as no surprise since his respondents all lived within a tight ethnic/geographical enclave.

The few studies that have been presented on Italian-American drinking behavior lend support to my hypothesis: Italian-American drinking behavior is directly related to generational membership, and the relationship is mainly associated with the process of acculturation. From the first generation through the succeeding ones, rapid change in drinking practices is suggested, with a decline in traditional patterns of wine drinking and an increase in the consumption of other alcoholic beverages that are more in line with the host society. This change is also suggested in the beliefs, attitudes, style, and context in which alcoholic beverages are consumed.

A STUDY OF ITALIAN-AMERICANS IN SAN FRANCISCO

To test the hypothesis of acculturated Italian-American drinking behavior, I conducted a study in San Francisco. The Italian community of San Francisco was selected for study because it is one of the most historically identifiable Italian communities in the United States. The area known as North Beach (originally bounded by the Embarcadero, Broadway, and Hyde Street) had its ethnographic peak in 1920, when an estimated 30,000 Italians lived there. Today, its geographical boundaries are not as definite as they were in the 1920s. There remain only clusters of Italian shops, bars, social clubs, delicatessens, restaurants, and business organizations. In addition to the Italian business section roughly scattered about Columbus Avenue, Green Street, Stockton Street, and Union Street, small neighborhood clusters still exist on Greenwich, Filbert, Lombard, and Chestnut streets.

According to recent population estimates from the Italian Consul General's office in San Francisco, there are approximately 100,000 persons of Italian parentage living in San Francisco, with some one million living in the entire Bay Area. Since the new immigration act of 1961, the average number of Italians entering the United States each year is 30,000. The consulate estimates that 10% of these (3,000 Italians) come to the San Francisco Bay Area. Approximately 66% of the Italian immigrants coming to the West Coast are from southern Italy. The new immigration laws do not require the immigrants to report their place of residence, so it is difficult to estimate how many new immigrants are making their homes in the North Beach area.

SAMPLE AND METHOD

Within the acculturation/generational change framework, a subsample of Italian-Americans was derived from available data collected from national surveys conducted by Cahalan and Cisin (1964–1967) and from San Francisco Bay Area surveys conducted by Cahalan, Knupfer, Clark, and Room in the same approximate time period.[3] I obtained the data from 6,159 questionnaire/interviews, which represent a combined random sample from three national and three San Francisco community surveys. In the combined sample, there were 252 Italian-Americans (54 in the first generation, 94 in the second generation, and 104 in the third generation). A comparison group of 502 third-generation British-American Protestants were also obtained from the combined sample.[4]

My analysis of the survey data showed that 75% of the drinking behavior measures conformed to acculturated drinking patterns and that in the majority of measures, the differences in drinking behavior between the Italian-American generational groups in comparison to British-American groups were maintained when controlling for pertinent demographic variables (i.e., age, sex, social position, and marital status). This supports my hypothesis that drinking behavior among Italian-Americans is directly related to their generational membership group and is not just an artifact of age or social position. This suggests that as they become more acculturated into American society, their drinking practices change significantly, with a concomitant increase in problem drinking.

In addition to using the interview and questionnaire data from the National and San Francisco community surveys discussed above, in 1975 I conducted key-informant interviews and made participant observation studies. Key informants came from first through third generations (25 were interviewed). A strategy I devised ensured that participant observation would give maximum exposure to situations or incidents that were directly involved or related to Italian-American drinking behavior

[3] For details on the methodology connected with how the sample was derived and how the data were analyzed, the reader is referred to Simboli (1976).

[4] Operationally defined for this study, generational membership means the following: First generation includes all male respondents who stated they were born in Italy; second generation, those who stated they were born in the United States and their fathers were born in Italy; third generation, those who stated they were born in the United States, their fathers were not born in Italy, but that most of their ancestors came from Italy.

or beliefs toward drinking. Such events among the Italian-Americans are most likely to be social gatherings or family affairs (e.g., weddings, baptisms, and picnics, and family dining at home, local restaurants, and bars). My previous knowledge of the Italian-American community and my ability to speak Italian speeded my acceptance as a researcher studying Italian-American drinking behavior. Most of the informant interviews were done formally, using the same or similar type questions given in the Cahalan *et al.* (1964–1967) community surveys.

In recording the data gathered during participant observation, I took notes that were as inclusive as possible, recording verbal communications that were elicited or overheard, and noting persons present, the time and place of the incident, and a general description of the context in which the incident occurred. My analysis on participant observation and informant interviews was performed in a fashion parallel to the analytical design used for the national survey data; that is, informants were placed in the proper generational group and drinking behavior incidents were coded. Drinking-behavior incidents were defined in terms of both physical and verbal expressions of drinking attitudes and beliefs related to the drinking of alcohol by an individual or group.

DRINKING BEHAVIOR AND BELIEFS

Findings from the San Francisco study show that Italian-Americans are anything but "dry." I encountered only one informant who abstained from the use of alcoholic beverage. She was an elderly lady (first-generation Italian) who claimed she never drank in her life, even though her Italian husband was an avid winemaker. All informants when questioned on this matter said that most Italians they knew (first, second, or third generation) had consumed or were consuming either beer, wine, liquor, or some combination of the three.

Most of the first-generation informants still consume wine on a daily basis, especially around the principal meal of the day, and still strongly believe that wine sustains their health when drunk in moderation. Some of the first-generation informants said that they used to make their own wine at home. Home winemaking was part of their family tradition brought from the "old country." In some cases the winemaking tradition was passed on to the sons of the immigrant generation, but for the most part the tradition died with the first generation. One informant, a second-generation commercial winemaker who had inherited his winery from his father, said that Prohibition was the main reason for the significant decline in second-generation home winemaking. He stated that even though the Volstead Act ("Prohibition") allowed each family to produce 200 gallons of wine, there was a stigma attached to making it. Part of the stigma was the identification of Italian-Americans as "wino," "grape-crusher," or "Dago-red." These ethnic slurs contributed to the decreasing popularity of winemaking. Also with the motivation to produce brandy illegally, the second generation saw winemaking as a secondary process to producing distilled spirits for the black market.

In most instances, second-and third-generation informants remember either their parents or grandparents drinking wine regularly at home. Some of the informants in the second generation drank wine at home when they were younger, but when they left home and started their own family they changed to beer and hard liquor. None of the third-generation informants mentioned that they regularly had wine in the home while growing up. All said they preferred beer or hard liquor to wine.

Bar-going is a highly structured social activity among Italian-Americans, especially those of the first generation. According to one key informant (first generation), serious drinking is done in the *cantina* (wineshop). The cantina is almost always patronized by men, who play various card games. One particular card game called "Padrone e Sotto Padrone" ("Boss and Under-Boss") can involve playing for large amounts of wine. For example, the winner becomes the *padrone* (boss) and the person who has the next highest score becomes the *sotto padrone* (second in command). The winner has his choice (with the permission of the sotto padrone) of who gets the drinks—if he chooses not to drink himself. However, problems can arise if the sotto padrone does not agree with the padrone in his choice; if they cannot agree the padrone must take the drinks himself. On occasion, if the game is long, the padrone may drink as much as three liters of wine.

There are many bars in the North Beach community, but only three or four could be classified as truly Italian—that is, run by Italians for Italians. The only one that came close to resembling a cantina was the Portofino, which had three card tables next to the bar. During time spent in the Portofino, especially on weekends, I saw the card tables in constant use by Italians (mostly first generation). Most of the drinks ordered during play were either wine or beer.

During the many times spent drinking at Italian bars in North Beach, I never saw anyone drunk or disorderly that I knew to be (or was identified as) Italian (first, second, or third generation). This was also true at Italian parties and social gatherings where alcohol was served (i.e., weddings, baptisms, parties, and picnics). In many of these events, alcoholic beverages were served whether children were present or not. In most cases when first- and second-generation families attended, younger children were present when alcohol was served, suggesting that the use of alcohol is still well integrated into the family structure.

I have already reviewed the health beliefs associated with the moderate use of wine and how these beliefs are manifested in the medicinal and normal use of wine by children. Addressing this aspect, a first-generation informant said that the belief that wine can safely be consumed by children is part of the socialization process whereby children experience alcohol with their parents in the family setting. A few of the second-generation informants said that they were introduced to wine and its health benefits at an early age through family home winemaking. In a few cases those in the second generation who drank later on in their teens, or who knew of others in the second generation who started drinking in high school, mentioned that their parents rarely drank wine at the dinner table. Third-generation informants mostly started drinking, mainly beer, in their late teens. They had no recollection of health beliefs associated with the use of wine by their grandparents.

Having a close relative with a serious drinking problem can disturb the equilibrium of family structure and affect family beliefs about drinking behavior. Only one informant (third generation) mentioned a close relative with a serious drinking problem, referring to her father as an alcoholic. Her family was very close, as are most Italian families, but after her younger sister died at an early age, her father starting drinking heavily and went on many binges. Although her father was perceived by her as an alcoholic, he has never had any troubles with the police, his job, or his friends. Another informant (first generation) volunteered that his father, a hard-working man, was a heavy drinker. The father drank wine daily and had liquor on festive occasions. According to the informant, "at times my father became violent when drunk and struck my mother." When probed further as to the extent of his father's drinking problem, the informant said that he did not consider that his father had even a mild drinking problem.

When informants were asked about their beliefs about drinking, that is whether drinking was either good or bad for one, healthy or not healthy, or how they or other members of their generation or other generations felt about drinking, most of the first generation gave indifferent or noncommital responses, such as the following response: "It's all right to drink and if you don't that's OK also." However, the informants of the second and third generation generally approved of drinking. Most of the first-generation and a few second-generation informants still expressed beliefs about good health and moderate wine drinking, but no third-generation informants drank wine regularly or expressed any health beliefs regarding its consumption. Regarding drunkenness, especially in the first and second generation, strong disapproval was expressed (e.g., "disgusting," "can't stand to see people intoxicated"). In the Italian-American subculture, drunkenness is usually perceived as an extreme degree of intoxication in which one is literally not responsible for his actions.

First-generation informant interviews revealed few instances of problem drinking for their generation. Relative to the degree of alcohol problems among close relatives, one first-generation informant mentioned that his father occasionally struck his mother when drunk, but he did not perceive this as a problem. Second-generation informant interviews also revealed few cases of problem drinking. Only one third-generation informant mentioned that she personally knew of several second- and third-generation Italian-Americans who had problems with alcohol; however, this perception may be relative as evidenced by the informant who did not view hitting a wife (his mother) as a problem.

In summary, community survey data, participant observation, and key informant interviews suggest that few Italian-Americans are abstainers. Alcohol consumption went from moderate drinking among first-generation Italians, to heavy drinking among second and third generations with some concern about occasional intoxication, especially at festive events or celebrations. Wine was the primary choice of the first generation and beer and hard liquor the primary choice of the second and third generations. Most of the first- and second-generation informants had relaxed attitudes toward drinking and believed wine to be good for one's health when drunk in moderation, whereas the third-generation informants generally did not. Informants in general did not perceive or recollect problem drinking in any generational group with the exception of one third-generation key informant who stated that many Italian-Americans in her generation had drinking problems.

It was found that not many of the first or second generation even perceived the third generation as being Italian in the ethnic sense, nor for the most part did the third generation participate in Italian community affairs. In addition, the survey questionnaire data collected in San Francisco sampled mostly third-generation Italians living outside the North Beach community. Thus, it is reasonable to assume that my own key informant interviews and participant observation among third-generation Italians concurs with the earlier survey findings of Cahalan et al. (1964–1967). As a matter of fact, these findings support the acculturation/problem drinking relationship mentioned earlier—drinking behavior among Italian-Americans is directly related to generational membership group and is not just an artifact of age or social position and with more acculturation into American society, drinking practices change significantly with a concomitant increase in problem drinking. I suggest that any third-generation Italian actively involved in Italian community affairs is less acculturated and therefore less prone to problem drinking.

CONCLUDING REMARKS AND IMPLICATIONS

Some of the questions raised by these findings run counter to alcohol theory and research findings. In any case, the conventional simplistic explanation of a direct relationship between alcohol consumption and problem drinking still demands more research especially in the area of ethnic differences. One of the main findings of this study is that Italian-American health beliefs concerning wine are an important aspect of their drinking behavior and that any future effort investigating Italian-Americans should look closely at how health beliefs govern their drinking behavior.

This study supports the evidence of the importance of the relationship between cultural patterning and drinking behavior; it points to the possibility of preventing drinking problems by changing the way drinking behavior is institutionalized and socialized in society. This is a tricky implication. One could easily conclude from the findings on Italian drinking behavior that moderate use of wine—beginning in childhood and integrated into the family circle—would protect against problem drinking. Obviously this approach is related to some sort of "inoculation" theory that assumes the smaller the percentage of abstainers in a population, the more likely the population is to be acquainted with alcohol effects on the nervous system and therefore the better able the population to deal with potential problem drinking situations. The fact is that the drinking behavior of non-acculturated Italian-Americans inherently involves a complex of cultural relationships, which includes personality, norms, beliefs, values, attitudes, and social structure. For example, it has been shown in studies on southern Italians in Italy and the United States that Italian family solidarity is a major cultural trait as compared with other family systems in the United States and many northern European countries (Banfield, 1958; Barzini, 1964; Covello, 1967; Gans, 1962; Campisi, 1948; Moss & Thompson, 1959). Most of these studies suggest that family honor and togetherness is highly valued in Italian culture. As suggested by Gambino (1974), the southern Italian family social structure (*l'ordine della famiglia*) was firmly transplanted in the United States. This structure involved social controls that dealt swiftly and unambiguously with any family member who displayed any behavior that would shame the family name. *L'Ordine della famiglia* also meant that any member of the family could be counted on for support in times of trouble. In sum, the Italian family structure may also be highly related to the low incidence of problem drinking among Italians.

With the many subcultures, institutions, groups, and other collectives in the United States, each with its own unique social structure, beliefs, and values, it is easy to see that alcohol "inoculation" may create a drinking problem when it otherwise would not occur. For example, the "inoculation" problem-drinking protection for Italian-Americans does not seem to work for the French. In France wine is also the national beverage and French children, like Italian children, are also introduced to wine at an early age—usually in the form of wine mixed with water (Anderson, 1968; Sadoun et al., 1965). Yet France is reputed to have the highest alcoholism rate in the world. The point is that until the interrelationships between cultural drinking norms and social structure are fully understood, it is dangerous to apply superficial drinking behavior models.

Another factor that may change the way drinking behavior is institutionalized and socialized in society is the time frame. If, as can be inferred from this study, it has taken approximately 75 years for Italian-Americans to assimilate the drinking behavior of the larger American culture, the theoretical implication is that it would take at least that long to accomplish the reverse, other things being equal. This may seem ridiculous (even in theory); however any social policy program aimed at preventing alcoholism by changing and/or institutionalizing drinking norms and values must be considered in long-range terms and will likely be costly just in terms of program length and needed resources.

The difficulties mentioned earlier lead to the common assumption that cultural factors related to drinking problems will be among the factors least apt to affect prevention programs that are aimed at changing social policy. Thus, the findings of ethnic differences in drinking behavior often lead to a stalemate when the implications for social policy are discussed (see, for example, Hiltner, 1967; and Stivers, 1971). This is not to say that we should give up trying to apply ethnic findings to changing social policy; data in the literature suggest that considerable changes in drinking behavior have happened in relatively short periods of time, following thoughtful changes in policy (Wilkinson, 1970; Room, 1971; Terris, 1967).

REFERENCES

Amfitheatrof, E. *Children of Columbus: An informal history of the Italians in the New World.* Boston: Little, Brown, 1973.

Anderson, B. G. How French children learn to drink. *Trans-Action*, 1968, 5, 20–22.

Bales, R. F. *The "fixation factor" in alcohol addiction*. Unpublished doctoral dissertation, Harvard University, 1944.

Banfield, E. C. *The moral basis of a backward society*. Chicago: Free Press, 1958.

Barzini, L. *The Italians*. London: Hamish Hamilton, 1964.

Blane, H. T. Acculturation and drinking in an Italian-American community. *Journal of Studies on Alcohol*, 1977, *38*, 1324–1346.

Campisi, P. J. Ethnic family patterns: The Italian family in the United States. *American Journal of Sociology*, 1948, *53*, 443–448.

Covello, L. *The social background of the Italo-American school child: A study of the southern Italian family mores and their effect on the school situation in Italy and America*. New York: Leiden and E. J. Brill, 1967.

Efron, V., & Keller, M. *Selected statistics on the consumption of alcohol . . . and on alcoholism*. New Brunswick, NJ: Rutgers Center of Alcohol Studies, 1972.

Foerster, R. F. *The Italian emigration of our times*. Cambridge, MA: Harvard University Press, 1919.

Gambino, R. *Blood of my blood: The dilemma of the Italian-Americans*. Garden City, NY: Doubleday, 1974.

Gans, H.J. *The urban villagers*. Glencoe, IL: The Free Press, 1962.

Greeley, A. M., & McCreedy, W. C. The transmission of cultural heritages: The case of the Irish and Italians. In N. Glaser & D. P. Moynihan (Eds.), *Ethnicity: Theory and experience*. Cambridge, MA: Harvard University Press, 1975.

Hiltner, S. Alcohol prevention and reality: Comment on the article by M. E. Chafetz. *Quarterly Journal of Studies of Alcohol*, 1967, *28*, 348–349.

Jellinek, E. M. Cultural difference in the meaning of alcoholism. In D. J. Pittman & C. R. Snyder (Eds.), *Society, culture, and drinking patterns*. New York: Wiley, 1962.

Jessor, R., Young, H. B., Young, E. B., & Tesi, G. Perceived opportunity, alienation and drinking behavior among Italian and American youth. *Journal of Personality and Social Psychology*, 1970, *15*, 215–222.

Lolli, G., Serianni, E., Golder, G. M., & Luzzatto-Fegiz, P. *Alcohol in Italian culture*. Glencoe, IL: The Free Press, 1958.

Lucia, S. P. *Wine as food and medicine*. New York: Blakiston, 1954.

Lucia, S. P. *A history of wine as therapy*. Philadelphia: Lippincott, 1963.

Moss, L. W., & Cappannari, S. C. Folklore and medicine in an Italian village. *Journal of American Folklore*, 1960, *73*, 95–102.

Moss, L., & Thompson, W. H. The south Italian family: Literature and observation. *Human Organization*, 1959, *18*, 35–41.

Room, R. *The effect of drinking laws on drinking behavior*. Paper presented at the annual meeting of the Society for the Study of Social Problems, Denver, Colorado, 1971.

Sadoun, R., Lolli, G., & Silverman, M. Drinking in French culture. *Monographs of the Rutgers Center of Alcohol Studies*, No. 5. New Brunswick, NJ: Rutgers Center of Alcohol Studies, 1965.

Schemerhorn, R. A. *These our people*. Boston: D. A. Heathland, 1949.

Silone, I. *Bread and wine*. New York: Atheneum, 1962.

Simboli, B. J. *Acculturated drinking practices and problem drinking among three generations of Italians in America*. Unpublished doctoral dissertation, University of California, Berkeley, 1976.

Snyder, C. R. *Alcohol and the Jews*, Glencoe, IL: The Free Press, 1958.

Stivers, R. *The bachelor group ethnic and Irish drinking*. Unpublished doctoral dissertation, Southern Illinois University, 1971.

Terris, M. Epidemiology of cirrhosis of the liver: National mortality data. *American Journal of Public Health*, 1967, 57, 2076–2088.

Vecoli, R. The Italian Americans. *The Center Magazine*, 1974, 7, 31–44.

Wilkinson, R. *The prevention of drinking problems: Alcohol control and culture influences*. New York: Oxford University Press, 1970.

Whitehead, P. C., & Harvey, C. Explaining alcoholism: An empirical test and reformation. *Journal of Health and Social Behavior*, 1974, 15, 57–65.

Williams, P. H. *South Italian folkways in Europe and America*. New York: Russell and Russell, 1938.

Williams, P. H., & Straus, R. Drinking patterns of Italians in New Haven: Utilization of the personal diary as a research technique. *Quarterly Journal in Studies of Alcohol*, 1950, 11, 586–629.

6

Polish-American Drinking
Continuity and Change

PAUL J. FREUND

INTRODUCTION

A growing body of research demonstrates that members of specific ethnic groups often have distinctly different patterns of alcohol use. Many of these studies have focused on and/or contrasted problem-drinking Irish with Jews, who have a low incidence of alcoholism (Ablon & Cunningham, 1981; Bales, 1962; Knupfer & Room, 1967; Strivers, 1976; Synder, 1958). Other ethnic groups studied include the Italians (Blane, 1977; Lolli, Serianni, Golder, & Luzzato-Fegiz, 1958; Simboli, 1976), Hispanics (Gordon, 1978; Paine, 1977; Trotter, 1982), Asians (Sue, Zana, & Ito, 1979; Wilson, McClearn & Johnson, 1978), and Armenians (Freund, 1980). However, with the exception of a brief mention in Thomas and Znaniecki's classic study of the Polish peasant (1918–1920) and a few scattered articles, largely concerned with other issues, (Taft, 1936; Zand, 1961; Finestone, 1967) there has been little research on Polish-American drinking. Even monographs on various Polish-American communities in Chi-

PAUL J. FREUND • Institute for African Studies, Community Health Research Unit, University of Zambia, Lusaka, Zambia. The research was conducted in 1979–80 while the author was a postdoctoral fellow in the NIAAA Program Social Science Research Training on Alcohol Grant No. T32 AA 07131, Dwight B. Heath, Program Director, Brown University, Providence, RI.

cago, Boston, Buffalo, and Detroit (Lopata, 1976; Morawska, 1977; Obidinski, 1968; Wrobel, 1979) contain at most an occasional reference to
drinking or the role of alcohol in the Polish community.

The lack of attention to Polish-American drinking behavior is due
partly to the context of drinking. Until recently, most drinking occurred
in social clubs, homes, or during ethnic events within the Polish community. Therefore, public awareness of drinking behavior or alcoholic
problems was limited. A large part of the popular image of Polish drinking behavior derived from second-hand information or jokes about Polish *picnics*, polka parties, and wedding receptions, which have the reputation of being heavy drinking affairs. As a consequence, Polish-
Americans became sensitive to the suggestion that they drank excessively or drank more than other ethnic groups.

At the time of my research, however, the issue of alcoholism and
alcohol-related problems was being addressed by Polish priests and community leaders.[1] In addition, the American Bicentennial celebration re-
vitalized the interest of Polish-Americans in their own ethnic heritage.
The combination of these factors enabled me to carry out my research
with little resistance. My interest in studying the interrelationship of
ethnicity and values sprang out of my earlier research among Armenian-
Americans, in whom there was a close association between a homogeneous, cultural value system and a low incidence of problem drinking.
A pilot survey I conducted in several Polish-American communities in
Rhode Island indicated that the study of Polish-American drinking and
values might also prove rewarding.

Although many social, demographic, and historical factors (e.g.,
acculturation, suburbanization, national prohibition, changing age structure) have influenced the current attitude of Rhode Island, Polish-American communities toward alcohol, there has also been a considerable
persistence of attitudes and behavior from eighteenth-century peasant
Poland, through immigration, up to the present. This chapter evaluates
the impact of these changes and focuses on attitudes and behaviors that
have continued within the Polish value system.

---

[1] There were no statistics available at the time of my research on the extent of alcoholism
or alcohol-related problems among Polish-Americans in Rhode Island. Subjective as-
sessments by alcohol counselors, social workers, and priests indicated that the alcohol
problem was increasing. However this increase may be only apparent due to the increasing awareness of alcoholism. Because my own research focused on sociocultural
and historical factors surrounding alcohol use I did not collect detailed epidemiological
data on alcoholism rates.

DATA COLLECTION

The methods used in the research, which was divided into four stages, combined historical ethnographic and sociological survey methods of investigation. The first stage was devoted to interviews conducted in each of the major Polish-American communities in Rhode Island, using an interview schedule of both direct and open-ended questions. The one-hundred item questionnaire was divided into four parts: (1) attitudes toward alcohol and frequency/occasion of use (e.g., frequency of alcohol use, day/week/month; occasion of use, bar, home, club; opinions regarding when and where alcohol should be used; beverage preference); (2) ethnic identification (e.g., membership in ethnic associations and Polish social clubs, attendance at ethnic events, opinions regarding Polish values); (3) socialization and learning in regard to alcohol (e.g., occasion and age of first alcohol use, recall of parents' attitude toward alcohol); (4) patterns of help-seeking behavior (e.g., where would one go for treatment of an alcohol problem, conviction regarding whether or not alcoholism is a disease). Informants were obtained through a procedure of "expanding social networks" whereby each person interviewed was asked for names of others who might be interested. Less formal interviews were also conducted with parish priests and with directors of the major Polish voluntary organizations.

The second stage of the research involved a historical investigation of Polish peasant customs and the early history of the Polish settlement in Rhode Island, derived from archival resources at the Rhode Island Historical Library.

The third stage combined participant observation and interviews of the directors of 16 Polish social clubs throughout the state. Each of the directors was asked a series of questions concerning the history of the club, current membership characteristics and attrition, benefits offered, hours of operation, and the nature of activities sponsored by the club.

The fourth stage utilized a mailed questionnaire to a sample of 100 persons randomly selected from referrals of previous contacts, lists provided by the Rhode Island Heritage Commission, and Polish church directories. The interview schedule was a shortened version of the instrument used in the first stage and similarly was designed to tap attitudes toward ethnicity and alcohol. Information was also obtained by observing various Polish-American social events throughout the research period.

THE SAMPLE

The formal sample consisted of 55 Polish-Americans interviewed during the first stage of research along with 47 persons who responded to the mailed survey (response rate 47%), for a total of 102 persons. The breakdown of the sample by sex, generation, and religion was as follows. Sex: male, 60; female, 42. Generation: first generation (foreign born of foreign parents), 12; second generation (American born of foreign parents), 46; third generation (American born, one or both parents American), 29; fourth generation (American born of American parents), 15. Religion: Roman Catholic, 79; Polish National Catholic Church,[2] 21; Methodist, 2.

THE AMERICAN ETHNIC SALOON

When the Polish immigrants arrived in the immigration wave of 1890–1916 and settled in the large industrial cities of the Northeast and Midwest,[3] they found at least one feature of American urban life that was reminiscent of the rural village. This feature, the ethnic saloon, was a thriving popular institution in cities throughout America. Zand (1961) notes that two or three saloons at every street intersection was not uncommon in Polish-American communities. In Rhode Island, the Polish-American saloons were concentrated near the factories and textile mills in the Central Falls area (12), Pawtucket (6), Warren (4), Woonsocket (3), Coventry (2), and sections of the city of Providence (8) that came to be known as "beer hell" among the clergy and by the local population.

It is easy to understand the popularity of the saloon when one considers their multiple functions. The saloon provided a place to rest

---

[2] The Polish National Catholic church, which is unique to the United States, has as its main distinguishing characteristics a lay board of control marriage and family rights of priests, the use of Polish during the mass, and rejection of the concept of papal infallibility.

[3] The vast majority of the Poles immigrating to Rhode Island came during the great waves of "economic immigration," 1890–1910. Any estimate of the actual number of Polish immigrants is problematic because of changing definitions of who was a Pole by the Census Bureau. Poland was partitioned three times in the eighteenth century by Russia, Prussia, and Austria and ceased to exist as an independent political entity from 1795 to 1918. Immigrants during this period were usually listed as from either Russia or Austria and sometimes grouped with Slavs on quotas. Polish immigrants to Rhode Island settled near the textile mills and factories in Central Falls, Pawtucket, Woonsocket, Warren, Providence, and Coventry. The 1976 population of the Rhode Island Polish-American community was estimated at 25,000 to 32,000 by the Rhode Island Heritage Commission.

and relax in relative luxury and warmth, after long boring hours of factory work. Moreover, the saloons offered a wide range of services and recreation generally unavailable elsewhere. This range included foreign newspapers, writing materials, toilet conveniences, communication with relatives and friends in the home country, a labor exchange, credit, pool and billiards, card games, and an occasional free lunch. The saloons of this period have been characterized as "poor men's clubs," designed primarily as sanctuaries for the male. Although Polish-American saloons had a "family entrance" it was rarely used. As Popham (1978) writes, a man "could relax in the company of other men, equally poor, similarly dressed, and just as glad to escape for a little time from their overcrowded homes" (p. 282).

Although the American ethnic saloon resembled the village tavern in some respects, it differed significantly in that it was not a center for community or family gatherings, nor did it play a significant part in the life of the community. In Poland the tavern (*Karczma*) was an integral part of village community affairs. It was here that contracts were sealed, village disputes settled, celebrations held, and marriages arranged. The village tavern was also a family gathering place and was not restricted to males.

There were, as Benet (1951) notes, a large number of holidays and festivals, all of which were acceptable and appropriate occasions for drinking. Seasonal changes and events were celebrated with vodka and whiskey. These included Spring (Green Holidays—*Zielone Swiatki*), the cutting of the first shocks, and harvest festivals. Religious holidays and various saint days were also observed with the use of alcohol. During *shrovetide* carnival (the final days before Lent) *Zapusty*, old men would capture single men who had been expected to marry during the preceding season and chain them to huge logs, which they were forced to pull. A man could buy his way out of this situation only by standing drinks at the village tavern. The feast of St. Martin was marked by offering the first wine of the season. On St. Stephen's day, the tavern served as a hiring center for the following season's farmhands. Agreements were sealed with vodka (*litkup*), which made them legally binding.

Life events, with the exception of funerals, were appropriate occasions for alcohol. This was particularly true for weddings, hence the folk expression: "Water for a Christening, vodka for a wedding, and tears for a death!" Each stage of the wedding, including the betrothal, the wedding ceremony, and the reception, was marked by alcohol. Christenings were also celebrated in the tavern and the child's godparent covered the cost of the liquor.

The specific attitudes of the Polish peasant toward alcohol are revealed in such literary sources as Reymont's (1925) series of novels about the Polish peasant. The fact that Polish peasants regarded alcohol as harmless and beneficient is evident in the following quotations:

> What's good for cattle may be bad for man, they say; what to drink is now and then not bad, but harm from vodka no one ever had. (*Autumn*, p. 251)

> To drink vodka is not a sin, if only at the proper season and with brothers; this is a wholesome thing, it cleanses the blood and drives away distempers. (*Autumn*, p. 227)

Drinking customs such as consuming large amounts at one sitting and proposing toasts (*Na Zdrowie*—for health) are expressed in these words:

> Who drinks at all, should drink one quart complete—likewise, who makes merry, should do it all Sunday long. But have you work to do? Man, do it with all your might, grudge not your force, but put forth all your strength. (*Autumn*, p. 227)

> It is a duty to drink to one's health. (*Autumn*, p. 10)

Although the American ethnic tavern reinforced some of these drinking customs and attitudes it never really incorporated features of the *Karczma* for the family and the community. There are a number of possible reasons for this. The urban saloon was first and foremost an American institution and secondarily an ethnic gathering place. In this new urban environment, the saloon became isolated from the community because immigrants drank with fellow workers and no longer socialized with family as they had in the village tavern. Further, as a consequence of adaptation to the urban environment, the ethnic saloon came to be regarded as a separate institution with values that differed from those of the ethnic community. Unlike social clubs, ethnic saloons were not established to preserve traditional customs or to perpetuate ethnic values. Therefore, they were situational and transitory, concerning only the immediate problems of adjustment for first-generation immigrants.

A number of studies from this period allude to the problem of alcoholism among Polish-Americans in Boston, Philadelphia, Cleveland, and Buffalo (Bushee, 1903; Parmelee, 1909; Taft, 1936), but none provide sufficient statistical data from which to draw meaningful conclusions. A further difficulty is that many of these studies grouped the Polish with Slavics, Austrians, Russians, and Lithuanians.

Thomas and Znaniecki's (1918–1920) classic study addressed the problem of alcoholism as a manifestation of the observed and predicted social disorganization of Polish immigrants freed from traditional social controls. They predicted that alcoholism would increase when people who were formerly dependent on the controls extant in a small, highly autocratic village escaped from these controls upon migration. Moreover, they argued that the reason why the immigrant wished to drink was that when drinking in company, social emotions were heightened, which substituted for the primary-group atmosphere of village life. The fact that the predicted moral disorganization of the Polish-American community never occurred is due to the existence of effective subsitutes for the traditional social controls (cf. Lopata, 1964, 1976). In addition, a historical process was gaining momentum that would have a significant impact on Polish-American attitudes, namely Prohibition.

PROHIBITION: POLISH-AMERICAN REACTIONS

The saloon came under increasing attack during the first two decades of the twentieth century by such highly organized temperance groups as the Anti-Saloon League and the Women's Christian Temperance Union (WCTU). As temperance sentiment gathered force, the saloon became the scapegoat for a wide variety of vices and social problems.

The temperance campaigns were ultimately successful and the Volstead Act, beginning a period of national prohibition, was signed into law in 1919. Prohibition was particularly resented by Polish-Americans because they saw the Amendment as an attack on their two most cherished cultural values, individualism and personal independence. Polish-language newspapers in Detroit and Chicago often ran editorials against Prohibition. Feelings were also strong among Polish-Americans in Rhode Island, as exemplified in the testimony of many first-generation men. They regarded prohibition as "unnatural" and felt that the Amendment was an attempt to make them second-class citizens. Moreover, they believed that the prohibitionists and the anti-saloon forces were really attempting to curtail personal liberty rather than to prohibit liquor. They resented having to conceal a glass of beer, vodka, or whiskey, which they felt was deserved after a hard day's work.

It is understandable why ethnic Americans viewed national prohibition differently than temperance proponents if we consider the values they espoused. Prohibitionists were predominantly Protestant, with nativistic and middle-class values, whereas the majority of ethnic groups were Catholic, urban, and working class (cf. Clark, 1976; Gusfield, 1963). Temperance groups in many urban areas engaged in campaigns to change the attitudes of the ethnic "drinking workman." For example, they tried to dissuade laborers of the belief that beer was necessary for the performance of manual labor. Seng (1973) notes that WCTU activists in

Chicago set out to change immigrant attitudes toward alcohol by point-ing out that the beer making and liquor distilling were products of a "wet" European environment and therefore un-American. The WCTU was also active in Rhode Island. However, they were unsuccessful in their attempt to begin a temperance education program in Polish-Amer-ican schools because of anti-Prohibition sentiment.

Although one of the consequences of Prohibition among other groups was increased drinking at home, Polish-American men, particularly of the first generation, continued to drink in social clubs. The social clubs, some of which were established before Prohibition, served as partial substitutes for the saloon. They continued to perform some of the same functions as the ethnic saloon, while also providing financial aid, in-surance benefits, and citizenship assistance to their members. Moreover, Polish social clubs differed from ethnic saloons in that they were orig-inally established to perpetuate cultural traditions.

THE ROLE OF THE SOCIAL CLUB

It has been argued that the major function of the social club is to protect the individual from the impersonal aspects of mass society. This was particularly true of the Polish-American social clubs I studied, which provided a comfortable, secure retreat within a familiar environment. Many of the patrons had known and worked with each other most of their lives. Patrons frequently reported that the reason they came to the club was that they felt safe and knew what to expect. In bars, however, "you never know who may come in and bother you." Patronized by "regulars" who were long-term residents of the community, each social club set its own norms as to the degree of inebriation or boisterousness tolerated. Old-timers were allowed a certain freedom that was not ac-ceptable in others. In the social club, control was exercised not so much by the bartender, but rather by the other patrons.

The drinking patterns that characterized first- and some second-generation men included standing rounds and the frequent drinking of toasts. Both these patterns were traditional Polish customs, which per-sisted in ethnic saloons and social clubs. Several club directors recalled how during the 1960s it was common for groups of six to ten men to come in after work, each buying the others a round of "boilermakers" (a glass of beer with a shot on the side) until each man had consumed as many as 10 shots of whiskey and 10 glasses of beer. Few Polish-American men continue this practice today. The majority of the men who engaged in this practice are now retired, and younger generations have either not joined social clubs or have different drinking patterns. The drinking of toasts, however, has persisted even among fourth-generation men.

The style of drinking by first- and second-generation men can be viewed as part of the status competition that Lopata (1976) describes as characteristic of Polish-American communities. One can gain status (ranking within the prestige hierarchy, as defined by the community) through education, job, material possessions, leadership in voluntary organizations, or behavior. For example, the practice of standing rounds was not only a way of establishing a reputation among one's peers, it served to build status within the ethnic community as well.

Social clubs have been particularly susceptible to alterations in the demographic structure of the Polish-American community. As a result of the suburbanization process and natural geographic dispersion, many clubs have lost as much as 30% of their membership. Recruitment continues to be a problem and the more varied clientele of neighborhood bars holds more attraction. Consequently, older members were over-represented in the membership of the social clubs, with several club directors estimating that over 60% of the members were age 65 or older. This age imbalance in some clubs was reflected in the socializing and drinking patterns. The old-timers, most of whom are now retired, came to the club in the late afternoon to have a few beers in a quiet atmosphere, to talk with old friends, and to play cards. Younger members, especially fourth generation Polish-Americans, came after work to relax, watch sports on television, play pool, and generally let off steam. There was evidence that some "old-timers" resented the younger members, which led to a reluctance to have them in the club. However, this rarely was expressed as hostility between the two groups. Many social clubs have actively solicited younger members by holding dances, parties, and other youth-oriented activities.

Nor were younger members joining the club for the reasons that had motivated previous generations to join. There was a decline in the listing of ethnic duty on the survey questionnaire ("because Polish-Americans should belong to their ethnic organizations") as one moves away from first-generation respondents. At the same time an increase in the number of social–recreational choices (e.g., "because of interesting social activities") were listed by the second to succeeding generations. This finding is similar to that reported by Morawska (1977), Lopata (1976), and Ostafin (1948). In spite of the frequent complaint by older members about a lack of interest in Polish ethnicity, fourth-generation respondents have demonstrated some renewed interest in Polish traditions and language. This is a promising trend for the continued existence of Polish-American social clubs.

Although social clubs will continue to adapt to changing social/demographic characteristics, they will undoubtedly maintain their important role in the social and cultural life of the Polish-American community.

AN ANALYSIS OF PRESENT ATTITUDES

The traditional attitude toward alcohol in peasant Poland was that it was a necessary part of socializing and a symbol of hospitality. Alcohol was a well-integrated feature of peasant life and was used for a wide variety of purposes, including sealing contracts, marking life events, and celebrating. At least some of these practices and attitudes have persisted among Polish-Americans. Thus, the offer of vodka as a mark of hospitality to guests, and the gift of a bottle of whiskey or vodka when visiting friends, was still practiced. Moreover, the old Polish custom of drinking alcohol after, rather than during a meal has continued among all generations.

There was, however, evidence of a generational shift in the attitude toward alcohol as a beneficient beverage. First-generation respondents tended to list more reasons in answer to the question "What are some good things that can be said about drinking?" than second- or third-generation respondents. This shift in attitude among later generations—the perception of alcohol in less positive terms—was evident in many of the answers to the questionnaire and during personal interviews as well.

The generations also differed in their recommendations as to where a person might seek assistance for an alcohol problem. The majority of first-generation respondents indicated close relatives or friends as a front-line resource for advice and help. Second and later generations all indicated an overwhelming preference for doctors or a self-help organization such as Alcoholics Anonymous (AA). This attitudinal shift away from family and friends toward more specialized resources was similar to the health-service utilization pattern of Polish-Americans in Baltimore described by Fandetti and Gelfand (1978).

Although respondents across all generations indicated an awareness and a general acceptance of the disease concept of alcohol, many qualified their answers with statements that moral weakness was the real, underlying cause of alcoholism.

Responses to questions concerning one's responsibility for actions while drunk and frequency of drunkenness were often phased in moral terms as well. Many noted that although people have the freedom to drink and to express their individuality, they were also responsible for their own behavior. A great deal of tolerance for heavy drinking and drunkenness was evident among Polish-Americans, as was true in peasant Poland, but limits were recognized, particularly when drinking caused the neglect of one's moral duty toward family or church.

Despite the tendency to see alcoholism as a moral issue, only a few indicated the priest as the first person to see for help. Many of the priests I interviewed noted that there was a reluctance for family members to seek help for suspected alcoholics because of the shame involved. Shame deterred many men from going to AA meetings in their own communities for fear that they would be seen by friends. Many preferred instead to travel considerable distances to attend meetings in other communities where they would be less likely to encounter Polish acquaintances. Shame, along with an attitude that has been described as a "rigid moralism" (Finestone, 1967), often made it difficult for recovering alcoholics to return to their families or regain acceptance in the Polish community.

Social class and occupation are also determinants of attitudes toward alcohol in the wider sense, and this was true in the Polish-American case. The majority of the Polish-Americans in my study were from a working-class background, which was reflected in many of their attitudes toward alcoholic beverages. Beer, vodka, or straight whiskey were regarded as the only appropriate drinks, whereas mixed drinks were viewed as "lady's drinks" and improper for a hard-working man.[4] Attitudes toward wine were more variable. Although wine was regarded as appropriate for family gatherings and celebrations, it was rarely consumed in social clubs or during ethnic events.

SOCIALIZATION AND EARLY LEARNING

Socialization in regard to alcohol use in nineteenth- and early-twentieth-century Poland was characterized by early introduction (age 8–10) by parents within a culture tolerant of the excessive use of alcohol. Because of the varying effects of acculturation, a new cultural environment, and changing attitudes, this pattern has changed. Today, introduction to the use of alcohol is typically delayed (average age 16) and

[4] It should be noted that attitudes toward the use of alcohol as a reward for work are not exclusive to the working class. Trice and Roman (1978) write that many middle-class Americans regard weekend and vacation drinking as earned rewards for work.

usually occurs outside the family and in the company of peers. This trend holds for second- through fourth-generation Polish-Americans. Another important factor in the formation of early attitudes toward alcohol among some second-generation individuals was the memory of excessive drinking by their fathers. In addition, second-generation Polish-Americans grew up during the years of national Prohibition, which also influenced their attitudes.

Out of all respondents, few recalled receiving explicit instructions about alcohol from either parent, a tendency seen with each subsequent generation. The majority of parents with teenage children believed that their children drank more than teenagers of previous generations. The most frequently cited reason by parents for increased drinking by teenagers was not peer pressure, but rather lack of respect or loss of parental control. The changes that have occurred in the authoritarian nature of the traditional Polish family have been well documented by Finestone (1967), Lopata (1976), Obidinski (1968), and Wrobel (1979), among others.

There are striking differences in opinions between the sexes with regard to alcohol and drinking among Polish-Americans. The majority of women interviewed observed that Polish males have a different attitude and tradition toward drinking than those of Polish females. This difference stems from a cultural tradition that regarded drinking as primarily a male activity. Polish men were supposed to seek the company of other men and to socialize in village taverns. Moreover, alcohol was associated with virility and manhood. Women were allowed to drink, but in most cases it was symbolic, such as the offer of vodka to mark acceptance of a marriage proposal. A woman was expected to show deference by turning her head when she accepted a drink from a man and was never to drink openly in the company of men. Although these traditions have changed with the increasing equality of male–female relationships and a democratic family structure, many of the attitudes have persisted. For example, most women interviewed felt that it was inappropriate to accompany their husbands to a social club. Moreover, they felt that the stigma attached to excessive drinking by women was particularly strong in the Polish-American community. These attitudes are also changing, particularly among the fourth generation, with increased marriage outside the community.

Religion also traditionally acted as a powerful social control. Although the survey found no direct association between the degree of church participation (i.e., church attendance, membership on the church board, church-related activities) and reported drinking frequency, churches did serve to raise awareness to alcohol-related problems. Parish priests delivered frequent sermons on the dangers of excessive alcohol use and tried to change attitudes in regard to drinking behavior. For example, some priests discouraged the use of alcohol during ethnic events held on church grounds. This issue sparked many heated discussions during meetings of social club directors, church boards, and voluntary organizations. The priests argued that drunkenness at ethnic celebrations presented a bad image of the Polish community to the increasing number of outsiders attending ethnic events. A counter argument was that alcohol was an important and necessary part of the Polish cultural heritage and religious and secular events could not be properly celebrated without alcohol. This conflict illustrates the complexities of ethnicity, alcohol use, religious authority, concepts of historical tradition, and changes in self-perception.

THE CURRENT ALCOHOL PROBLEM IN POLAND

One of the most interesting findings to come out of the investigation of Polish-American attitudes was a widespread awareness of the alcoholism problem in Poland. The majority of those interviewed knew about the alcohol problem in Poland through a variety of sources, including Polish newspapers, relatives, sermons by priests, or conversations with recent immigrants. In addition to increasing the awareness of alcoholism among Polish-Americans, the Polish situation is interesting because of the interpretation of the cause by Polish-Americans. Most Polish-Americans attributed the alcoholism problem to an oppressive political regime in Poland. In addition, another frequent interpretation was that drinking in Poland was an expression of personal freedom. These interpretations are significant because they reveal the close interplay between cultural values and attitudes toward alcohol among all Polish-Americans.

It is equally interesting to look at the explanations proposed by Polish researchers for the rapid rise in alcohol consumption and alcoholism in Poland. Polish sociologists associate the problem with rapid urbanization, which has loosened old social structures that functioned as instruments of social control, in the absence of any functional new structures (Kryczka, 1979). At least partial support for the urbanization theory has come from several pilot studies conducted in new Polish urban centers (Zakrzewski, 1971). Although urbanization may be a factor, it is not the sole cause of the current alcoholism problem. For example, studies of attitudes toward alcohol among inhabitants of village communities have shown that there has been little change since the turn of the century (Gorski & Moczarski, 1972). Although there were changes in attitudes toward alcohol, particularly since the second world war, many atittudes are similar to those of Polish immigrants in the United States during the late nineteenth century and early twentieth century. Moreover, a consistent finding in attitude surveys in Poland and in consumption data studies is that rural dwellers consume more alcohol than city dwellers (Falewicz, 1968; Swiecicki, 1968). Although this is the reverse of what we would expect in industrializing economies, it is not surprising when viewed within the framework of peasant-rural cultural values. Rural immigrants brought with them to urban environments traditions of heavy drinking and attitudes that viewed alcohol as acceptable and harmless. As in the Polish-American case, the traditions and attitudes individuals bring to the new environment are important factors in explaining current drinking behavior. In addition, the fact that Polish attitudes have remained virtually unchanged confirms the strength of Polish cultural values and helps explain the persistence of these attitudes and values among Polish-Americans.

CONCLUSION

This case study of Polish-American drinking substantiates the relationship between ethnicity, values, and alcohol use. I have argued that Polish-American attitudes and practices are rooted in Polish peasant values and customs, which despite the varying effects of acculturation, historical events, and other social changes, have persisted. One of the themes of the study was that these attitudes and practices continued because they were closely linked to cultural values. The cultural belief that alcohol is an integral part of the Polish cultural heritage and that drinking is an expression of personal freedom and individualism is deeply held. Finally, the study has shown the importance of considering sociocultural factors in understanding drinking behavior. The study of customs, practices, and attitudes surrounding alcohol in an ethnic group can tell us a great deal about a wide range of social behavior, including attitudes toward American culture, ethnic self-perception, changing socialization patterns, and inter-generational changes. Although these relationships are often complex, unraveling them can be an interesting and valuable exercise.

REFERENCES

Ablon, J., & Cunningham, W. Implications of cultural patterning for the delivery of alcoholism services. *Journal of Studies on Alcohol*, 1981 (Suppl. 9), 185–206.

Bales, R. F. Attitudes toward drinking in the Irish culture. In P. J. Pittman & C. R. Snyder (Eds.), *Society, culture and drinking patterns*. New York: Wiley, 1962.

Benet, S. *Song, dance and customs of peasant Poland*. New York: Roy Publishers, 1951.

Blane, H. T., Acculturation and drinking in an Italian-American community. *Journal of Studies on Alcohol*, 1977, *38*: 1324–1346.

Bushee, F. A. Ethnic factors in the population of Boston. *Publications of the American Economic Association*, 1903, *4*, 293–477.

Clark, N. H. *Deliver us from evil; an interpretation of American Prohibition*. New York: Norton, 1976.

Falewicz, J. K. Uwarunkowania spotycia napojow alkohologych: Wyniki: ankiety ogolnopolskiej z 1968. (Conditions of alcoholic beverage consumption: Results of a nationwide survey in 1968.) *Problemy Alkoholizmu*, 1968, *20*, 1–10.

Fandetti, D. V., & Gelfand, D. E. Attitudes toward symptoms and services in the ethnic family and neighborhood. *American Journal of Orthopsychiatry*, 1978, *48*, 477–481.

Finestone, H. Reformation and recidivism among Italian and Polish criminal offenders. *American Journal of Sociology*, 1967, *72*, 575–588.

Freund, P. J. Armenian-American drinking patterns. *Alcohol, Health and Research World*, 1980, *5*, 47–50.

Gordon, A. J. Hispanic drinking after migration: The Dominican case. *Medical Anthropology*, 1978, *2*, 61–84.

Gorski, J., & Moczarski, K. *Alkohol w Kulturze: Obyczaju*. (Alcohol in culture and custom.) Warsaw: Wiedza Powszechna, 1972.

Gusfield, J. R. *Symbolic crusade: Status politics and the American Temperance Movement*. Urbana: University of Illinois Press, 1963.

Knupfer, G., & Room, R. Drinking patterns and attitudes of Irish, Jewish and white Protestant American men. *Quarterly Journal of Studies on Alcohol*, 1967, *28*, 676–699.

Kryczka, P. Some phenomena of social pathology in Poland. *The Polish Sociological Bulletin*, 1979, *2*, 101–109.

Lolli, G. E., Serianni, E., Golder, G. M., & Luzzato-Fegiz, P. *Alcohol in Italian culture: Food and wine in relation to sobriety among Italians and Italian-Americans*. New Brunswick, NJ: Rutgers Center of Alcohol Studies, 1958.

Lopata, H. Z. The Function of Voluntary Associations in an ethnic community: Polonia. In E. W. Burgess & D. J. Bogue (Eds.), *Contributions to Urban Sociology*. Chicago: University of Chicago Press, 1964.

Lopata, H. Z. *Polish-Americans: Status competition in an ethnic community*. Englewood Cliffs, NJ: Prentice-Hall, 1976.

Morawska, E. T. *The maintenance of ethnicity: Case study of the Polish-American community in Greater Boston*. San Francisco: R&E Research Associates, 1977.

Obidinski, E. *Ethnic to status group: A study of Polish-Americans in Buffalo*. Doctoral Dissertation, State University of New York, Buffalo, 1968.

Ostafin, P. A. *The Polish community in transition: A study of group interactions as a function of symbioses and common definitions*. Doctoral dissertation, University of Michigan, Ann Arbor, 1948.

Paine, H. J. Attitudes and patterns of alcohol use among Mexican-Americans: Implications for service delivery. *Journal of Studies on Alcohol*, 1977, *38*, 544–553.

Parmalee, N. *Inebriety in Boston*. New York: Eagle Press, 1909.

Popham, R. E. The social history of the tavern. In Y. Israel *et al.* (Eds.), *Research advances in alcohol and drug problems*, Vol. 4. New York: Plenum Press, 1978.

Reymont, L. *The Peasants: Autumn, Winter, Spring, Summer* (4 vols.), translated by Michael H. Dziewicki, New York: Knopf, 1925.

Seng, D. I. *Ethnic, racial and class attitudes toward national prohibition, 1920–1933*. Master's thesis, DePaul University, Chicago, 1973.

Simboli, B. J. *Acculturated drinking practices and problem drinking among three generations of Italians in America*. DPH Dissertation, University of California, San Francisco, 1976.

Strivers, R. *The hair of the dog: Irish drinking and American stereotype*. University Park: Pennsylvania State University Press, 1976.

Sue, S., Zane, N., & Ito, J. Alcohol drinking patterns among Asians and Caucasian Americans. *Journal of Cross-Cultural Psychology*, 1979, *10*, 41–56.

Synder, C. R. *Alcohol and the Jews*. Carbondale: Southern Illinois University Press, 1958.

Swiecicki, A. Consumption of alcoholic drinks by pupils and students of primary, secondary and high schools in Polish cities. *British Journal of Addiction*, 1967, *62*, 357–366.

Swiecicki, A. Ksztaltowanie sie Spozycia napojow alkoholowych od XVII w do dz: (Consumption of alcoholic beverages in Poland from the 18th century to the present day.) *Problemy Alkoholizmu*, 1968, *16*, 4–7.

Taft, P. Nationality and crime. *American Journal of Sociology*, 1936, *1*, 724–736.

Thomas, W. I., & Znaniecki, F. *The Polish peasant in Europe and America*. Boston: Richard G. Badger, 1918–1920.

Trice, H. M., & Roman, P. M. *Spirits and demons at work: Alcohol and other drugs on the job*, 2nd Ed. New York: New York State School of Industrial and Labor Relations, Cornell University, 1978.

Trotter, R. T. Ethnic and sexual patterns of alcohol use: Anglo and Mexican-American college students. *Adolescence*, 1982, *17*, 305–325.

Wilson, J. R., McClearn, G. E., & Johnson, R. C., Ethnic variation in use and effects of alcohol. *Drug and Alcohol Dependance*, 1978, *3*, 147–151.

Wrobel, P. *Our way: Family, parish and neighborhood in a Polish-American community*. Notre Dame, IN: University of Notre Dame Press, 1979.

Zakrzewski, P. Spozycie alkoholv w rejonie uprezemyslawianym oraz niektore zwiazane z tym problemy. (Alcohol drinking in a region becoming industrialized and some associated problems.) *Problemy Alkoholizmv*, 1971, *19*, 1–5.

Zand, H. S. Polish-American leisureways. *Polish-American Studies*, 1961, *18*, 34–36.

7

Jewish-Americans and Alcohol
Processes of Avoidance and Definition

BARRY GLASSNER and BRUCE BERG

One overriding reason has been given to examine the drinking patterns and attitudes of Jews: they present low rates of alcoholism and alcohol abuse. Although Jewish populations evidence a high percentage of adult drinkers (Riley & Marder, 1947; Mulford, 1964; Cahalan & Cisin, 1968; HEW, 1972; Levy, 1973), they have low alcoholism and alcohol problem rates (Snyder [1958], 1978; King, 1961; Chafetz & Demone, 1962; Roberts & Myers, 1967; Room, 1968; HEW, 1972; Lowenthal, Wald, & Klein, 1975; Schmidt & Popham, 1976; Greeley & McCready, 1978; Glassner & Berg, 1980). The usual statistic given in recent years is that more than 7% of the adult United States population are alcoholics, but less than 1% of the adult Jewish population is so classified (Unkovic, Adler, & Miller, 1975, 1977). Some researchers argue for a similarly low rate of alcohol abuse in Israel (King, 1961), and historical analysis suggests that alcohol problems have been rare in Jewish communities for the last 2,500 years (Keller, 1970).

Our own work[1] has concentrated upon these questions: Which aspects of Jewish-American life contribute to the avoidance of alcohol

---

[1] Much of the present paper is drawn from Glassner and Berg (1980, 1984).

BARRY GLASSNER • Department of Sociology, Syracuse University, Syracuse, New York 13210. BRUCE BERG • School of Criminology, Florida State University, Tallahassee, Florida 32306.

problems? and In what ways do Jews think about alcoholism and alcoholics? We addressed these issues through an empirical interview-based study of a Jewish community in New York State. In this chapter, we review the central findings from this study. First, we suggest four social processes that appear to contribute to the low rate of alcoholism among Jews. These involve (1) defining alcohol problems as something that does not afflict Jews; (2) early learning of actions and symbolism that integrate drinking into everyday activities; (3) interacting primarily with persons who hold to one's own drinking norms; and (4) utilizing a set of techniques to avoid social pressures toward excessive drinking. In the second part of the chapter, we report differences among groups of Jews in their definitions of alcoholism. The Orthodox tend to offer disease notions, the Conservatives are ambiguous in their definitions, and the Reform and nonpracticing talk of alcoholism in terms of psychological weaknesses.

METHOD

A stratified random sample was developed in a city located in central New York State that is consistently listed among the ideal northeastern cities for market and social scientific research because it offers "mean demographics" on such variables as population size, ethnic composition, and age and income distributions. In other words, the study community is potentially representative of many American communities.

The sample was developed over a six-month period by compiling lists from religious, social, university, and fraternal organizations (including one from the Jewish Welfare Federation, which is reliably claimed to include the names of 95% of all Jews living in the area, including many unaffiliated or nonpracticing Jews). After expending considerable effort in telephone conversations (during which time the topic of drinking was never mentioned), we achieved an acceptance rate of 91%.

The sample of 88 is equally divided into these groups: 24 young adults, 32 middle age, and 32 older; 44 male and 44 female; and 16 Orthodox, 24 Conservative, 24 Reform, and 24 nonpracticing subjective denominational affiliations. This was done in order to facilitate comparisons among subgroups.[2] Based upon such indicators as income and education levels, the sample is similar to Jews nationally (see Glassner & Berg, 1980, p. 650).

---

[2] The two cells for young Orthodox were not filled, however, because hundreds of telephone calls did not produce sufficient names.

Interview sessions were extensive, based upon a schedule that combined short-answer demographic questions with structured open-ended questions. The goal was to encourage respondents to discuss the issues fully. Many of the 144 questions were follow-up queries to a series of "How," Why," "Could you tell me more," and similar probes. Essentially, our interview schedule used Snyder's (1958) format as a developmental reference, in order to permit direct comparisons with his findings. However, it went into greater detail on secular topics. The interview was standardized to the extent that questions were consistently asked using the same wording and in the same order from one session to the next.

Ordinarily, interviews lasted two to three hours and ranged from one and one-half to three and one-half hours. Each interview was transcribed verbatim and coded by at least two independent coders. Only those responses about which the coders agreed and about which the respondent did not contradict himself or herself during the course of the interview are reported below as supporting our conclusions. There is considerable evidence within almost every transcript that the interviewer established good rapport and had open, detailed discussion.

PROTECTIONS FROM ALCOHOLISM

From the transcript material, we have identified four processes that seem to protect Jews from alcohol problems:

ALCOHOL PROBLEMS AS NON-JEWISH

There appears to be a widespread belief among members of the sample that alcoholism is something that does not happen to Jews.[3] A substantial majority of the respondents indicated this belief, even though we did not ask any questions about it. In most cases the respondents directly stated this association. An Orthodox counselor in his 60s reported:

> Nobody ever drank to get drunk. I mean that wasn't, isn't a Jewish concept. Liquor and wine is part of Jewish, you know, holiday and tradition. More sociability at parties.

A Reform college professor in his 40s stated:

> [A Christian friend] drinks a lot. I don't think he's an alcoholic, but his idea

---

[3] The Yiddish expression is *shikker vie a goy* (drunk as a Gentile), usually quoted by our respondents as *shikker as a goy*.

of socializing is to drink, so I buy a bottle of Scotch to have in the house, just in case, because he likes J&B. So when he comes over, I'll always offer him a drink. He and his wife are very drink-oriented people. They drink, and I associate that with non-Jewishness myself. I always associate Jewish tradition with food, as most people do, I think.

A Reform housewife in her 30s has long held this association:

We were exposed to liquor plenty. Actually, my father worked for a beer company. It sounds like a stupid generalization, but non-Jewish people drink more heavily than Jewish people. That's a generalization I've been brought up with . . . and I still think it's true.

In some cases, the association is made, but in less direct statements. A Reform housewife in her 50s noted:

I've been a guest at [a local Jewish County Club]. . . . You know, they never spend money on their bar there like they do in most Gentile country clubs.

Our finding that Jews define alcoholism as an out-group characteristic is a reconfirmation of the finding within earlier studies (Snyder, 1958, Ch. 5; Kramer & Leventman, 1961). Indeed, Zimberg (1976) goes so far as to claim that "the sociocultural attitudes equating Jewish identify with sobriety are perhaps the major factors accounting for low alcoholism in Jews."

MODERATION PRACTICES FROM CHILDHOOD

Snyder (1958) argued that Jewish sobriety "depends upon the continuity and vitality of the Orthodox religious traditions." He emphasized the importance of Orthodox religious affiliation *per se* for his respondents, but within his work one also sees that it would be possible to extend the implications of the protective characteristics of Orthodoxy. He argued, for instance, that "the thread of orthodox life may be woven into many regional fabrics without losing its sobering influence." Our data suggest that such protection has been extended even further into the contemporary "secularized" community. Through religious and ceremonial usage of alcohol, Jews continue to learn "prescriptive" (Mizruchi & Perruchi, 1970) drinking norms. In contrast to the expectations of those authors (Zimberg, 1976; Malzberg, 1940; Bales, 1944) who expected the loss of traditional protections with the loss of Orthodoxy, we have detected a good deal of tenacity and adaptability of traditions and their latent symbolism.

Children continue to have many opportunities to learn to associate drinking of alcohol primarily with special (dare we say "sacred") occasions (see Knupfer & Room, 1967, p. 690). On the last page of his book,

Snyder (1978) listed several conditions we find are currently met in new ways: "Where drinking is an integral part of the socialization process, where it is interrelated with the central moral symbolism and is repeatedly practiced in the rites of a group, the phenomenon of alcoholism is conspicuous by its absence."

Sobriety amid drinking continues to be the norm in Jewish homes. Our interviews suggest that Jews perpetuate this association and its connection with ritual, not only by affiliation with religious life, but also by restricting drinking to special secular occasions and by cataloguing drinking as a symbolic part of festive eating.

Almost without exception, our interviewees reported that drinking in their homes during childhood was predictable, and drunkenness was never condoned or was condoned only on very rare occasions. A Conservative secretary in her 60s said:

> Parents don't usually put a drink in front of themselves or their children. Mine didn't, and I wouldn't with my kids. Some kids will say, "You do it, why can't I do it?" We never did drink except for a special occasion.

A nonpracticing medical researcher in her 30s noted:

> It was a ritual in our daily life. It was a family routine. [My father] would sit and read the paper for a little while and have the drink and then we would have dinner and my mother would have a drink with him.

Nearly one-half the sample could not remember their first drink, but recall that all their early drinking was during childhood in the home as part of religious ceremonies. Of the remainder of the sample, 89% recall that their first drinks were recreational with their families, in the home or at the synagogue, and occurred before the age of 13. Thus, only about 5% of the sample recall their first drinks as outside the family and at a time later than childhood.

Our sample thus meets the conditions that several studies have found result in low adult alcohol problem rates: parents who agree about drinking norms for themselves and their children; formative drinking experiences within the family; and controlled drinking at an early age (Wilkinson, 1970; McGonegal, 1972). Jews are sheltered from what Pittman (1967) calls the ambivalent drinking culture of the United States, which consists of conflicting attitudes toward alcohol consumption. The Jewish family maintains instead what Pittman calls a permissive drinking culture in which drinking is permitted, but excessive drinking is not.

Not only is Jewish drinking limited primarily to predictable family, social and religious occasions, but also the drinking is closely associated with eating (cf. Hill, 1977). Most of the interviewees alluded to the relative importance of food over drink, and the absence of drinking without eating. A nonpracticing businessman in his 40s described:

> I can remember more my mother arranging platters of cold cuts and potato salad or baking because she was having people [to the house] than I can ever remember her serving drinks. The emphasis was always on the food, and the liquor was deemphasized. If it was there it was because there were people who expected it to be there.

A Conservative schoolteacher in her 30s observed:

> Drinking was just never entered as an activity. It's part of a thing like eating. It's OK with foods. It goes, to me, it goes along with eating.

The moderating effect of associating drinking with eating was first noted by Feldman (1923) in a comparison of the contrasting practices of English Gentiles and English Jews. In recent decades, several studies (reviewed in Plaut, 1967 and Zucker, 1976) have suggested that groups characterized by widespread drinking but few alcohol problems have integrated their drinking into a variety of basic activities. In the case of Jews, this integration is taken a significant step further, toward making drinking a symbolic act. Drinking serves as a symbolic punctuation mark that helps to separate certain positive events (religious services, weddings, dinners, etc.) from all other events (Mandelbaum, 1965). By being a symbolizer, the drinking takes on a symbolic character itself.

INSULATION BY PEERS

If drinking patterns learned in childhood are to be maintained, and if the self-definition of Jews as moderate drinkers is to remain believable, Jews must collectively reiterate moderate drinking patterns in adulthood. This is accomplished in part through family practices, as we have already noted. However, the serious threat to moderate drinking occurs outside the home, at parties and other public occasions, when considerable drinking is often encouraged or expected.

In large part, the ability of our respondents to avoid excessive drinking contexts seemed to be due to their socializing mainly with other Jews. Thus, 76% of the sample report that all or nearly all their friends are Jewish, although most say that their work associates are primarily non-Jews. The respondents are not, as a group, ideologically opposed to integrating. When asked if it mattered one way or the other whether they spent recreational time with Jews or with non-Jews, 64% answered negatively, 25% affirmatively, and 14% (including two individuals who answered affirmatively) described types of situations in which it would or would not matter.

A gravitation toward Jewish things and people (90% see "Jewishness" as important in their lives) protects Jews from excessive drinking in obvious and not so obvious ways. First, one is unlikely to find others who would encourage, cooperate with, or even condone excessive drinking. Thus, any rationalization for extra drinking becomes difficult. This is particularly evident in those cases when Jews drink primarily among other Jews. The process goes further, however, to selection of non-Jews who drink like Jews (cf. Alexander & Campbell, 1970). A nonpracticing college professor in his 50s said:

> Most people at the party were not Jewish, actually. This one guy was making a real ass of himself. He'd had too much to drink and it made everybody uncomfortable. I guess our friends just are not heavy drinkers. I mean, we carry drinks around at parties, but when this guy got drunk he really stood out. . . . I think he eventually got the message, because he was one of the first to leave.

Socializing with Jews and other moderate drinkers makes inconvenient or unavailable the use of alcohol to various practical ends; in contrast to other cultural groups, such as the Oaxacans of Mexico, in which intoxication provides persons with license to shout insults and attend social gatherings without invitation (Dennis, 1975), Jews who become intoxicated are likely to be scorned or pitied.

A net result of the moderate drinking environment is that Jews maintain images of normal and preferred social interactions over their life spans that do not include excessive drinking. This point is evident, for example, in the case of an Orthodox businessman in his 40s who answered the question, "When did people drink when you were growing up?" with, "My family just didn't drink. People we are with now just don't seem to drink." Or the Conservative male college student who answered a question about parents' drinking patterns with, "My parents rarely drink at all. Myself, I went through the drinking phase with my friends, but we've turned the page now."

Group avoidance is thus a dynamic process. By modeling consistent group drinking practices and norms in adulthood, Jews perpetuate the drinking associations into the next generation. Also because unique alcohol consumption practices serve to emphasize ethnic differences in this multiethnic nation (Cisin, 1978), adult drinking patterns contribute not only to perpetuation of the first process (defining drunkenness as non-Jewish), but also to group boundary maintenance.

AVOIDANCE REPERTOIRE

Despite the social interaction patterns previously described, drinking histories collected in the interviews reveal that at least 46% of the respondents found themselves in social situations during their adulthood where they were pressured to drink more than they wished to drink. Here they utilized what we have termed "avoidance repertoire," or interaction techniques appropriate to the social context, which permitted them to control their alcohol consumption. We recorded several exchanges similar to the following discussion with a nonpracticing businessman in his 40s:

> *Q:* Does it matter if others drink with you?
> *A:* No, it seems to matter more whether I drink with them.
> *Q:* What do you mean?
> *A:* My not drinking tends to make more people uncomfortable than their drinking makes me.

The drinking histories, symbolic associations, and uniformity of group opinion we have noted thus far seem to result in these respondents feeling they can offer an assertive "No" when encouraged to drink more than they wish. When we asked respondents how they avoid excessive drinking during various pressure situations, many were curious why we would suspect that such a situation might prove to be a problem. They perceive their avoidance repertoire as nonaction. A Conservative housewife in her 40s said:

> I don't care one way or another whether I have a drink. If everybody is drinking and I feel like having a drink I'll have a drink. If everybody is drinking and I don't want a drink, I don't drink.

Nevertheless, upon further probing, many respondents do reveal a variety of techniques used in such avoidance. Several report that they begin a party with a mixed drink, but refill the glass throughout the night with only the mixer (e.g., moving from gin and ginger ale to just ginger ale). Other respondents report that their family as a unit will act to prevent unwanted drinking. Some tell of plans whereby the wife will reprimand the husband for drinking too much at a party, when actually he is still "nursing" his first or second drink. Other couples develop reputations as nondrinkers by making jokes or by avoiding such situations.

Other respondents report specific avoidance repertoires they had worked out for particular situations. For example, a Conservative photographer in his 40s tells of the problems he used to experience at weddings, where sometimes he was "on the verge of being hit because they felt I wasn't showing proper respect for the bride and groom by toasting them." This photographer reports that since those early years he has consistently utilized a series of "convincing lines" to indicate that he cannot drink while he is working. He couples this with taking more pictures at times when others are involved in heavy drinking. "Being Jewish, your reflexes and everything else sharpen you, and that expedience staves off any future attempts for people to push drinks on you," he explained.

In summary, this research suggests four processes that protect Jews from alcohol abuse: (1) defining alcoholism as something that happens to non-Jews; (2) subscribing to practices and symbolism learned early, which integrate sober drinking into daily life; (3) restricting most primary relations to others with similar drinking norms and practices; and (4) developing a repertoire of avoidance techniques.

DEFINITIONS OF ALCOHOLISM

The four social processes just discussed operate to lessen the possibility of Jews developing alcohol problems. These appear to be present among all subgroups of Jews. In contrast, we find marked differences among the affiliation categories of Jews regarding *perceptions* of alcoholism and alcoholics. Let us look at how each affiliation group defines alcoholism.

ORTHODOX

Of the available definitions of alcoholism, we suspected Orthodox Jews would select a highly moralistic version that blamed secular society. Alcoholism has long been considered a moral weakness, and only in the past few decades has a major effort been made to counter this view by educating the public that alcoholism is a disease (Conrad & Schneider, 1980; Ries, 1971). Since the Orthodox are the most traditional Jewish affiliate group and the least accepting of their fellow Jews' accommodation to American ways, one would anticipate that they would be the least likely to accept such "new-fangled" notions as disease definitions of alcoholism. Instead, our findings suggest that 63% of the Orthodox (compared to 25% of Reform and nonpracticing) offer disease definitions of alcoholism in response to our open-ended question, "What do you think alcoholism is?"

A teacher in her 30s replied:

It's a kind of sickness. It can be cured, at least in the beginning.

An engineer in his 40s responded:

I think it is a disease. People that are alcoholics, a lot of them, can't, they need some sort of professional help, plus family help together.

As these quotations attest, Orthodox respondents frequently talked about the alcoholic being cured, whether by conventional medicine or by a combination of medicine and determination.

One half of the Orthodox respondents say they have never known anyone who was even accused of drinking too much, and the persons cited by the other half of the sample as possible alcoholics were in every case very distant from the respondent (e.g., an aunt's second husband and "someone I once met in college"). None of the Orthodox respondents reported knowing a Jew who was considered to have a drinking problem.

In short, the Orthodox respondents appear to have stayed a great distance away from alcoholism and to have picked up a readily available definition of the phenomenon. The respondents' disease definition is externalistic and simplifying. It treats alcoholism as something that happens to someone, much like a bacterial infection. The conviction is probably not held very deeply. Indeed because the Orthodox find little practical need for defining alcoholism, many respondents had not consciously thought of the issue before we asked about it.

CONSERVATIVE

Alone among the affiliate groups, Conservative respondents rambled in their discussions about the key issues of concern in our study. Asked to define alcoholism, the median Conservative response is 80 words long, compared to a median of 45 words for the remainder of the sample. About 33% of the Conservatives' definitions could not be coded into a single category, compared to 5% of the remaining sample. Transcripts of the Conservatives' responses frequently display hesitations and ambiguities when compared to comments made by members of other affiliation groups. A writer in his 30s (who did not falter in response to other questions) answered:

> I don't think it is a disease. I think that it is too much drinking. You know, I don't think, you know, every day, with somebody, somebody very well. It's nobody's, it's not anybody's fault that it's a disease. I think it's a character weakness of a person, I think that when people say it's a disease, I think it's crazy.

A repairman in his 40s said:

> I think it's a, um, it's a need to drink and to consume alcohol, on a regular basis. Not just a need to drink, but it's, it's habit forming to the point where you can't do without it. And, it doesn't necessarily mean that a person has to get drunk. But it's being hooked on something that, uh, you need to use as a crutch, or, uh, filler for other things.

The latter excerpt includes the sole consistent characteristic within the Conservative respondents' interpretations of alcoholism: the idea that drinking is habit forming. Forty-six percent of the Conservatives (compared to 10% of the other respondents) include habit as a defining characteristic: "It's habit forming. It's a drug. Your body demands a certain amount. . . ." "You get the habit and up and use more and more. . . ." "There is a compulsive need, even when nobody else is around. . . ." This emphasis on habit is the only common feature of the Conservatives' responses.

The idea of alcoholism as a response to life's troubles appears in Conservatives' discussions, but seldom in the pseudo-psychological form we will note below for the Reform and nonpracticing. Among Conservatives, this image is enmeshed with other impressions. For a student in his 20s, it is a characteristic of an illness:

> Alcoholism is a sickness, I think, it is a, you have to seek refuge through alcohol. You really can't deal with the world as it is or you are having trouble facing your own reality. An escape.

Thus, the Conservatives are midway between Orthodox and Reform and nonpracticing in several respects. As a subjective affiliation, Conservatism is viewed as less religious than Orthodoxy and more religious than Reform or nonpracticing. Conservatives generally see themselves and are considered by other Jews to balance traditional Jewish life-styles and values with involvement in American society. Sklar (1972, p. 207) has called Conservatism a "halfway house" between Reform and Orthodox, "a type of Judaism which, while not Orthodox, derives from traditional sources; while not completely Reform, it is sufficiently advanced so as to be 'Modern.' " His research suggests that a difficulty for the Conservative movement has been its lack of a distinct body of religious and secular beliefs and practices (pp. 199–245).

Their confusion and emphasis regarding definitions of alcoholism coincide with these positions. Conservatives are divided into those (29%) who accept the disease definitions prevalent among the Orthodox and others (33%) who—like the Reform and the nonpracticing—adopt psychological versions. But the largest group of Conservatives (38%) is unclear about how to define the phenomenon. The major commonality in Conservatives' definitions—habit—is itself a notion midway between the disease definitions of the Orthodox and the psychologistic definitions of the Reform and nonpracticing. In common usage, "habit" is a vague term implying something that happens to a person's mind and body and hence is unintentional or outside the person's control, but also something that involves a person's own acts.

Conservatives as a group are in the middle of a continuum also with regard to their experiences with alleged alcoholics. One third indicate they have never known an alcoholic (compared to 50% of the Orthodox, and 7% of the Reform and nonpracticing). Of those who have known someone suspected of alcoholism, one third have known a Jewish alcoholic. Most of the alleged alcoholics with whom the Conservatives are familiar are distant acquaintances, but unlike the Orthodox, some of the suspected alcoholics are personal friends or in-laws.

REFORM AND NONPRACTICING

In contrast, the Reform and nonpracticing interviewees tended to define alcoholism in psychological terms as a psychological dependence and weakness.

A housewife in her 50s said that an alcoholic is:

> Somebody who would need a drink to face the day, or face anything in the day. Just to get himself going, or herself going. In other words, would just really need it all through the day. To get up, to feel relaxed, to feel self-confident.

A college student in his 20s called it:

> A dependency on alcohol. In other words, it's not just going out and having a drink, but needing that drink to relax. If you can't relax without that drink, you're an alcoholic, whether it is one drink or a bottle.

Seventy-one percent of the Reform and nonpracticing respondents defined alcoholism as a dependency with psychological overtones.

Reform Jews have accommodated to and been active in American society for many years, a state of affairs that has provided them with social involvements different from those of more traditional Jews. One result is that Reform and nonpracticing respondents differ fom the Orthodox not only in how they view alcoholism, but concurrently, in their contact with alcoholism and drinking. The dramatic difference is demonstrated by the fact that 93% report they have known an alleged alcoholic and 37% say they have known a Jewish alcoholic. Unlike the Orthodox, many of the presumed alcoholics are close acquaintances, including aunts and uncles, parents' friends, in-laws, parents, and personal friends. The difference is also apparent in the role drinking occupies in the respondents' daily lives. Reform and nonpracticing Jews regularly engage in social drinking. Most respondents say they drink every couple of days, either at business or at social gatherings, whereas Orthodox Jews report nonceremonial drinking as rare.

The Reform and nonpracticing respondents consider drinking to be a part of social life in two primary ways. First they see alcohol as a natural ingredient in social gatherings.

For instance, a retired manufacturer in his 70s says he drinks:

> To make the people that we are with more comfortable when they are at our house. The other way around, when I'm out, I don't want them to feel that I am not social.

A business owner in his 30s says:

> It is very social to drink. It's a common ground either for discussion or as a starting point to being together.

These sorts of comments are totally unlike those of the Orthodox respondents, as is the case with the second social version of drinking in which Reform and nonpracticing Jews talk of drinking as useful in its own right. Many respondents described the importance of drinking with business associates, but for several others the secular symbolism of drinking went still further.

A college professor in his 40s explained:

> I think the reason why I keep a store of liquor in the house and why I put wine on the table [is because] it's sort of a syndrome of, well, we're doing the elegant thing . . . it was always the thing to do if you were rich. Since none of us were rich, this is why I think we do it now.

SUMMARY

The study of Jewish drinking patterns and beliefs continues to provide important information about a variety of alcohol-related issues. We have suggested four protective processes that appear to contribute to the avoidance of alcohol abuse by American Jews. In addition, we discuss some significant variations among subgroups of Jews in beliefs about alcoholism and alcoholics.

It will be particularly instructive if further research can focus upon changes in drinking patterns and beliefs among Jews, as these occur in contrasting types of communities or over different historical periods. In addition, comparative research with other ethnic groups might reveal patterns of protective processes and beliefs similar to those we have described for American Jews.

REFERENCES

Alexander, D. N., & Campbell, E. Q. Normative milieux and social behaviors. In G. L. Maddox (Ed.), *The domesticated drug*. New Haven, CT: College and University Press, 1970.

Bales, R. F. The *"fixation factor" in alcohol addiction*. Unpublished doctoral dissertation, Harvard University, 1944.

Cahalan, D., & Cisin, I. H. American drinking practices. *Quarterly Journal of Studies on Alcohol*, 1968, *29*, 142.

Chafetz, M. E., & Demone H. W. *Alcoholism and society*. New York: Oxford University Press, 1962.

Cisin, I. H. Formal and informal social controls over drinking. In L. Ewing & B. Rouse (Eds.), *Drinking: Alcohol in American society*. Chicago: Nelson-Hall, 1978, 145–158.

Conrad, P., & Schneider, J. *Deviance and medicalization*. St. Louis: Mosby, 1980.

Dennis, P. The role of the drunk in an Oaxacan village. *American Anthropologist*, 1975, *77*, 856–863.

Feldman, W. M. Racial aspects of alcoholism. *British Journal of Inebriety*, 1923, *21*, 1–15.

Glassner, B., & Berg B. How Jews avoid alcohol problems. *American Sociological Review*, 1980, *45*, 647–664.

Glassner, B., & Berg, B. How Jews define alcoholism. *Journal of Studies on Alcohol*, 1984, *45*, 16–25.

Greeley, A. M., & McCready, W. C. *Societal influences on drinking behavior*. NORC paper presented to the International Medical Advisory Conference of the Brewing Associations, Toronto, October, 1978.

HEW. *Alcohol and health*. First Special Report to the U. S. Congress from the Secretary of Health, Education and Welfare. Washington, DC: U. S. Government Printing Office, 1972.

Hill, T. M. Survey of Jewish drinking patterns. *Military Chaplain's Review*, 1977, 65–77.

Keller, M. The great Jewish drink mystery. *British Journal of Addiction*, 1970, *64*, 287–295.

King, A. R. The alcohol problem in Israel. *Quarterly Journal of Studies on Alcohol*, 1961, *22*, 321–324.

Knupfer, G., & Room, R. Drinking patterns and attitudes of Irish, Jewish and white Protestant American men. *Quarterly Journal of Studies on Alcohol*, 1967, *28*, 676–699.

Kramer, J. R., & Leventman, S. *Children of the gilded ghetto*. New Haven, CT: Yale University Press, 1961.

Levy, L. Drug use on campus. *Drug Forum*, 1973, *2*, 141–171.

Lowenthal, U., Wald, D., & Klein, H. Hospitalization of alcoholics and the therapeutic community. *Harefuah*, 1975, *89*, 316–320.

McGonegal, J. The role of sanction in drinking behavior. *Quarterly Journal of Studies on Alcohol*, 1972, *33*, 692–697.

Malzberg, B. *Social and biological aspects of mental disease*. Utica, NY: State Hospital Press, 1940.

Mandelbaum, D. G. Alcohol and culture. *Current Anthropology*, 1965, *6*, 281–288.

Mizruchi, E. H., & Perruchi, R. Prescription, proscription and permissiveness: Aspects of norms and deviant drinking behavior. In G. Maddox (Ed.), *The domesticated drug*. New Haven, CT: College and University Press, 1970.

Mulford, H. A. Drinking and deviant behavior, USA 1963. *Quarterly Journal of Studies on Alcohol*, 1964, *25*, 674.

Pittman, D. J. International overview: Social and cultural factors in drinking patterns, pathological and non-pathological. In D. J. Pittman (Ed.), *Alcoholism*. New York: Harper & Row, 1967.

Plaut, T. F. *Alcohol problems: A report to the nation by the cooperative commission on the study of alcoholism*. New York: Oxford University Press, 1967.

Ries, J. Public acceptance of the disease concept of alcoholism. *Journal of Health and Social Behavior*, 1971, *18*, 338–344.

Riley, J. W., & Marder, C. The social pattern of alcoholic drinking. *Quarterly Journal of Studies on Alcohol*, 1947, *8*, 265–273.

Roberts, B. H., & Myers, J. K. Religion, national origin, immigration and mental illness. In S. I. Weinberg (Ed.), *Sociology of mental disorders*. Chicago: Aldine, 1967.

Room, R. Cultural contingencies of alcoholism. *Journal of Health and Social Behavior*, 1968, *8*, 99–113.

Schmidt, W., & Popham, R. E. Impressions of Jewish alcoholics. *Journal of Studies on Alcohol*, 1976, *37*, 931–939.

Sklar, M. *Conservative Judaism*. New York: Schocken, 1972.

Snyder, D. *Alcohol and the Jews*. New Haven, CT: Yale University Press, 1958.

Unkovic, C. M., Adler, R. J., & Miller, S. E. A contemporary study of Jewish alcoholism. *The Alcohol Digest*, 1975, *9*, vi–xiii.

Wilkinson, R. *The prevention of drinking problems*. New York: Oxford University Press, 1970.

Zimberg, S. *Socio-psychiatric perspective on Jewish alcohol abuse*. Paper presented to the Task Force on Alcoholism of the Commission on Synagogue Relations, New York, March 1976.

Zucker, R. A. Parental influences on the drinking patterns of their children. In M. Greenblatt & M. Schuckit (Eds.), *Alcoholism problems in women and children*. New York: Grune & Stratton, 1976, 211–238.

8

Historical Meanings of Irish-American Drinking

RICHARD STIVERS

Irish-Americans have for a long time had one of the highest rates of alcoholism of any ethnic group in the United States, a country that itself has had an extremely high rate of alcoholism (See Appendix). The story of Irish-American drinking begins in Ireland, but does not end there as some have thought. Robert Bales (1944, 1962), for instance, in an otherwise erudite work thought that Irish attitudes toward drinking should be similar to those of Irish-Americans. That is one reason why he doubted statistics that indicated a higher rate of alcoholism among Irish-Americans than among the Irish in Ireland (Bales, 1944, p. 34). In my earlier work (Stivers, 1976), I addressed the statistical discrepancy between the two groups and attempted to show that the cultural attitudes toward drinking were dissimilar and moreover that the structural context of these attitudes were distinct.

Before examining the historical context of Irish drinking, I will discuss my sources of data and methodology. The data include demographic statistics, historical documents, including literature and drama, and secondary historical sources. Data on drinking include statistics on

This chapter is a revised and edited version of material drawn from the author's *A Hair of the Dog: Irish Drinking and American Stereotype*, University Park: Penn State Press, 1976 and his "Irish Ethnicity and Alcohol Use," *Medical Anthropology*, 1978, 2, 121–135.

RICHARD STIVERS • Department of Sociology, Anthropology and Social Work, Illinois State University, Normal, Illinois 61761.

heavy drinking, such as the arrest rate for drunkenness and the death rate from cirrhosis of the liver, and observations of drinking behavior. Data on the context of drinking, which are drawn from a great variety of historical sources, are more important than the data on drinking behavior as such. The reason for this is the interpretive method I have employed.

This method involves situating the behavior in its cultural or meaningful context. It represents an attempt to "read" the symbolism of the behavior. The cultural context of an action is specifically determined through its institutional setting. But, and this is the most difficult part, the institutional setting of the action needs to be related to its other institutional settings.

For example, in this chapter, I interpret Irish-American drinking in its religious, familial, political, and economic contexts—not as separate contexts, but rather as interrelated contexts. This method, moreover, involves taking social structure into account as a source of meaning without reducing meaning to social structure.

IRISH DRINKING IN THE NINETEENTH CENTURY

In the late eighteenth and early nineteenth centuries, drinking behavior and attitudes were similar in England, Scotland, and Ireland and much of northern Europe. Drinking rituals punctuated the workday of every occupation and profession. Moreover, drinking had become part of virtually every rite of passage (Dunlop, 1839).

European temperance movements dramatically changed this—making hard drinking disreputable for the middle classes and the "respectable" working classes. The temperance movement not only started earlier in Ireland, it also declined earlier. For all intents and purposes, the Irish movement was over by the end of the Great Famine in 1848. In other European countries, hard drinking eventually became taboo to all but the lower classes and the leisured upper classes. Ireland, however, had no substantial Catholic middle-class movement and consequently the old drinking patterns returned, but this time in a different socioeconomic context (Stivers, 1976, pp. 33–50).

The changed context included British land reforms that gave Irish Catholics the opportunity to own land. An emergent norm of primogeniture permitted families to consolidate their holdings. The Great Famine had taught the Irish that the minute subdivision of land was economically unfeasible (Connell, 1968, pp. 114–115). Because marriage was now tied to the ownership of land, a large class of unmarried men and women was created, some of which would emigrate. But even those privileged few who did marry were forced to do so at a later age because parents tried to hold onto their farms as long as possible. Consequently, a pattern of few and late marriages emerged (Arensberg & Kimball, 1968; Connell, 1968).

An emphasis on chastity and related prohibitions against birth control and abortion compounded the problem. Illegitimacy was not only sinful, it threatened the new family farm economy in which marriage and land ownership were inexorably linked. The church's official policy advocating early marriage for all (save those with a religious vocation) contradicted this norm. The peasant clergy, however, condoned the pattern of few and late marriages (Connell, 1968, pp. 124–131). The infamous Irish "puritanism," which segregated the sexes prior to marriage, was as much an economic phenomenon as it was a moral one (Kennedy, 1973, p. 172). Emerging out of the family farm economy and a strict sexual morality was the institution of the bachelor group.

Its folklore centered on the theme of freedom from responsibility as expressed in sports activities, fighting, storytelling, and hard drinking. The rite of passage from boyhood to manhood was signaled by taking a drink in the company of bachelor group members in the local pub. Male identity was thus intimately bound to an ethic of hard drinking. The group became a means of controlling bachelors by diverting their attention from the responsibility of marriage, family, and farm toward the non-responsibility of leisure activities (McNabb, 1964, pp. 224–236). Hence it was crucial that married men also take part in "bachelor" group activities; otherwise this contradiction of Irish society would have been evident. In this sense, the bachelor group ethic of hard drinking constituted a cultural remission, that is, a tacitly approved way of circumventing agrarian asceticism and Irish puritanism and of bridging the gap between the ideal of farm ownership and marriage and the bleak reality most men faced.

However much hard drinking was approved of, drunkenness was not condoned. The Church defined intemperance as a mortal sin; therefore, high bachelor group status went to the man who could hold his own after a full night's drinking. Most of the clergy supported moderate drinking rather than total abstinence; hence drunkenness was rather narrowly defined. The Irish ambivalence toward drinking was expressed in the sympathetic saying, "Drink is a good man's failing." This folk saying represented an unconscious recognition that the hard drinker was a bulwark of the community and its stem-family farm economy.

His sacrifice of his traditional inheritance and marriage permitted farms to remain intact. Hence it was better to permit hard drinking and even occasional drunkenness than to risk the possibility of early marriage or illegitimacy, both of which threatened the entire socioeconomic order. The bachelor group ethic of hard drinking, with the tacit approval of the clergy and the participation of married men, served to bridge the gap between the haves and have-nots of Irish social structure (Stivers, 1976, pp. 86–98).

IRISH-AMERICAN DRINKING FROM THE MID-NINETEENTH THROUGH THE EARLY-TWENTIETH CENTURY

The Irish were the first of the great wave of western European immigrants to the United States. That they faced great problems of settlement, unemployment, and poverty is unquestioned (Handlin, 1959; Ernst, 1949; Potter, 1960). Many other immigrant groups, however, faced similar problems; yet they did not have similar high rates of alcoholism. Therefore, one must look beyond the argument that the stress of migration and resettlement led to drunkenness and alcoholism.

THE IRISH-AMERICAN FAMILY

The Irish-American family is comparable to its counterpart in Ireland. Irish-Americans were for the most part reluctant to marry (Woods & Kennedy, 1962, p. 145), and they had the lowest marriage rate of any ethnic group (Bagenal, 1882, p. 534). Those who did marry married late. In a study in Buffalo, Mattis (1975, p. 11) discovered that in the period from 1877 to 1882 the average age at marriage for men was 35 and for women 31 when both partners were Irish and 32 for men and 27 for women when only one partner was Irish. No other ethnic group approximated this late age at marriage.

Fragmentary evidence in casual observation and fiction points to an austere husband–wife relationship and an overly protective mother–son relationship. These relationships are similar to Irish family relationships (Stivers, 1976, pp. 107–111). Stephen Crane's novel *Maggie* depicts an Irish-American family in which the husband and wife lead almost mutually exclusive existences. Each is master of his or her domain. The husband controls the domains of work and leisure, whereas the wife is in charge of the household, childrearing, and religion. There is a tendency for the mother to manipulate the children, especially the sons, by appealing to their guilt and sympathy (Greeley, 1972, p. 110).

The heavy emphasis on chastity, especially for women, was sustained in America. Although the second generation may not have been as strict in adhering to the norm as the first generation (Bushee, 1903, p. 114), Irish women still had a reputation for being chaste (Woods, 1898, p. 172). Chastity was extolled above every other virtue (Fontaine, 1940, pp. 25–33). The end result was a psychological segregation of the sexes.

THE ECONOMICS AND POLITICS OF DRINK

The Irish, like the Italians and the Jews later, tended to settle in the large urban areas of New York, Boston, Philadelphia, and Chicago. It was in the ethnic slum that Irish drinking assumed a new context and thus a new meaning.

The drink trade played an enormous part in the life of the Irish-American neighborhood. The *New York Tribune* reported in 1854 that up to three fourths of that city's saloons were run by Catholics who made up only one fourth of its population (Potter, 1960, p. 518). At the same time the overwhelming majority of grogshops in Boston were operated by the Irish (Handlin, 1959, p. 121). For some of the Irish, the liquor business was a means of economic advancement; for others it was a way of eking out a living.

Together with the church, the saloon was the focal point of community life. It was sometimes a hiring hall and always a campaign stop for politicians. Irish politicians often owned a saloon; concomitantly, a saloonkeeper was fully expected to enter politics. The saloonkeeper/politician often used drink as the asking price for jobs, especially political appointments.

Irish-Americans had a reputation for being the best and most readily organized political group. This was due in no small measure to the great prevalence of street gangs among the Irish boys (Bushee, 1903, p. 151). A vital link between the younger street gang and the older ward political machine was the continued participation of some older men in various street gangs. Woods (1898) distinguished between the respectable and disreputable gangs. The disreputable gangs, which outnumbered the respectable ones, spent much of their time drinking and fighting, whereas the respectable gangs formed social clubs to advance their political and economic goals. The various street gangs and social clubs provided the rudimentary organization for higher level political organizations.

Over the gangs and social clubs stood the "machine club." Membership included all those men in the ward who had important political appointments. Its leader was the ward boss; under him were "lieutenants" and "heelers." The lieutenants were political aids who implemented high-level political decisions; the heelers did the leg work and put pressure on people to vote for their candidate. Leaders of street gangs and social clubs also belonged to the machine club (Woods, 1898, pp. 118–122).

The ward boss and his machine club performed many services for the community. Food, clothing, jobs, and repairs were provided for those in need. The ward boss took care of his own; they in turn were expected to vote for him and his candidates (Woods, 1898, pp. 126–127).

A brief comparison of Irish and Irish-American neighborhoods is crucial to my theory. The Catholic religion and the priest maintained greater control over everyday life in Ireland than in America. At this time Ireland was not politically autonomous; consequently political discussion often centered on anti-British themes, including the repeal of the union between Great Britain and Ireland. Because the Catholic Church in Ireland was identified with the movement to gain greater freedom from Protestant England, the local priest was given an enormous amount of authority. Often the only educated person in the village, the priest became an expert on virtually everything. Moreover, he looked after people's material needs (Connell, 1968, pp. 144–145).

By contrast, in the Irish-American neighborhood, the priest's power covered mainly religious and familial concerns—the private sector; the political boss's power encompassed politics and business—the public sector. Therefore, politics and the political boss had supplanted some of the duties of religion and the priest. To a certain extent the ward boss replaced the priest as the dominant figure in the Irish-American neighborhood, at least for men (Stivers, 1976, p. 114).

The boss and his heelers had reputations as hard drinkers. The heeler's reputation was well deserved; the boss's was based more on past accomplishments. The demands of the ward boss's duties militated against much actual hard drinking (Glazer & Moynihan, 1963, pp. 226–28). Nonetheless, he was able to maintain his image as a hard drinker for several reasons: his working-class origins and earlier hard drinking; his association with hard-drinking heelers; and his pity for the drunkard whose family he looked after. Paramount, however, was the desire of his supporters to believe that their leader personified Irish manhood. The boss's hard drinking was the projected image his supporters had of themselves. The street gang and machine club then were the counterpart of the Irish bachelor group.

ADJUSTMENT, IDENTITY, AND HARD DRINKING

To say that hard drinking remained part of the group identity of Irish-American men is not enough. Under the impact of immigration, the psychological significance of group identity was intensified. Even though immigrants sometimes came as members of families, the act of migration was essentially individual (Handlin, 1951, p. 38). That is the ties of the old world community would never be reestablished, to the same degree, in America. To the isolated individual under the stress of migration and resettlement, cultural prejudice and discrimination, one's former group identity became vastly more important: it became an exaggerated way of differentiating oneself from others (Handlin, 1951, p. 285). One's ethnic identity could be transformed into a sort of secular religion, more specifically, a political religion (Herberg, 1960, pp. 6–16).

Hard drinking made one more Irish; it distinguished one from members of other ethnic groups. Hence in the affirmation of a drinking-oriented life-style, one was nationalistic. Ultimately drink—even more than it had in Ireland—acquired spiritual value; it became sacred. No one better understood this than Stephen Crane (1969), who capatured the essence of Irish-American Drinking in *George's Mother*. In this short novel, Crane depicts the initiation of a young man into the moral community of drinkers. Crane wrote that his hero's newfound friends "drank reverently." Protagonist George Kelcey's reaction was that

> he was all at once an enthusiast, as if he were at a festival of religion. He felt that there was something fine and thrilling in this affair, isolated from a stern world, and from which the laughter arose like incense (p. 146).

Drink, as a spiritual value, as something sacred, was part of a larger religion of Irish nationalism and Irish life-style. It was the exact point at which nationalism and life-style were intertwined. In Ireland drinking had been a consequence of communal conviviality; now in America, it had become the mystical means of community, creating an imaginary community to fill the void where a real one had once stood. Drinking became the occasion for recalling memories of Ireland. It was compensation for the grim reality of Irish-American life.

In the transformation of hard drinking from the bachelor group context to the machine club and street gang context, a subtle change had occurred: hard drinking was now symbolic of *Irish* identity. Previously, in Ireland, it was only symbolic of *male* identity.

Whereas hard drinking had been culturally remitted in Ireland, it was now under the province of direct moral control. Remissions and controls both go to make up the moral demands of a culture, but there is an important difference between them. Hard drinking as cultural remission was a release from the dominant symbolism of religious devotionalism/puritanism and was tolerated by the Catholic church in Ireland as long as it fell short of drunkenness. This of course served as a check on the hard drinking and encouraged drinkers at least to simulate sobriety. But in America, Irish drinking as a cult embodied nationalistic self-consciousness and individualistic life-style. Drinking became a religious phenomenon, under direct moral control, and as such implied strict obligations to oneself and others. Hard drinking had all the appearances of a religious obligation—the obligation to be Irish and to promote one's Irishness. The implication was that the more one drank, the more Irish one became. The habitual drunkard was at least a religious true believer and at worst a religious fanatic. On the social level, hard drinking was encouraged by saloonkeeprs and liquor dealers for economic reasons and by politicians for political reasons in their quest for a power base (Woods, 1898, pp. 128–138).

But as religion always creates its extremists, so also did the political religion of Irish drinking. Hard drinking as Irish group identity provided the opportunity for excess for the troubled individual. Alcoholic drinking, because it is so difficult to distinguish from hard drinking, was an "invisible" adaptation to stress. An addicted drinker blends in with a community of hard drinkers (Stivers, 1976, p. 130).

"Momism" and "Bossism"

The psychological implications of the Irish-American family and the ward political machine for drinking need to be explored. Erikson's (1963) concepts of "momism" and "bossism" are especially appropriate here.

Momism is the tendency of the mother to assume the roles of both mother and father. This places pressure on her to rear her children in a rigid, moralistic manner. Erikson (1963) argues that the American frontier helped create a strict division of labor between mother and father. Mother became "cultural censor, religious conscience, esthetic arbiter, and teacher." Mother became "Mom" only to the extent that Father became "Pop," the father who abdicates his dominant place in the family, especially in education and culture.

Puritanism was the critical force in the creation of "Mom." Continual migration, unchecked immigration, urbanization, class mobility, and female emancipation, in short, rapid social change, placed puritanism on the defensive (Erikson, 1963, p. 292). Erikson (1963) describes what happened:

> Puritanism, beyond defining sexual sin for full-blooded and strong-willed people, gradually extended itself to the total sphere of bodily living, compromising all sensuality—including marital relationships—and spreading its frigidity over the tasks of pregnancy, childbirth, nursing and training. The result was that men were born who failed to learn from their mothers to love the goodness of sensuality before they learned to hate its sinful uses. Instead of hating sin, they learned to mistrust life. Many became puritans without faith or zest. (pp. 292–293)

Momism is a readily apparent phenomenon among the Irish in Ireland and in America. A puritanical sexual code was upheld by the mother in the Irish family; in fact sin in general came close to being equated with sexual sin in particular. The Irish mother often functioned as an autocrat in the home, although she was intermittently gentle with and overprotective toward her children. The Irish and Irish-American pattern of few and late marriages was reinforced by this kind of puritanism.

Paradoxically there is a tie between the Irish mother and the priest. The mother acted as a representative of the Church in the family. Erikson (1963) observes that with the transformation of puritanism into its rigid, defensive versions, "the church community becomes a frigid and punitive Mom" (p. 319). No wonder then that the Irish priest was sometimes referred to as "she." The priest sometimes acted just like one's mother in the rigid enforcement of the moral code. The Irish mother often desired that at least one son become a priest. The priest was the mother's ideal of manhood for her son; the boss, in contrast, was the father's ideal.

The "boss" is the male counterpart to "Mom." The boss is a self-made autocrat who is manipulative in his personal relationships for political purposes and who bends or violates the law on behalf of his supporters and himself (Erikson, 1963). Because the Irish-American machine boss did so much to help those in his own ethnic group, his criminal and quasi-criminal offenses appeared to be a consequence of his great zeal to help his own. His self-interest was equated with the interests of Irish-Americans. In an attempt to improve the position of Irish-Americans in a highly competitive society, the boss turned his electorate into a machine, thereby rendering it irresponsible.

Both momism and bossism leave their victims overly dependent and irresponsible, that is, unable to choose for themselves. Erikson (1968) has discussed the psychological consequences of momism and bossism:

> Where the resulting self-definition, for personal or for collective reasons, becomes too difficult, *a sense of role confusion* results: the young person counterpoints rather than synthesizes his sexual, ethnic, occupational, and ty-

pological alternatives and is often driven definitely and totally for one side or the other. (p. 87)

In the Irish-American family, the mother held up to her son an ideal—the saintly Irish priest. Contradicting this, adult males preferred to the sons the hero of the political boss, the epitome of a man of the world, a force to be reckoned with, a hard drinker. Irish-American literature indicates that some mothers bitterly complained about their husband's or brother's drunkenness to their son. Concurrently, hard drinking was defined as positive identity by the son's male friends and as negative identity by his mother. Irish-American men often were placed in the position of having to choose between their mother and their friends.

Of course, some mothers did not contrast the alternatives of priest and boss as much as others, thereby affording their sons the opportunity to synthesize the two roles. But those mothers who had stifled their son's ability to synthesize the contrasting roles or even choose for themselves which role to assume inadvertently confirmed their sons as drunkards through their harping about drink's pernicious consequences. Erikson (1968, p. 174) has observed in clinical settings that a negative identity can provide the surest sense of identity when two positive identity choices cancel each other out. If the positive identities of saintly priest and hard-drinking boss cancel each other out, the negative identity of habitual drunkard becomes both real and desirable. Therefore, it was the mother's inability to distinguish between drinking and drunkenness and her inability to tolerate her son's drinking in any form that contributed to her son's choice of the negative identity of drunkard.

The Stereotype of the Irish-American as Drunkard

The stereotype of the Irish-American as drunkard reinforced and intensified the Irish-American attitudes toward drink. The Irish must be numbered among the ethnic groups most discriminated against in American history (Ernst, 1949, p. 66). The Irish were often compared to blacks in the nineteenth century: dirty, lazy, rowdy, and ill-mannered (Curtis, 1971, p. 101). But above all, the Irish were thought to be drunkards (Clark, 1973, p. 103; Wittke, 1956, p. 262). The British had long regarded the Irish in Ireland as drunkards; therefore, it is no surprise that East Coast Establishment British-Americans so regarded the Irish immigrants.

The stereotype of the Irish-American as drunkard became institutionalized in a plethora of ways: (1) employment practices—contractors and employers often paid the Irish part of their wages in rotgut whiskey and forced them to drink heavily in order to extort their wages or make them work harder (Potter, 1960, p. 320); (2) job discrimination—the Irish were often denied work because their reputed drunkenness made them appear unreliable (Ernst, 1949, p. 67); (2) police discrimination—the Irish were arrested for drunkenness while members of other ethnic groups were left alone, and disorderly conduct was thought always to be the result of drunkenness (Lane, 1967, p. 76; Potter, 1960, p. 526); (4) mass media—the stereotype was transmitted visually and verbally in the form of cartoons, newspaper sketches, and stage plays (Stivers, 1976, p. 142–150).

"Stage Irish" refers to an Irish caricature that always included the actor's carrying and drinking from a bottle of whiskey. The stage portrayal of Irish drinking was invariably humorous, but the depictions in newspaper cartoon and editorial were quite vicious. One 1867 Thomas Nast cartoon depicts Irish-Americans beating up respectable citizens and policemen. The Irish-Americans have rum and whiskey bottles in their pockets (Stivers, 1976, p. 144). These contrasting portraits of Irish-American drinking will be referred to as the happy drunk image and the habitual drunkard image.

The happy drunk harms no one save himself (which is de-emphasized). He is affable and lovable when drunk, a comic figure speaking nonesense one moment, waxing eloquent the next. He is great entertainment. At worst, he simply collapses at the end of the evening. Examples include songs such as "The History of Paddy Denny's Wife and His Pig" and "Tim Finnegan's Wake" (Wittke, 1956, p. 245).

The habitual drunkard, by contrast, is a menace to everyone, including himself. His drinking frequently leads to crime and brawls for himself and loss of income and peace of mind for his family. Finally, he becomes brutalized by alcohol. Thomas Nast, Fredrick Opper, and Joseph Keppler were among the most famous and most vicious American caricaturists of Irish-Americans in the late nineteenth century (Curtis, 1971, Ch. 5).

Before 1890, the American stereotype of the Irish, especially in New England, had accentuated the dire consequences of drinking, although the happy drunk image competed for attention at the same time. After 1890, the stage Irishman, with his more humorous, positive traits, was accepted universally, even by social scientists (Saveth, 1948, pp. 56–57; Solomon, 1956, p. 71). In summary, prior to 1850, the negative image (habitual drunkard) predominated; from 1850 to 1890, the negative and positive images competed for recognition; after 1890, the positive image (happy drunk) predominated (Solomon, 1956, pp. 152–155).

The main reason for the gradual ascendancy of the happy drunk image over the habitual drunkard image appears to be the begrudging acceptance of Irish-Americans by "native" Americans (British-Americans). Solomon (1956, p. 152–155) has demonstrated that after 1890 American fear and hatred turned toward the more recent immigrants from southern and eastern Europe (e.g., Italians, Poles) whose customs appeared more alien and threatening than those of the earlier immigrants (e.g., Irish, Germans, English, French, and Scandinavians). Those desiring to restrict the numbers and kinds of immigrants (restrictionists) contrasted the old with the new immigrant as a model for distinguishing between those who were capable of being Americanized and those who were not. Moreover, by 1890, many Irish-Americans had proved to be hard-working, patriotic citizens and had become modestly successful. They could be pointed to as examples of what American tolerance and goodwill could do for the "right" kind of immigrant.

It is one thing to show that prejudice exists and still another to demonstrate its effect on its subject. Negative identities, such as that of habitual drunkard, can be internalized when a group faces discrimination, but concomitantly desires acceptance from the discriminating group. Erikson (1968) speaks to this point:

> Therapeutic as well as reformist efforts verify the sad truth that in any system based on suppression, exclusion, and exploitation, the suppressed, excluded, and exploited unconsciously accept the evil image they are made to represent by those who are dominant. (p. 59)

But it was more than merely accepting an image, for the Irish wanted desperately to be part of America. Irish leaders accepted almost all the dominant ideals of Americans, such as hard work, frugality, perserverance, and success (Potter, 1960, p. 282). Irish-Americans made Ireland's history over in the ideal image of America and exaggerated their former group traits by caricaturing their origins (Wittke, 1956, p. 161; Shannon, 1963, pp. 132–133). Irish-American culture had become more American than Irish.

One does not have to infer unconscious processes, however, to show how readily Irish-Americans accepted the positive image of the happy drunk. Beer (1926, p. 152) has argued that the stage-Irish caricature had become "sacred with Americans." The Irish working class never resented the comic Irishman (Shannon, 1963, pp. 142–143; Ernst, 1949, p. 148). Both Irish and Americans attended the plays featuring stage Irishmen and read comic strips such as "Happy Hooligan," which promoted caricatured Irish characters.

Therefore, the stage Irishman was acceptable to the "native" American and the Irish-American because it enabled both groups to forget what had been done to the Irish: the prejudice, discrimination, and exploitation. A farcical Irishman was palatable to native Americans because he was not a menace and remained beneath them. It was likewise amenable to the Irish because it was simultaneously a source of their distinctiveness and of their qualified acceptance into American society. It was as though native Americans could accept only those ethnic groups who had been typed, pigeonholed, and thus pacified; and the ethnic groups who desired so much to be accepted were willing to take whatever was proffered. The stage-Irish drunk was a bond between Americans and the Irish.

At this point we can see how the stereotype of the Irish-American as drunkard reinforced and intensified the Irish-American ethic of hard drinking. Drinking had become a spiritual value in the new religion of Irish-American nationalism. But if one were Irish when drinking, it followed that one proved most Irish when one had the most to drink. The drinking was now open ended. To drink excessively symbolized one's Irishness, created a community among co-drinkers that had been lost, and was a bond of acceptance into American society. On the psychological level, those Irish men whose identity choices (priest *versus* boss) were most contrasted and whose mothers most railed against the evil of drink were the most likely to become alcoholic.

IRISH-AMERICAN DRINKING IN THE MID-TWENTIETH CENTURY: CONSUMERISM AND THE RISE OF THE "NEW ETHNICITY"

The new ethnicity of the 1960s refers to the renewed ethnic consciousness among white ethnic groups following on the heels of the black civil rights movement. During this time, consumerism became a spiritual value for many Americans.

For several decades prior to the rise of the "new ethnicity" in the late 1960s, there was a lack of interest in ethnicity among social scientists and a prevailing sense that the American Melting Pot had largely assimilated immigrants, at least white immigrants. Glazer and Moynihan (1963) attempted to show how illusory this complacency about ethnicity was, as they skillfully argued that many immigrants and their offspring in the urban centers were still living a ghetto existence. But assimilation does not necessarily imply acculturation. By this I mean that one can simultaneously be discriminated against in the social structure (unassimilated) and be more or less culturally integrated (acculturated).

This is not to suggest that Irish identity was weakening for the first- and second-generation Irish-American ghetto dweller. But for third- and fourth-generation Irish-Americans who had become more assimilated, Irish identity was less decisive and perhaps for some had become almost an afterthought.

For the third- and fourth-generation Irish-American and those not confined to a ghetto environment, certain changes in American culture were becoming as much a motivation for heavy drinking as their Irish-American identity. For those most removed physically and socially from the time of immigration, wider cultural changes were perhaps the paramount influence. This was difficult to perceive, for the pattern of heavy drinking behaviorally remained the same, but its meaning was being altered. What follows is an interpretation of this process based upon well-documented changes in American culture.

IRISH-AMERICAN DRINKING IN THE CONTEXT OF CONSUMERISM

Before looking at the impact of the "new ethnicity" upon the meaning of Irish-American drinking, we should first look at consumerism as a spiritual value. The concepts *sacred* and *secular* are important in this regard. Simply, *sacred* will refer to that to which man attributes greatest spiritual value and according to which meaning and orientation are gauged. Sociologically, the sacred becomes a basis for culture and acts to integrate the individual into the group in establishing primordial ties among individuals. *Secular* on the other hand, is that to which no unusual significance is attributed. It is the safe or neutral part of life, unprotected by stringent religious regulations (Caillois, 1959).

As qualities, what is sacred and what is secular vary from culture to culture across time. Thus, they are dynamic, not static, qualities. Often secularization refers to that which once was considered sacred but now has become or is being secularized; similarly sacralization refers to that which at one time was deemed secular but now is in the process of being made sacred.

If drinking was sacralized in relation to Irish-American identity in the nineteenth century, it was becoming secularized in the twentieth century as ethnicity became less important to Irish-Americans. But if, however, drinking was becoming secularized in regard to Irish-American identity, it was simultaneously become sacralized in regard to consumption. The term here means defining life's meaning in terms of goods and services to be consumed and turning even what is spiritual into something to be consumed. The more Irish-Americans became economically successful, the more they could "afford" to be drawn into a life-style devoted to consumerism.

To elaborate on this point, one could look to one of the more profound changes in twentieth-century America, a transition from a work ethic to a fun ethic. Martha Wolfenstein (1951) documented this transition with respect to childrearing practices, and David Riesman (1961) placed the change in a larger societal and cultural setting. This is not to suggest that work ceased to be important to people or that the work ethic dissipated altogether. But increasingly work came to be less an end in itself, a self-contained ethic, and more a means of self-fulfillment.

There has been a movement away from production and toward consumption in American culture. In Riesman's (1961, p. 189) felicitous phrase, "popular culture is in essence a tutor in consumption." Daniel Boorstin (1973, pp. 89–164) has shown that the roots of a consumption ethic go back to the nineteenth century.

The myth of progress, so deeply adhered to over the past two centuries, initially provided a hope that mankind could through technology and political reform create the good society and good life in this world (Becker, 1932). But as the world wars and other atrocities of this century made it increasingly difficult to believe in a future utopia, more and more hope was pinned to experiencing the good life now. And this meant consumption.

Indeed, the ethic of fun so fundamental to the "religion" of consumption threatens even to eclipse the work ethic. Recently, William F. Buckley commented on public opinion polls that indicated that a majority of young people in the United States, England, and Germany when given the choice would rather not work than work. Studs Terkel's *Working* (1975) indicates the spiritual vacuum most people experience in their work today. Consequently we turn to leisure, sports, recreation, vacations, etc., to find fulfillment.

As Ellul (1975, Ch. 3) has indicated, technology and the consumer goods and services it provides are sacred today. Technology is what guarantees our future, solves our problems, and offers us the good life. Consumer goods and services, as manifestations of technology, provide a "sense of the sacred" in a plethora of ways: For one, it is a sports car; for another, a stereo; for still another, a painting. And for certain Irish-Americans, it was alcohol.

Given a cultural predisposition to view alcohol as symbolic of Irish-American identity, it is very likely that in the age of consumerism some Irish-Americans would choose alcohol to experience the good life. Thus, heavy drinking as a symbol of Irish-American identity and heavy drinking as a form of consumerism would be synthesized. This subtle change would be imperceptible, however, because the overt behavior pattern of heavy drinking would have remained the same. In sum, I am hypothesizing that the closer we get to the 1960s, the larger the role consumerism as a spiritual value plays in Irish-American drinking.

Another factor may have hastened the sacralization of Irish-American drinking in relation to consumerism: the stereotype of the Irish-American as drunkard. Heavy drinking was a symbol of Irish-American identity. What this did, of course, was to cause many to equate Irishman with drunkard. The American historian Carl Wittke (1956, p. 262) remarked that for a time in the second half of the nineteenth century "stage Irishman and drunkard were practically synonymous." One effect among Irish-Americans who accepted the stereotype was to make drinking even more important than it had been before. Now it was equivalent to Irish. It would not take much in this situation for the drinking to become even more important than one's Irish-American status, to become an end in itself. Thus, at first one drank heavily because one was Irish, whereas now one was Irish because one drank heavily. Perhaps this subtle shift in emphasis from Irishness to drinking can be gleaned from the recent confession of a third-generation Irish-American, "Drinking and the people I drink with are sacred to me." Yet he drank more with non-Irish than with Irish friends, and drinking was mentioned first. Truly it was the drinking that had become sacred to him (Stivers, 1978, p. 130).

FROM THE OLD TO THE NEW ETHNICITY

What are the implications of the new ethnicity for Irish-American drinking? Help in answering this question has been provided by Marty (1976). In his *A Nation of Behavers,* he has a chapter on ethnic religion, in which, to avoid using perjorative terms, he distinguishes between Ethnicity A and Ethnicity B. Ethnicity A refers to the cultural situation of a community of immigrants or of an ethnic neighborhood still largely dominated by immigrants, in which identification with consciousness of one's ethnic heritage is "inescapable, automatic, and reflexive" (Marty, 1976, p. 164). That is, because of discrimination and prejudice, a group is isolated, forced in upon itself, and in compensation, sacralizes its group identity. Ethnicity B, on the other hand, is "escapable, intentional, and reflective" (Marty, 1976, p. 164). It involves a conscious decision by third- and fourth-generation ethnics to reassert their common origin and heritage. Certainly Ethnicity A aptly describes the plight of Irish immigrants in the nineteenth century. And as well, Ethnicity B fits the situation of those of the third and fourth generation who are intensely self-conscious of their ethnic identity (Marty, 1976, pp. 164–169).

The final word is not on the percentage of Irish-Americans who have become new ethnics. But let us assume for the moment that at least a sizable minority of later generation Irish-Americans has become intensely ethnic. Should this cause a resurgence of heavy drinking as symbolic of Irish-American identity? It is necessary before answering this question to explore the larger meaning of Ethnicity B (the new ethnicity).

The new ethnicity appears to be a form of religion that was anticipated by Will Herberg (1960) who remarks

> It is "peace of mind" that most Americans expect of religion. "Peace of mind" is today easily the most popular gospel that goes under the name of religion; in one way or another it invades and permeates all other forms of contemporary religiousity. It works well with the drift toward other-direction characteristic of large sections of American society, since both see in adjustment the supreme good life. What is desired, and what is promised, is the conquest of insecurity and anxiety, the overcoming of inner conflict, the shedding of guilt and fear, the translation of the self to the painless paradise of "normality" and "adjustment!" Religion, in short, is a spiritual anodyne designed to allay the pains and vexations of existence. (p. 267)

For third- and fourth-generation Irish-Americans outside the confines of a closed ethnic community, the new ethnicity can readily become a religiosity of self-assurance. Suddenly one's heritage, one's roots, one's ethnic identity take on inordinate importance in an effort to shore up one's sagging self-esteem. Especially suspect are the recent converts who are devout Irish-Americans, but who know almost nothing about Irish history and their own heritage (Greeley, 1972). They have become consumers of their ethnic heritage. Thomas Luckmann (1967) noted this merger of consumerism and private religion:

> The individual, originally socialized into one of the "versions" (church denominations) may continue to be "loyal" to it to a certain extent, in later life. Yet, with the pervasiveness of the consumer orientation and the sense of autonomy, the individual is more likely to confront the culture and sacred cosmos as a "buyer." One religion is defined as a "private affair" the individual may choose from the assortment of the "ultimate meanings" as he sees fit—guided only by the preferences that are determined by his social biography. (pp. 98–99)

Consumerism and private religions, then, like the new ethnicity, are mutually reinforcing. To select one's own religion is a form of consumerism, and consumerism is form of private religion. To consume one's ethnic identity and ethnic heritage is but a "denomination" within the larger religion of consumption.

Insofar as the stereotype of the Irish-American as drunkard has not

completely dissipated, the new ethnicity presents an obvious danger to
Irish-Americans: once again heavy drinking is identified with being Irish
and is sacred. One might expect that the new ethnicity for Irish-Americans would further intensify the tendency toward heavy drinking that
consumerism itself has sustained in this century (Stivers, 1978).

CONCLUSION

In conclusion, I suggest that heavy drinking among Irish-Americans
has had three meanings:

1. Heavy drinking was symbolic of Irish-American identity in the
   nineteenth century;
2. Heavy drinking next was secularized/sacralized as a form of consumerism for Irish-Americans in the twentieth century; and finally
3. Heavy drinking becomes a means of consuming one's ethnic
   identity and heritage.

Although the pattern of heavy drinking has remained constant, its meaning has changed significantly. To understand this process it is necessary
to understand the transformation of Irish ethnicity as part of an overall
national change from a country of immigrants, with their variegated
communities and neighborhoods, to a mass society of consumers whose
outward expressions of what they consume constitute a large part of
their ethnic distinctiveness.

APPENDIX

The various measures of heavy drinking indicate that Irish-Americans have had
for each statistical index the highest rate among the ethnic groups tested for the
period 1860–1950. Indices of heavy drinking include the arrest rate for drunkenness, the death rate for cirrhosis of the liver, the death rate for acute alcoholism, and the admission rate for alcoholic psychoses. Tables A-1 and A-2
illustrate the Irish-American propensity for heavy drinking. More statistical data
on the subject can be found in Stivers (1976, Ch. 1).
    One must exercise caution in interpreting these data. Convictions for drun-
kenness and admissions for alcoholic psychoses are categories that sometimes
are interpreted differently from period to period and from control agent to control
agent. Thus one is not always sure what is being measured—the behavior of
the drinker or the perception of the control agent, which is sometimes a prej-
udiced perception. Irish-Americans were one of the most discriminated ethnic

TABLE A-1.
Convictions for Drunkenness and
Disorderly Conduct by Nativity in New
York City, Courts of Special Sessions, 1859[a]

Nativity	Rate of drunkenness and disorderly conduct per 100,000 population
Ireland	3,234
Scotland	1,548
England	1,190
France	967
Canada	650
Germany	641
United States	383

[a] Data from *Immigrant Life in New York City, 1825–1863* (p. 204) by R. Ernst, 1949, New York: King's Crown.

TABLE A-2.
Average Annual Number of Foreign White First Admissions with
Alcoholic Psychoses to all Hospitals for Mental Disease in New York State,
1929–1931, per 100,000 Population, Age 25 and over, Classified by Country
of Birth and Sex[a]

Country of birth	Males	Females	Total
Ireland	50.1	11.0	30.5
Scandinavia	15.1	0.6	7.9
Italy	8.6	—	4.3
Germany	7.0	0.4	3.8
England	5.4	4.4	4.8
Native whites of native parents	7.7	1.5	4.6

[a] Data from *Social and Biological Aspects of Mental Disease* (p. 203) by B. Malzberg, 1940, Utica, NY: State Hospitals Press.

groups in American history. Even when one takes this into account, because the Irish-American rates for most indices of heavy drinking are so much greater than those of other ethnic groups, I think that one should still conclude that the Irish-American propensity for heavy drinking was great indeed.

REFERENCES

Arensberg, C., & Kimball, S. *Family and community in Ireland.* Cambridge, MA: Harvard University Press, 1968.

Bagenal, P. Y. *The American Irish and their influence on Irish politics.* London: Kegan Paul, Trench, 1882.

Bales, R. F. *The "fixation factor" in alcohol addiction: An hypothesis derived from a comparative study of Irish and Jewish social norms.* Unpublished doctoral dissertation. Harvard University, 1944.

Bales, R. F. Attitudes toward drinking in the Irish culture. In D. Pittman & C. Snyder (Eds.), *Society, culture, and drinking patterns.* New York: Wiley, 1962.

Becker, C. *The heavenly city of the eighteenth century philosophers.* New Haven, CT: Yale University Press, 1932.

Beer, T. *The mauve decade.* New York: Knopf, 1926.

Boorstin, D. *The Americans: The democratic experience.* New York: Knopf, 1973.

Bushee, F. Ethnic factors in the population of Boston. *Publications of the American Economic Association,* 1903, 4.

Caillois, R. *Man and the sacred.* New York: The Free Press, 1959.

Clark, D. *The Irish in Philadelphia.* Philadelphia: Temple University Press, 1973.

Connell, K. H. *Irish peasant society.* Oxford: Claredon Press, 1968.

Crane, S. *The works of Stephen Crane,* Vol. 1, *Bowery Tales.* Charlottesville: University Press of Virginia, 1969.

Curtis, L. P. *Apes and angels.* Washington, DC: Smithsonian Institution Press, 1971.

Dunlop, J. *A philosophy of artificial and compulsory drinking usages in Great Britain and Ireland.* London: Houlston and Stoneman, 1839.

Ellul, J. *The new demons.* New York: Seabury, 1975.

Erikson, E. *Childhood and society.* New York: Norton, 1963.

Erikson, E. *Identity: Youth and crisis.* New York: Norton, 1968.

Ernst, R. *Immigrant life in New York City, 1825–1863.* New York: King's Crown, 1949.

Fontaine, E. de la. Cultural and psychological implications in case work treatment with Irish clients. In *Cultural problems in social case work.* New York: Family Welfare Association of America, 1940.

Glazer, N., & Moynihan, D. P. *Beyond the melting pot.* Cambridge, MA: MIT Press, 1963.

Greeley, A. *That most distressful nation.* New York: Quadrangel, 1972.

Handlin, O. *The uprooted.* New York: Grosset & Dunlap, 1951.

Handlin, O. *Boston's immigrants.* Cambridge, MA: Harvard University Press, 1959.

Herberg, W. *Protestant, Catholic, and Jew.* Garden City, NY: Anchor, 1960.

Kennedy, R. *The Irish.* Berkeley: University of California Press, 1973.

Lane, R. *Policing the city: Boston, 1822–1855.* Cambridge, MA: Harvard University Press, 1967.

Luckmann, T. *The invisible religion.* New York: Macmillan, 1967.

Malzberg, B. *Social and biological aspects of mental disease.* Utica, NY: State Hospitals Press, 1940.

McNabb, P. Social structure. In J. Newman (Ed.), *The Limerick rural survey, 1958–1964.* Tipperary, Ireland: Muinter Na Tire Rural Publications, 1964.

Marty, M. *A nation of behavers.* Chicago: University of Chicago Press, 1976.

Mattis, M. *Irish mobility in Buffalo, New York, 1855–1875.* Paper presented at the Midwest Sociological Meetings, 1975.

Potter, G. *To the golden door.* Boston: Little, Brown, 1960.

Riesman, D. *The lonely crowd.* New Haven, CT: Yale University Press, 1961.

Saveth, A. *American historians and European immigrants.* New York: Columbia University Press, 1948.

Shannon, W. V. *The American Irish.* New York: Macmillan, 1963.

Solomon, B. *Ancestors and immigrants.* Cambridge, MA: Harvard University Press, 1956.

Stivers, R. *A hair of the dog.* University Park: Penn State Press, 1976.
Stivers, R. Irish ethnicity and alcohol use. *Medical Anthropology,* 1978, *2*, 121–135.
Terkel, S. *Working.* New York: Avon, 1975.
Wittke, C. *The Irish in America.* Baton Rouge: Louisiana State University Press, 1956.
Wolfenstein, M. The emergence of fun morality. *Journal of Social Issues.* 1951, *7*, 15–25.
Woods, R. A., *The city wilderness.* Boston, MA: Houghton Mifflin, 1898.
Woods, R. A., & Kennedy, A. J. *The zone of emergence.* Cambridge, MA: Harvard University Press, 1962.

9

Appalachia
The Effects of Cultural Values on the Production and Consumption of Alcohol

GRACE TONEY EDWARDS

INTRODUCTION

Steely eyes, glinting rifles, and clear white moonshine—the terms are synonymous with a stereotyped view of Appalachia. Although the stereotype is not without basis, it falls far short of a complete picture of the region. The rifle-toting, jug-swigging mountaineer, popularized by comic strips and prime-time television shows, is a caricature growing out of the eastern highlands' reputation for producing much of the nation's illegally distilled spirits. But this mountaineer's counterpart is the strait-laced, churchgoing father who works from nine to five at the local city hall. Or it is the young woman who attends college in her hometown and then goes back to teach in the high school she graduated from. Or it is the coal miner who works the hoot owl shift so that he can care for his young son while his wife takes a secretarial course.

The possible illustrations rapidly multiply, for Appalachia is geographically large and culturally diverse. The mountain range to which it clings stretches from Maine to Mississippi, but officially the area dubbed

GRACE TONEY EDWARDS • Radford University, Appalachian Studies Program, Radford, Virginia 24142.

Appalachia begins in the north with southwestern New York State (*1979 Annual Report Appalachian Regional Commission,* 1980). It pulls in a goodly portion of western Pennsylvania, a bit of Maryland and Ohio, all of West Virginia, a sizable chunk of western Virginia, eastern Kentucky, east Tennessee, and western North Carolina, smaller pieces of northwestern South Carolina, north Georgia, northern Alabama, and finally, northeastern Mississippi. Thirteen states in all are represented in this long slice of territory that includes not only mountains, but broad plateaus, river valleys, and foothills as well. Representative of the region's diversity are its sizable cities, small towns, and vast stretches of sparsely populated peaks and valleys; its coal mines, steel mills, and textile factories; its ski resorts, tourist attractions, wealthy second-home developments, and pockets of poverty. Again, the illustrations multiply in order to show that Appalachia is pluralistic with many cultures at work; but most of the people, at least the native inhabitants, have some connection to the folk culture that traces its roots to the very earliest settlement of the region. The generalizations posited here then are based largely on values emerging from the dominant folk culture, as I know it in southwestern Virginia, West Virginia, eastern Kentucky, east Tennessee, and western North Carolina.

Before I begin to describe and generalize, perhaps the establishment of my own credentials is important. In Appalachia, almost any sort of transaction between strangers is prefaced by an account of family ties and place of residence. On both counts I am legitimately Appalachian. As a native of the foothills of western North Carolina, I come from generations of mountain-born and bred Scotch-Irish and English stock. Growing up in the shadow of the Carolina mountains, I had no inkling that my own cultural heritage would eventually become my prime subject of study. However, as a student and eventually teacher at Appalachian State University in Boone, North Carolina, I grew more and more fascinated with mountain culture and mountain people. Finally at the University of Virginia I formalized that fascination by earning a Ph.D. in English with a concentration in Appalachian folklore and literature.

Pursuit of a job where I could cultivate my interest brought me to Radford University in Radford, Virginia, located in the New River Valley between the Blue Ridge and Alleghany ranges. Here, while developing a program in Appalachian Studies, my work has taken me from the fiction of Gurney Norman to the sociological facts of John Stephenson, from field work with traditional musicians to a seminar on alcohol abuse with medical professionals. My observations in this chapter are based on a wide array of experiences: personal knowledge, clinical contact, studies on alcohol production and consumption, creative literature, folklore both as practiced and collected/analyzed, numerous conversations with various professionals and grassroots Appalachians. Most of what I say is a synthesis of these diverse influences, which coalesce here to form my own understanding of alcohol use in Appalachia.

SETTLEMENT OF THE APPALACHIAN REGION

Now what about that stereotype of the mountain moonshiner? How has Appalachia become so intricately linked with the illicit production of alcohol? Is it the mountaineer who consumes the vast quantities of corn liquor that have allegedly flowed from his mountains? What is the value system that nurtures the moonshiner and the non-moonshiner alike?

To begin to answer these questions, we can reconstruct a series of vignettes depicting the settlement of the area now known as Appalachia. By the early 1700s, emigrants from the British Isles and other selected western European countries had begun to move into the mountainous regions of the New World. They were forced into the highlands because the fertile flat lands to the east had already been claimed by their predecessors. The Appalachian Mountains were at that time the western frontier, the first great geographical barrier to westward expansion. Though wresting a living from the hilly, rocky land promised a hard future for the newcomers, they accepted the challenge and even seemed to appreciate the isolation of their new homeland.[1]

Perhaps the character of these particular settlers required a degree of autonomy provided by the tall peaks and hidden coves. Preponderantly Scotch-Irish, they were proud, independent, and fiercely protective of their individual rights. These traits had grown in them over a period of some two thousand years of having to survive under extremely adverse conditions. The group now known as the Scotch-Irish had at its nucleus a large remnant of the Celtic population, King Arthur's descendants. Pushed from place to place by various invaders of Britain, the Celts and other ethnic splinters eventually found themselves in the hills of the Scottish border country. From there, at the request of the British crown, they immigrated to northern Ireland in the early 1600s. In short

---

[1] I am indebted to Dr. Cratis Williams, professor emeritus of Appalachian State University in Boone, North Carolina for most of the historical information given here. It is gleaned from the references cited under his name at the end of the chapter as well as from his numerous lectures, and conversations with him, over the years.

order this resourceful group proved its excellence in farming, in the woolen and linen trades, in the establishment of a public education system for both boys and girls (Williams, 1972).

By the early 1700s, the Scotch-Irish had succeeded so well at what they were sent to Ireland to do that the established British order considered them a threat. Thus, excessive curbs were placed on the producers' markets and, for good measure, their religious zeal as well. By the 1730s, the Scotch-Irish, once again in search of a home, were immigrating in droves to America (Williams, 1972, 1980).

Sometimes traveling south from Pennsylvania by way of the great valley of Virginia, sometimes traveling due east across the coastal plains and piedmont sections of the Carolinas and Virginia, they moved into the mountains, bearing their calfskin copies of the classics and the English poets. They reestablished their culture, which was strong enough to subsume most of the other ethnic groups they encountered. Except for a few special communities, such as the Mennonites or the German Baptists, the English, French, and German settlers allowed their cultures to be absorbed by the Scotch-Irish. As a consequence the dominant culture of the Appalachian Mountains is still Scotch-Irish, and today Scotch-Irish blood is believed to run through the veins of approximately 60% of all native Appalachians (Williams, 1980, Videotape).

EARLY DISTILLATION AND CONSUMPTION OF ALCOHOL

One of the significant legacies of that culture was the ability to distill spirits from grains, fruits, and vegetables. No people could lay claim to a finer alcoholic product than the Scots and the Irish. Historical accounts show that whiskey-making flourished in Scotland as early as 1494 (Carr, 1972). In the hands of the Scots it had undergone an evolutionary change from the use of mead (a combination of honey and other substances) as its basic ingredient to the use of grain. The Irish, too, discovered the value of grain to produce a spirited beverage competitive with that of the Scots. With more than two hundred years experience behind them when they immigrated to America, the Scotch-Irish probably set up their homemade pot stills as soon as they had harvested their first crops to feed them. They assumed the right to convert their excess foodstuffs into liquid form and they vehemently opposed any attempt to regulate or tax the fruits of their home distilleries.

Some of the Scotch-Irish still living in the British Isles in the late 1700s and early 1800s had been subject to excise taxes and even prohibition of small-gallonage stills. Hence, they had become proficient at plying their trade undercover. In America little restriction hampered the distiller until 1791 when a new nation flexed its growing muscle by passing a law that established an excise tax on the "domestic manufacture of alcoholic spirits" (Carr, 1972, p. 18). Outrage over the tax mounted to such a fury in Pennsylvania that the Whiskey Rebellion resulted, and subsequently long years of animosity between the whiskey-maker and the revenuer.

With their hardy approach to life tempered by a Calvinistic philosophy of fatalism about what lay ahead, the mountaineers multiplied rapidly, for they believed in large families. They named their children Homer, Virgil, Cicero, Helen, Penelope, and Beatrice after the classic writers and heroes they admired. They adhered to the patriarchal structure of the family, with the father firmly established as the authoritative head of the household. Williams (1972) notes that "the woman accepted her role as being part and parcel of her lord and master. She deferred to his wishes" (p. 1); yet in her own quiet, resilient, and determined way she managed to effect certain changes to her own advantage without her husband even realizing his demure wife's strength.

In 1905, Emma Bell Miles, a resident of Walden's Ridge, Tennessee and a pioneer in cultural interpretation in the highlands, observed the differences in male and female roles. Although alcohol use was not a major focus, she noted its restriction to the male. In her book, *The Spirit of the Mountains* (1905, 1975), she wrote:

> [The man's activities] are the adventures of which future ballads will be sung. He is tempted to eagle flights across the valleys. For him is the excitement of fighting and journeying, trading, *drinking*, and hunting, of wild rides and nights of danger. To the woman, in place of these, are long nights of watching by the sick, or of waiting in dreary discomfort the uncertain result of an expedition in search of provender or game. . . . The woman belongs to the race, to the old people. He is a part of the young nation. His first songs are yodels. Then he learns dance tunes, and songs of hunting and fighting and *drinking*, and couplets of terse, quaint fun. It is over the loom and the knitting that the old ballads are dreamily, endlessly crooned. (pp. 68–69; my italics)

Twice in this passage Miles associates the male with drinking and drinking with excitement, boisterous fun, activities of youth or youth-seekers. Historically, in Appalachia, drinking has indeed been the male's domain, and it has been reserved for special times, festive occasions, celebrative events. At the risk of excessive generalizing, I submit that even today the same conventions hold somewhat true in the region, particularly among the more traditional, more conservative segment of the population. Older women are not prone to drink publicly, or if they do, they are discreet and restrained. This is not to say that no female alcoholics exist in Appalachia, for they do; but according to Dr. Robert Mullaly (1982), formerly of St. Albans Psychiatric Hospital in Radford, Virginia, their numbers appear to be much smaller than their male counterparts, and they are most often closet drinkers, in an almost literal definition of that term.

Despite her reluctance to imbibe publicly and for pure pleasure, the Appalachian woman has long been associated with the production and consumption of alcohol. Before individual distilling became a covert activity, she helped her spouse convert his excess corn and rye into whiskey; she may have gathered the apples or peaches for a run of brandy. (See Jess Carr's *The Moonshiners*, 1977, and Sherwood Anderson's *Kit Brandon*, 1936, for two fictional views on woman's role in whiskey-making and whiskey transporting.) When the winter winds swept the eastern ridges, she ladled out many a teaspoonful of "cough medicine" to herself and the children, probably a thick syrup made from honey and whiskey or maybe from rock candy dissolved in corn liquor. Perhaps she concocted a mixture of honey, vinegar, and whiskey to give to Granny when she complained of rheumatism. Maybe she soaked a bit of rag in the potent liquid to place on the throbbing tooth of a child. And she may have even rubbed the teething baby's tender gums with a cloth dabbed in brandy. The medicine chest of the early Appalachian home was seldom without some form of spirits. It is not uncommon today to find many of the old remedies calling for alcohol still in use, along with self-prescribed patent medicines from the drugstore.

BIRTH OF THE MOONSHINER

Rarely does the Appalachian family openly exhibit its stock of alcoholic beverages, whether their use be medicinal or otherwise. Changes in distilling laws and in moral attitudes gradually forced production and consumption into hiding. With the imposition of excise taxes on home distilleries and other restrictions that eventually found their way into the mountains, the moonshiner was born (Carr, 1972). Rather than bow to laws he thought unjust, the Scotch-Irish mountaineer took his still to the caves and hollers. By the last quarter of the nineteenth century, he had discovered a means of turning a profit on his corn crop. Living high in the mountains and far away from any sizable market, he found transportation of his whole grain practically impossible. But by converting that corn into liquid, he could take it out on a mule's back and make more money to boot. As Jess Carr puts it, "twenty bushels of corn might bring ten dollars; turned into corn liquor, it would bring seventy-five dollars" (1972, pp. 72–73). No enterprising mountaineer could resist that kind of deal, but he angrily resented the government interference that reduced his profits. Harking back to that Scotch-Irish independence, he believed it his right to do with his crops what he would. Isolated and autonomous in his mountain home, he had little interest in the national welfare; his concerns for himself and his family were much more immediate.

With the turn of the century, the temperance movement, which had been simmering for one hundred years, began to boil. As legal avenues for production and sale of alcohol were closed off one by one, the moonshiners simply stoked their fires and bottled greater and greater quantities of their corn. Prohibition, depression, and two world wars further fueled the production of illegal whiskey. Unfortunately, shortcut methods motivated by greed, competition, and fear of revenuers' raids began to replace the centuries-old techniques and recipes that had resulted in quality whiskey. Rotgut, popskull (vernacular names for inferior whiskey), and lead poisoning were among the results of moonshining in the southern mountains in the twentieth century. Franklin County, Virginia, lying along the crest of the Blue Ridge, gained the reputation of being the moonshine capital of the world. According to Jess Carr (1972), Sherwood Anderson reported in *Liberty* magazine in 1935 that some 70,448 pounds of yeast had been sold in Franklin County over a four-year period. Other figures revealed that single families had used as much as 5,000 pounds of sugar in one month. More than 600,000 non-gurgling five-gallon cans were shipped into the county by rail. Anderson deduced that if all those containers went out again full of moonshine, Franklin County produced approximately 3,501,115 gallons of illicit alcohol during that four-year period (p. 116).

Who drank it all? How much was for the home folks? Not much, actually, for the whole county had a population then of only 24,000 (Carr, 1972, p. 115), which would have amounted to almost 146 gallons of moonshine for every man, woman, and child. Most of the brew was exported to urban centers, in both the South and the North. In fact, even had the home folks been able to consume the products of their mountains, many of them would have refrained, for they were by then well within the grip of a Bible-Belt moralism that dubbed drinking a sin (Carr, 1972).

MOUNTAIN VALUES

The change in attitude toward drinking alcoholic beverages had begun as early as 1800 with the rumblings of temperance advocates. When the great revival swept through the mountains a short time later, it left in its wake a preponderance of Baptists and Methodists who adhered to their own brand of fundamental Calvinist theology. One day the preachers were denouncing strong drink from their pulpits, despite the fact that only a short while ago they had accepted as wages a jug of premium corn whiskey. Suddenly, those who drank were doomed to a sinner's hell. Wives and mothers became the watchdog guardians of their men's souls. One account after another in the first half of this century relates the mother's plea to stay away from "that stuff." Dave Couch, a confessed moonshiner from Harlan County, Kentucky, figures he made and sold more than 10,000 gallons of whiskey (Roberts, 1980). "But" he says,

> my money never did do me no good I got out of whiskey. It came easy and it went easy. And I believe today the reason I never got nothing out of whiskey making was 'cause my mother was all the time again'it and she told me never to do it and she bagged me not to. But I went ahead, after she left here, going again' her will. And I think that is the reason the money I got out of whiskey never done me no good. (p. 59)

Couch's story about his mother's influence on his life, even as an adult, directs our attention toward what is perhaps the single most important value among Appalachian people: familism. To place that value in its proper context, let us look first at a grouping of traits previously attributed to the Scotch-Irish settlers: individualism, self-reliance, and pride. The strong desire for privacy and autonomy, the desire to govern themselves according to their own principles, the desire to be beholden to no one have endured in some measure across the years to present-day mountaineers. Loyal Jones (1975) tells a story demonstrating self-reliance and pride that comes out of the 1960 blizzard in the North Carolina mountains.

Most of Ashe County had been buried in up to sixty inches of snow for weeks and weeks. Finally, toward the end of March a pair of workers from the Red Cross managed to get in to a small house several miles distant from any neighbors. They clambered onto the porch and knocked on the door, to be greeted by a little old lady who was visibly surprised to see visitors.

With a broad smile one of the men said, "Hello, we're from the Red Cross."

She replied, "Well, I don't believe I'm a-goin' to be able to help you'ns any this year. It's been a right hard winter" (p. 509).

CLINICAL IMPLICATIONS

With this sort of orientation, obviously any problem that emerges will be subject first to a self-conceived, self-directed resolution. If that fails, the problem may be deemed unsolvable and hence relegated to the fatalistic category of "just the way things are." The alcohol abuser may excuse his actions on these grounds. He is not prone to take his problem to someone else. His pride and self-reliance prevent that. Even with the strong family bonds that exist, the problem drinker will rarely seek help from a family member. He either denies his problem or becomes resigned to living with it. Only when the situation becomes desperate and the individual is no longer in control, will the family intervene.

If the alcoholic is a churchgoer and his guilt becomes too great, he may eventually confide in his pastor, but not because he wants professional help. The Reverend Joe Burton of the First Baptist Church, Radford, Virginia contends that the problem drinker does not look on the preacher as a professional; rather he comes to him out of a "rescue and supportive concept." His drinking is a sin; he needs to confess it and apologize for it; he questions where he stands in relation to God. He seeks reassurance of his place in the Kingdom of Heaven, and yet he can rarely accept that reassurance, for as Burton (1983) puts it, "the gospel of grace is not understood." The drinker wallows, then, in his guilt and ultimately may fall back on his fatalistic philosophy of what will be, will be.

In the event that outside assistance is offered or even voluntarily sought, the Appalachian native looks for personalism in his relationship with the help-giver. He sees every activity, every association as a person-to-person or one-to-one commitment. He has no inclination to abstract or generalize situations. He wants personal, individualized attention. After all, in his day-to-day life he accepts the people he meets for what they are; he assesses them on the basis of their conversation with him, their friendliness, their *personal* traits, rather than on accomplishments, credentials, titles, appearances, or other criteria. Therefore, any help extended, whether in the form of goods or services, is likely to be rebuffed, rejected, or thwarted unless a personal relationship is first established. If one goes to a mountaineer's home, for example, to discuss business, he does not immediately plunge into the affair at hand. Instead he talks first about the weather, the family, the hound dog, the garden—whatever topics seem important to the native's life. He finally approaches the business in an indirect manner such as "Do you reckon a feller might be able to buy a piece of land around here anywhere?" (For a fictional view of this trait see Mary Lee Settle's, 1956, 1960, 1973, 1980, 1982 novels, particularly the Beulah Quintet, which includes *Prisons, O Beulah Land, Know Nothing, The Scapegoat,* and *The Killing Ground.*)

Appalachians strive to be agreeable to friends and strangers alike. They dislike direct confrontations on any matter. Therefore, in order to avoid disagreement and confrontation, they may commit themselves to a pact or an action that they never intend to keep. They may promise to come to a meeting on a certain date or to bring a family member in for conference, but that action never comes to pass because the Appalachian feels no personal relationship with or commitment from the individual initiating the action.

Further, this acceptance of people as they are, acceptance of faults and all, may lead to a harboring within the family of the member who drinks too much. To the kin, that's "just Daddy," and oftentimes there's nothing to do about it except "stay out of his way when he's at his worst." Or perhaps a compassionate family member perceives his relative's inability to help himself and intercedes. Gurney Norman (1977), a contemporary author native to Grundy, Virginia, writes in a collection of short stories called *Kinfolks* of Uncle Delbert's drinking binges, which grow longer and longer. His college-age nephew, who was initiated years ago into his own experience with alcohol by this same uncle, aches to see Delbert destroying himself. Finally, he goes to try to help and succeeds in squelching that particular binge, but the reader suspects the solution is only temporary.

A young married woman in Pulaski, Virginia, reports that her father-in-law regularly goes on a drinking spree about every eight months. Consuming multiple cases of beer and bottles of wine and hard liquor, he drinks for days and nights on end. Attempts to reason with him are futile, although he well knows what the result will be. Suggestions that he look for help from a hospital or counselor are rebuffed by a firm, "I don't need no help. I can quit when I want to." Finally, when he has drunk himself into a stupor, his sons deliver him to the nearby VA hospital. After a drying-out period, he comes home and goes about his normal routine for another eight months. Then the cycle starts anew.

The daughter-in-law relates the story very matter-of-factly without any indication of judgment on her part. It is as if she acknowledges his right to go his own way and follow his own will as long as he is able.

Then, when he is no longer in control, the family steps in to obtain treatment of the symptoms. But never at any point does she theorize on the cause of his behavior. She never questions why, but only what to do when the situation becomes desperate. Her reaction is perhaps explained by that ingrained belief that his business is *his* business. And he is still the father of her husband's family with all the due respect that that position entails.

FAMILY TIES

I have already begun to discuss the influence and significance of family. Kinship in Appalachia has long been a bond that sanctions the actions of family members, regardless of what they are. An old mountain saying captures the essence of this trait in these words: "They're still my family, my kin, no matter how far they wander."

From the beginning settlements in the mountains, the family unit was the nurturing body. Because of the isolation, the long distances between farms, and the sparse population, that had to be the case. The extended family often provided marriage partners among the relatives because choices were so few. Cousins married each other; pairs of brothers married pairs of sisters, creating "double cousins" in their offspring; so that soon, in a small community, everybody was related to everybody. That situation is still prevalent in rural mountain communities in which the migration patterns have been fairly static (Howell, 1981).

The family, then, becomes a clan with intense loyalty toward one another. Even when family members do not like each other very much, they are tradition bound to defend and protect their own from outsiders and alien institutions, such as schools, government offices, and social and health care agencies. With the family acting to sabotage "suspect" outsiders, conflict within the kinship group is diminished, since it is the only refuge against the suspect agencies (Lewis, Kobak, & Johnson, 1978).

As a refuge, however, the family may become overly protective of its children. On one hand, it may hold them so close that they feel guilty about leaving home or even developing new ideas. On the other hand, the family is indeed a sanctuary in time of great need. If the coal miner gets to the point where he "can't take it anymore," he can go home to his parents (or grandparents, or uncle, or aunt, or even cousin) to be cared for until he has strength to go back to the mines—or elsewhere. If severe illness strikes, a grown child can go home to his parents or ask them to come to him for the succor he needs (Lewis, *et al.*, 1978). A marriage may even be temporarily dissolved until the crisis has passed, because one of the mates seeks refuge among his blood kin against whatever physical, mental, or emotional trauma he is experiencing.

As already indicated, the kinship group may extend to encompass the neighborhood or community. Social interaction tends to involve family rather than unrelated neighbors. Family activities are generally spontaneous and nonspecific, except for large family reunions that bring in out-of-town relatives (Howell, 1981). These typically occur among the immediate family on every major holiday and on every birthday. Among the larger clan of distant cousins and assorted appendages, the reunion is most likely an annual affair. If the bottle or fruit jar passes on these occasions, it is rarely passed openly. Rather, the men folk slip out from time to time to the smokehouse or the carport, returning in a more convivial mood than before. The women are scarcely ignorant of what is going on, but they pretend not to notice. Once in awhile, a more brazen female will dare to go out, too, for her own little nip of cheer. Something about the "forbidden fruits" idea makes the drinking all the more exciting and pleasurable. Perhaps it is also this notion that raises the old drunk to a stock figure in the Appalachian storyteller's repertory of jokes.

If an adult child breaks the family code, as in the case of divorce perhaps or marriage to a divorcee, he may find himself temporarily ostracized, though hardly ever on a long-term basis. In such an event, he is likely to seek refuge in a community or church group, or perhaps a helping agency, if he finds the group members person-oriented. The new group becomes his "family," even if for a short duration, and provides the nurture he needs.

This example may be extended to the alcoholic (a term, incidentally, most Appalachians would never apply to someone they know and love). Typically, as I have indicated before, the alcoholic is not ostracized by the family. His problem is ignored, evaded, or perhaps even accepted and dealt with. But it does tend to stay within the family or within the extended support group of community kin. If on a rare occasion the alcoholic himself moves outside the family for help, without the group's knowledge or consent, he may find that he is cut off from his refuge. Consequently, he looks for a replacement. His new "family" may be the helping agency he has turned to, which could work to his disadvantage, if the new family functions in the same way as the old, who inadvertently enabled him to deny his problem. On the other hand, unless the helping agency provides the individual, person-oriented approach he craves, the drinker is not likely to accept the proffered help.

To compound the problem, the Appalachian native may suffer the negative effects of biculturalism—being and acting one way in public and another way at home. Schools and other institutions often create dualities in the mountain people by imposing a set of mainstream standards and behavior patterns that simply do not blend with those inherited from the subculture. Most learn to deal with this discordance on the institutions' terms, but the strain can be great. In a talk given to sociologists in Baton Rouge, Louisiana, Mike Smathers noted that the educated mountaineer can take two routes in his approach to this problem: cultural schizophrenia or cultural transvestitism (Lewis *et al.*, 1978, p. 134). Whichever he chooses, at least he can rationally analyze his position. The uneducated mountaineer, on the other hand, may simply bounce from one culture to another without ever knowing why he feels so uncomfortable, so insecure, so torn by the differences in the environments he must traverse. Imagine the intensified identity crisis if the mountain native has a drinking problem as well.

LOVE OF PLACE

One way to combat the identity crisis that biculturalism may promote is to stay at home, and the Appalachian mountaineer has a strong inclination to do just that. The love of place or love for the land of one's birth in the most specific meaning of that term has long been a bond that keeps Appalachians at home, or if they must leave, it keeps them fretting to get back home. Usually "home" means the actual ancestral homeplace, although sometimes the term is broadened to encompass the area around the homeplace, perhaps a five- to ten-mile radius, in which relatives live. Such devotion to place surely goes back to the settlement of the region when the land provided the only living to be had. Through farming, hunting, fishing, and trapping, the Scotch-Irish and other ethnic groups managed to carve out a livelihood in a region noted for its severity of climate and terrain.

Whether it is the beauty of the homeland or the family traditions or the roots that go so deeply into the rocky mountain soil, Appalachians are place bound to the extent that few are happy outside their native environment. Seventy or more years ago as the coal camps began to spring up in central Appalachia, the men left their little farms to go to the mines to earn what they had scarcely known in their lives—cash money. But at first they did not take their families. The wives and children stayed home to maintain the farm and homeplace. The men lived in boarding houses and went home Sundays to visit. Eventually, however, miners found that they had to follow the coal, which meant moving from camp to camp, often ranging too far afield to go home on Sundays. Gradually they began taking their families to the camps, closing up the homeplace, but going back every chance they got—sometimes on Sundays, certainly during layoffs. It is significant to note that large, company-owned coal camp churches often had no attached cemeteries. The appropriate resting place was considered to be "back home"—often in a family cemetery right on the homeplace or perhaps in a home church cemetery (Howell, 1981).

During the decades of the 1940s, 1950s, and 1960s, out-migration from the Appalachian area was excessively large because of the search for employment. Many natives of southwestern Virginia, West Virginia, eastern Tennessee, and eastern Kentucky went north to the cities of Cincinnati, Columbus, Toledo, and Detroit. They usually found work in the factories and automobile plants, and they made friends, but most had left their hearts in the highlands. (See Harriette Arnow's *The Doll-maker*, 1954, for a poignant fictional portrayal of this theme.) Most talked of that future "dream time" when they would go back "home" to live. In the interim they went back often to visit—some driving one thousand miles every weekend. The dreams of returning to stay have not always come true, but the migration patterns have reversed themselves. Now in-migration is the more common practice, with both returning natives and outsiders flocking to the Southern Highlands.

Some alcohol counselors believe, in fact, that the in-migration of non-natives has produced more alcohol abuse problems in Appalachia than the moonshiners ever did among their own kind (All in the Family: Seminar on Alcohol Abuse, 1982). Such a statement is impossible to document of course, but the region has never been noted for excessive alcoholism among its inhabitants. In certain high-stress jobs, such as coal mining, and in certain high-poverty areas, alcohol abuse is fairly prevalent. The coal miners, however, have the advantage of the support of their labor unions and good insurance policies for hospital and psy-chiatric care. They are more likely to seek treatment for their problems today than would have been the case even 10 years ago. "And yet," the counselors say, "sustaining treatment or retaining positive results is so difficult. The mountain man is elusive. You think you've helped him, but then he starts fretting to go back *home*, and one day he's gone, and you never see him again. How do you know what's happened to him?"

Well, the chances are if he's home, he's content. The poet, Jim Wayne Miller, explains the attraction of the place, the mountain land, in his poem "Brier Sermon." He also recognizes how the various traits that bind Appalachians to the family and home can create conflict when posed against mainstream American culture. But Miller offers a solution even more fundamental than treatment for alcohol abuse. I quote from his collection entitled *The Mountains Have Come Closer* (1980).[2] This passage is delivered by a self-appointed preacher on a street corner in a small mountain town. He is talking to his fellow Appalachian mountaineers as they pass him and occasionally stop to listen.

Let me go back a little, let me tell you
how we got in this fix in the first place.
Our people settled in these mountains
and lived pretty much left to themselves.
When we got back in touch we started seeing
we had to catch up with the others.
And people came in telling us,
You've got to run, you've got to catch up.

Buddy, we've run so fast
we've run off and left ourselves
We've run off and left the best part of ourselves.
And here's something peculiar:
running we met people on the road
coming from where we were headed,
wild-eyed people, running away from something.
We said, What'll you have? and it turned out
they were running away from what we were running after.
They were on their way to sit a spell with us.
We had something they wanted.
When they got here, a lot of us weren't to home.
We'd already run off and left ourselves.
So they set to picking up
all the things we'd already cast off—
our songs and stories, our whole way of life.
We couldn't see the treasures in our house,
but they could, and they picked up what we'd abandoned.

You say, Preacher, you must be touched, that's
foolishness.
How can anybody run off and leave himself?
I say, Don't ask me. You're the one who's done it.

You've kept the worst
and thrown away the best.
You've stayed the same where you ought to have changed,
changed where you ought to have stayed the same.
Wouldn't you like to know what to throw away
what to keep
what to be ashamed of
what to be proud of?
Wouldn't you like to know
how to change and stay the same?

You must be born again.

---

[2] From *The Mountains Have Come Closer* (pp. 58–59) by Jim Wayne Miller, 1980, Boone, N. C.: Appalachian Consortium Press. Copyright 1980 by the Appalachian Consortium Press. Reprinted by permission.

Born, that is, Miller might have added, into the knowledge of who your father is, who you are, and why you are as you are.

REFERENCES

All in the Family: Seminar on alcohol abuse. St. Albans Psychiatric Hospital, Radford, Virginia. 24 September 1982.

Anderson, S. *Kit Brandon.* New York: Charles Scribner's Sons, 1936.

Appalachian Regional Commission. *1979 Annual report.* Washington, D.C.: Author, 1980.

Arnow, H. *The dollmaker.* New York: Avon Books, 1954.

Burton, J. Telephone interview with the pastor of the First Baptist Church, Radford, VA. 22 November 1983.

Carr, J. *The second oldest profession: An informal history of moonshining in America.* Radford, VA: Commonwealth Press, 1972.

Carr, J. *The moonshiners.* Nashville, TN: Aurora Publishers, 1977.

Howell, B. *A survey of folklife along the Big South Fork of the Cumberland River.* Knoxville: University of Tennessee Department of Anthropology Report of Investigations No. 30, 1981.

Jones, L. Appalachian values. In R. J. Higgs & A. Manning (Eds.), *Voices from the hills.* New York: Frederick Ungar, 1975.

Lewis, H. M., Kobak, S. E., & Johnson, L. Family, religion and colonialism in central Appalachia or Bury my rifle at Big Stone Gap. In H. M. Lewis, L. Johnson, & D. Askins (Eds.), *Colonialism in modern America: The Appalachian case.* Boone, NC: The Appalachian Consortium Press, 1978.

Miles, E. B. *The spirit of the mountains.* Knoxville: University of Tennessee Press, 1975. (A facsimile edition; first published in 1905.)

Miller, J. W. *The mountains have come closer.* Boone, NC: Appalachian Consortium Press, 1980.

Mullaly, R. Conversations with director of the alcohol abuse program at St. Albans Psychiatric Hospital, Radford, VA, July, August 1982.

Norman, G. *Kinfolks: The Wilgus stories.* Frankfort, KY: Gnonom Press: 1977.

Roberts, L. *Sang Branch settlers.* Pikeville, KY: Pikeville College Press of the Appalachian Studies Center, 1980.

Settle, M. L. *Prisons.* New York: Ballantine Books, 1973.

Settle, M. L. *O Beulah Land.* New York: Ballantine Books, 1956.

Settle, M. L. *Know nothing.* New York: Ballantine Books, 1960.

Settle, M. L. *The scapegoat.* New York: Random House, 1980.

Settle, M. L. *The killing ground.* New York: Bantam Books, 1982.

Williams, C. Heritage of Appalachia. *News from the Appalachian State ITT project,* 1972, Winter, 1–3.

Williams, C. *Appalachian language and culture.* Videotaped lecture given at Southwest Virginia Community College, Richlands, VA, 1980.

III

Case Studies
Black Americans

10

Ambiguity in Black Drinking Norms

An Ethnohistorical Interpretation

DENISE HERD

INTRODUCTION

Ambivalence regarding the use of alcoholic beverages is sometimes described as the key feature of American drinking patterns (Pittman, 1967). Alcohol use is esteemed as "one of the pleasures of life," yet it also evokes quite negative sentiments in others.

Among American blacks, the ambiguity around beverage alcohol seems particularly heightened. Two entirely disparate images have emerged regarding black drinking. The first, drawing on popular stereotypes and anthropological studies of ghetto life (Hannerz, 1970; Liebow, 1967; Lewis, 1955), characterizes drinking and drunkenness as prominent features of black culture. Lewis' (1955) description of drinking patterns in a small Southern town is exemplarary of this viewpoint:

> The pervasiveness of whiskey drinking among the adult males and lower-status women; the indulgent and tolerant attitude toward it in all quarters

DENISE HERD • Alcohol Research Group, Institute of Epidemiology and Behavioral Medicine, Medical Research Institute of San Francisco, Berkeley, California 94709. This study was supported by a National Institute on Alcohol Abuse and Alcoholism (NIAAA) Alcohol Research Center Grant (AA-05595) to the Alcohol Research Group.

except among the females, who perhaps feel strongest about its effects upon family income and expenditures, status, and personal relations; the frequency with which whiskey appears as a factor in non-martial sex play, "touchy" behavior, and "having a good time," whether attending baseball games, hunting, or idling; and the lack of elaborate rationalizations or justifications for its use or abuse—all these indicate a near-successful integration with other aspects of the culture. (p. 24)

In contrast, other studies suggest that there is a considerable reservoir of anti-alcohol sentiment among American blacks. Borker, Hembry, and Herd's (1980) description of black drinking patterns in a California coastal city revealed that alcohol use is subject to many negative sanctions; that there is not a cultural consensus regarding its value; and that abstention or infrequent drinking is an accepted pattern among urban residents. Affiliation with fundamentalist Protestant religions, which is widespread among working- and lower-class blacks, is associated with ambivalent or hostile attitudes toward alcohol use. These sentiments were directly reflected in statements by informants of the study:

I just don't think that it would be good to come home every day and take a drink. . . . I don't really think that it [alcohol] has a proper use, not really, when you . . . think about it. . . . Why do you want to do something that alters your personality or to make you relax, or make you feel freer to make you drop you inhibitions or whatever? Why do you need something to do that for you? You can do it yourself. So I don't think it really has a proper use because it should be about a natural high really. . . . Well, I guess I don't really think alcohol is that acceptable. . . . I think alcohol is dangerous and I have found myself saying, "I don't want to drink anything tonight," because maybe I had something to drink last night. Alcohol is not good; it's not good to have it in your system.

A previous ethnographic study of a lower income housing project (Sterne & Pittman, 1972) supported the prevalence of this attitude. The authors suggested that although alcohol was "near-successfully" integrated into black culture, liquor was negatively regarded and subject to ambivalent norms even among informants who were regular drinkers. They concluded that "consensus regarding alcohol use and consistency between drinking practices and attitudes is incomplete" (p. 653).

Support for both perspectives—for example, that black culture supports attitudes for patterns of heavy drinking *and* for abstaining—can be gleaned from survey data and social indicators of drinking patterns and problems. For example, surveys of drinking patterns have consistently found that black women are heavily abstinent, yet also manifest much higher rates of heavy drinking than white women (Bailey, Haberman, & Alksne, 1965; Cahalan, Cisin, & Crossley, 1969). White men and black men resemble each other in self-reported drinking patterns

(Bailey *et al.*, 1965; Cahalan *et al.*, 1969; Clark & Midanik, 1982), but blacks experience much higher rates of physiological complications related to chronic heavy alcohol consumption (Herd, 1985a; Pottern, Morris, Blot, Ziegler, & Fraumeri, 1981; Rogers, Goldkind, & Goldkind, 1982).

This chapter examines the problem of ambiguity and cultural conflict in drinking norms among present-day blacks from an ethnohistorical perspective. It argues that ambivalence around alcohol use in the black population, as in the general American population, is the product of complex historical changes. For example, it is generally acknowledged that the changes associated with the nineteenth-century temperance and prohibition movements resulted in conflicting images of beverage alcohol in American culture. The data presented here suggest that Black American attitudes have been affected in a similar manner. Black Americans were deeply involved in the anti-alcohol movements of the nineteenth century—spearheading a major black temperance movement and a network of "dry" social and religious organizations. These movements appear to have initiated very dramatic changes in the social meaning of alcohol beverage consumption in the black population. This transition can be summarized as follows:

In the early part of the nineteenth century, beverage alcohol was highly esteemed and integrated into all aspects of sociocultural life. Following their mass involvement in the antebellum temperance movement, blacks abstained from alcohol because it was a symbol of slavery and oppression. They upheld norms of moderation and abstinence throughout the nineteenth century until they were confronted with the racism and political hostility of the Southern prohibition movement. In reaction to these changes, alcohol re-emerged as a valued social commodity among blacks in the twentieth century, while temperance sentiments became privatized and relegated to the religious and domestic sphere.

These abrupt ideological shifts appear to have set the stage for the "ambivalent" attitudes of contemporary blacks. The prevalence of abstinence beliefs among today's blacks appears to be rooted in the black temperance and religious reform movement of the nineteenth century. The rise of a nightclub culture associated with heavy drinking is a newer phenomenon that overlays "dry" social values. The recent transition from "dry" to "wet" cultural norms may account for the polarization of drinking norms and the lack of integration of drinking into many cultural domains of black life.

Support for this hypothesis is evident in the trends of cirrhosis of the liver in the black population (Herd, 1985a; Malin, Kaelber, Sorenson,

Dadovrian, & Munch, 1980). These data reveal profound generational differences in rates of cirrhosis, with recent cohorts—those maturing during the rise of the nightclub culture in the 1920s and 1930s—showing several times the mortality level of those reared in the temperance-oriented milieu of the nineteenth century.

Here we examine the changes in the social meaning of beverage alcohol use among North American blacks and how they are related to changes in the larger sociopolitical systems affecting alcohol control. In so doing, the analysis is sensitive both to the structure of normative values operating within black culture and to the system of external social and political controls emanating from the larger society.

PRE-NINETEENTH-CENTURY IDEOLOGICAL CONCEPTIONS OF BEVERAGE ALCOHOL

The West Africans brought to the American colonies were members of cultures with a long tradition of using fermented grains and palm sap. Most of the non-Islamic groups in traditional Africa possess an extensive beer brewing culture closely associated with their agrarian subsistence patterns. The thick native beer is highly valued as a "food, a medium of exchange and a sacred liquid." In some groups, like the West African Kofyar who "make, drink, talk and think about beer," beer brewing is a focus of cultural concern and activity in much the same way as are cows in African pastoral societies (Netting, 1979). Even in societies without this focus, beer occupies a central role in their religious, social, and economic life. Beer is socially and symbolically important in the ancestral ceremonies and rites of origin; weddings, funerals, and births; work agreements; legal adjudication processes; and dances, feasts, and other social occasions that characterize peasant life (Heath, 1975).

The consumption of native-brewed beer or palm wine in pre-colonial times was apparently not associated with high rates of social complications (Cherrington, 1929, p. 1985). Brewing and drinking beer were regarded as collective social activities that were regulated according to norms of reciprocity and etiquette. This tradition placed a high value on moderate drinking and strongly disapproved of socially disruptive behavior (Umunna, 1967).

African symbolic and social orientations to alcoholic beverages persisted to some degree in the social life of Afro-Americans in the American colonies. The continuation of these norms was probably facilitated by the fact that the Euro-American colonists shared a similar ideology about the nature of beverage alcohol. Ales, brews, and ciders were made from a variety of grains and fruits; drink was considered a "good creature of God" and was especially esteemed for its nutritional and health-stimulating properties. Alcoholic beverages were used freely in all forms of social intercourse—they were an indispensable part of the numerous meetings, weddings, funerals, and work parties that took place in colonial life (Rorabaugh, 1979; Aaron & Musto, 1979).

Accounts from the nineteenth century describe social customs maintained by slaves that show marked similarity to rituals practiced in Africa (Johnson, 1937; Smith, 1838; Stampp, 1961). Similar accounts suggest that beverage alcohol was an integral part of the slaves' social and public life. New England slaves elected their own governers with an elaborate inauguration ceremony accompanied by much 'lection cake and 'lection beer. Liquor consumption was also a part of the numerous "house raisings, church raisings, apple parings, maple sugarings and corn huskings" that slaves participated in (Greene, 1942).

Black drinking in colonial America did not appear to be accompanied by high rates of drunkenness or social disorganization (Herd, 1985b). In contrast, excessive drinking was regarded as a significant problem among white settlers (Genovese, 1976; Rorabaugh, 1979) and Native Americans. The absence of major alcohol-related problems among blacks can probably be traced to their prior cultural experience with alcoholic beverages and the persistence of informal sanctions against disruptive drinking behavior.

BLACKS IN THE ANTEBELLUM NORTH: ALCOHOL, THE ENSLAVER

The beginning of the nineteenth century witnessed profound changes in Black American and general American orientation to alcohol use. The image of liquor as the "good creature" of the colonial era began losing ground to one that regarded alcohol as an overpowering demon capable of destroying mind and body. This transformation in world view can be traced to the onset of several waves of temperance and religious reform. The rise of these movements is attributed both to profound changes in the social structure and to the increased availability of spiritous liquor during the late eighteenth and the early nineteenth centuries.

Concern about drinking, particularly among the lower classes, became more prevalent in the aftermath of the American Revolutionary War, as traditional measures of class control and control of drinking establishments began to break down (Rorabaugh, 1979). At the same time, the market was glutted with cheap spirits and liquor consumption increased dramatically (Aaron & Musto, 1979).

Economic, familial, and political changes heightened the focus on temperance. The expansion of industrial capitalism, the changing character of the middle-class family, and increasing interethnic conflict made temperance a center stage issue in the early nineteenth century (cf. Levine, 1978; Gusfield, 1966).

Black Americans, probably more than other Americans of the same social status, became eager participants in the wave of temperance reform. In doing so, they were responding not only to the cultural context of the larger movement, but also to their specific sociopolitical circumstances. Black concerns about temperance revolved around the overriding issue of slavery and grew out of the close relationship between the anti-slavery and anti-liquor movements.

The temperance and abolition movements were intimately connected: both were politically aligned with Northern political and economic interests. Both were drawn from the reformist Protestant churches—the Quakers, Presbyterians, and Congregationalist denominations—that denounced slavery; and the two movements drew from the same middle- and upper-class constituency. Leading temperance figures were also noted anti-slavery activists. Dr. Benjamin Rush, who published the first temperance tract in 1785, *An Enquiry into the Effects of Ardent Spirits upon the Human Body and Mind*, was also a prominent leader of the abolition movement. Similarly, Anthony Benezet, the "best known early opponent of spiritous liquors" (Rorabaugh, 1979, p. 36), played a major role in the early anti-slavery movement.

The black church movement, which began at the turn of the eighteenth century, was one of the first avenues for promoting anti-alcohol sentiments among blacks. The churches were organized with the assistance of Benjamin Rush, Anthony Benezet, and other reformers who urged blacks to incorporate principles of middle-class morality and temperance into their governing body (George, 1973). The constitutions of the newly organized black churches carried specific provisions against the use of spiritous liquors. For example, a resolution by the second largest black church, the African Methodist Episcopal Zion Church, declared "strong drink to be a monster of frightful mein, and made its use a violation of faith and religion" (Dancy, 1866; p. 426, cited in Cheagle, 1969, p. 14). These churches continued to issue anti-drinking statements at regular conventions throughout the nineteenth century, thus serving to continually reinforce temperance and abstinence among blacks.

A mass movement for temperance among blacks began in earnest in the 1830s when both the general temperance movement and the anti-slavery movements gained momentum.

The ideological thrust of the black temperance movement focused on blacks' primary concern for emancipation, political equality, and social betterment. Prominent black abolitionists, such as Frederick Douglass, expanded on the rhetoric of the larger movement that alcohol was a powerfully addicting substance that would "enslave" its victims (Douglass, 1892). He argued that sobriety paved the way to freedom, since it was only the "sober" and alert slave who could plan and execute his escape from slavery.

> To enslave men, successfully and safely, it is necessary to have their minds occupied with thoughts and aspirations short of the liberty of which they are deprived. . . . It was about as well to be a slave to master, as to be a slave to whiskey and rum. When a slave was drunk, the slaveholder had no fear that he would plan an insurrection; no fear that he would escape to the north. It was the sober, thinking slave who was dangerous, and needed the vigilance of his master to keep him a slave. (pp. 147–148)

Temperance workers drew vivid parallels between enslavement to a master and bondage to alcohol. They warned the free black that no sooner would he "put the intoxicating cup to his lips than he would give back the lash to the slave driver" (p. 47) (Cheagle, 1969). The pledge for black freedmen in most temperance societies stated that (Temperance Tract for Freedman; cited in Cheagle, 1969):

> Being mercifully redeemed from human slavery, we do pledge ourselves never to be brought into the slavery of the bottle, therefore we will not drink the drunkard's drink: whiskey, gin, beer, nor rum, nor anything that makes drunk come. (p. 29)

The popularity of the temperance movement among blacks was indicated by the success of the Colored American Temperance Society. Formed in Philadelphia in the early 1830s on the expressed principle "of entire abstinence from the use of ardent spirits," the society had a mushrooming growth, reporting 23 branches in 18 cities within a year's time (Quarles, 1969, p. 94). By 1837, Pennsylvania blacks had several temperance societies, including a proliferation of Daughters of Temperance Unions, with a total membership of 1,500.

Black Ohio communities were particularly zealous in their support of the temperance movement. In 1840, over one-quarter of Cincinnati's black population belonged to either the adult society of 450 members or the youth branch numbered at 180. Black opposition made it impossible for other blacks to sell intoxicating liquor openly (Quarles, 1969).

Temperance activity also flourished in New England and in the

Eastern seaboard states. Between 1829 and 1838, temperance societies were organized in New Haven, Middleton, and Hartford. The prominent black clergyman and anti-slavery leader Jehiel C. Beman founded and coordinated the Connecticut State Temperance Society of Colored People. The members of the society pledged to "abstain from the use of all fermented liquors that produced intoxication, as well as all distilled spirits, and not to use them as common drink, nor traffic in them, neither furnish them for other persons thus to use" (*The Liberator*, 30 July 1836, cited in Cheagle, 1969, p. 33). Adopting an even stronger anti-liquor stand at subsequent conventions, the organization became known as the Connecticut Total Abstinence Society in 1836. In 1854, the society endorsed the Maine Law for prohibition (Cheagle, 1969).

A similar network of "colored" temperance societies were formed in New York, Massachusetts, and New Jersey. Black newspapers such as the *Northern Star, Freedman's Journal*, and the *Colored American* circulated temperance literature to the black populations in the North and the South. Blacks also organized a missionary movement to further the cause of temperance among Africans and colonized blacks in other regions (Cheagle, 1969).

By the late 1840s, the strong connection between temperance and abolition and the deep infusion of temperance ideology into black social life had substantially changed the role of alcohol in the culture. Whereas alcohol use had been integrated into nearly every aspect of social, economic, and political life during the African and the colonial periods, the temperance movement relegated alcohol use to a minor public role. Although blacks undoubtedly continued to use liquor, drinking was frowned upon in religious, political, or social settings. The Protestant church, with its pervasive influence in all areas of black life, continued to discourage the use of alcohol. The church and its related umbrella of social, political, and self-help activities served to reinforce anti-drinking sentiments. This legacy of temperance continued to permeate black thought for the remaining decades of the nineteenth century.

BLACK TEMPERANCE AND SOCIAL IMPROVEMENT UNDER RECONSTRUCTION

Formal temperance activity among blacks and in the society at large had waned during the intense period of sectional conflict during the Civil War. However, temperance reform began to emerge as a significant social force in the 1870s and 1880s. Women, especially members of the Womens' Christian Temperance Union (WCTU), played a central role in the movement of this period. Ideologically the movement was consistent with its abolitionist legacy. It was favorable to social reforms that would aid black freedmen and immigrants (Blocker, 1976). The movement was dominated by the middle classes, who regarded the anti-liquor campaigns as a means for promoting a change in social values and personal (especially familial) relations (Levine, 1980).

Blacks of this period were highly supportive of the temperance movement, which was tied to their struggle to attain political equality and maintain social acceptance.

Alcohol was seen as a barrier to needed economic and political gains, as well as hindering the adoption of proper moral and religious values. The noted educator and minister J. C. Price urged blacks to "fight liquor because it interfered with their new homes, new education, new citizenship, new church, and new responsibilities as free colored men!" (Isaac, 1965, p. 36). In planning a resettlement territory for displaced slaves, the Negro Convention of 1882 petitioned Congress to "prohibit the sale of all intoxicating beverages in the new territory" (Aptheker, 1951, p. 685).

The specific focus of the temperance movement reflected its primary constituents—the large network of black womens' clubs and the Protestant church. The period witnessed a blossoming of womens' self-help organizations devoted to improving domestic life and a vigorous program of social reform and welfare work. Temperance was regarded as a progressive social cause in both these areas. First, temperance values would improve the general health, welfare, and economic well-being of the population. Second, it would strengthen the moral fiber of families and young people.

Black women supported the temperance cause directly through "colored" departments of the WCTU. By 1919, there were "colored" WCTU chapters in 26 states. Temperance issues also received support from the broader women's movement. A National Association of Colored Women's Clubs was organized in 1896 under the leadership of women who had distinguished themselves in WCTU work, such as Lucy Thurman and M. A. McCurdy. At conventions of the Association, specific resolutions were issued in support of the WCTU (Aptheker, 1951), as well as statements calling for the shutdown of dram shops and liquor establishments (Kletzing & Crogman, 1897).

The Protestant church was probably the key institution promoting temperance philosophy in the black population. Temperance and abstinence were codified in the religious dogma and were also promoted as a means of self-improvement and of reform of social conditions. The vigorous anti-alcohol stance of the black church is illustrated by the resolutions adopted by the African Methodist Episcopal Church at its general conference (1896 or 1897) (Kletzing & Crogman, 1897).

> **Resolved:** 1. That we discourage the manufacture, sale and use of all alcoholic and malt liquors.
> 2. That we discourage the use of tobacco by our ministers and people.
> 3. That we discourage the use of opium and snuff.
> 4. That we endorse the great prohibition movement in this country, also work done by the Woman's Christian Temperance Union, and will use all honorable means to suppress the evils growing out of intemperance.
> 5. That it shall be a crime for any minister or member of the A. M. E. Church to fight against temperance, and if convicted of this crime he shall lose his place in the conference and the church. (pp. 176–177)

The church was an important source of temperance support, which went beyond the religious domain. Black churches were the center of social life—including political, educational, charitable, and even recreational activities. Besides a vast network of specific church organizations and missionary societies, black religious institutions sponsored colleges, hospitals, infirmaries, training schools, old folks' homes, orphanages, and rescue and mutual aid societies (cf. Frazier, 1966; DuBois, 1899; Kletzing & Crogman, 1897). These organizations internalized the implicit moral ideology of the church, which included pervasive anti-alcohol sentiments.

In the wake of strong public support for temperance, alcohol problems in the black population of this period appeared to be extremely low and were far outstripped by higher rates of drunkenness and alcoholism among whites. Koren's (1899) exhaustive analysis, "Relations of the Negroes to the Liquor Problem," concluded that chronic drunkenness was so rare among blacks that they were thought to be physiologically immune from alcoholism.

Koren's study was supported by both sociological (DuBois, 1899) and vital statistics data (U. S. Census Office, 1886, p. lxvii) of the period.

BLACKS, PROHIBITION, AND THE RISE OF WHITE SUPREMACY

By the turn of the century, the anti-liquor movement underwent fundamental changes that adversely affected the aspirations of blacks seeking social assimilation through temperance politics. The abolitionist and liberal social policies of the earlier movement were transformed into very hostile positions on black sociopolitical equality. Supporters of prohibition openly campaigned for "white supremacy," black political disfranchisement, and racial segregation. They argued that the black vote was hopelessly corrupted by liquor and that blacks themselves were liquor-crazed, volatile, and sexually depraved (Herd, 1983).

The changing image of blacks in the anti-liquor movement was the result of larger societal problems. Blacks became the scapegoats for the political factionalism, economic strife, and labor unrest that plagued the South at the turn of the century. The movement for white supremacy served to suppress radical movements among lower-class whites, to eliminate blacks and dissident whites from the voting rolls, and to divert attention away from economic reform. The racist themes of the prohibition movement both reflected and supported these repressive policies (Herd, 1983b).

Prohibition emerged as one of the primary rationales for the disfranchisement campaigns. Despite their long-standing support of the temperance movement, blacks were stereotyped as irresponsible voters who would be easily swayed by liquor to vote against prohibition measures. Politicians argued that the South would not go dry unless the black vote was eliminated.

The position of Southern prohibitionists on black voting rights is illustrated by the prohibition appeal to the Mississippi Convention of 1890 (Wharton, 1947):

> There are seventy-five counties in Mississippi, and forty of them are dry. These dry counties are in the white section. The thirty-five wet counties are mostly in the black belt and are kept wet by the negro vote. . . . What are you here for, if not to maintain white supremacy, especially when a majority of the whites stand for a great principle of public morals and public safety? We especially appeal to the delegates from the black counties. Gentlemen, can you in reason expect the white counties to stand by you and uphold white supremacy for you, while you discard the doctrine which has been our common safety, in the interest of the saloon keepers of your counties? How long is it to be expected that white solidarity can be maintained, if the negro is to be brought forward to arbitrate this great question in the interest of a minority of the whites, and they mostly foreign born, and not in sympathy with our institutions? (p. 208)

Similar statements echoed in political and legislative campaigns throughout the South. In Tennessee the "State-Wide" prohibition organization of the Democratic Party meeting in 1909 pushed for the enforcement of prohibition, urged disfranchisement of blacks, and resolved that "State-Widers" should enter the Democratic primary in 1910 (Isaac, 1965, p. 182).

In Atlanta's gubernatorial battle between Hoke Smith and Tom Watson, prohibition and disenfranchisement were the two principal tenets of Smith's appeal (Walton, 1969, p. 68). His platform was based upon the need to "reform" the suffrage through devices that would totally exclude blacks from all elections "without disfranchising a single white man" (Crowe, 1968, p. 238). The Atlanta prohibitionist the Reverend Sam Jones "often called for the expulsion from the polls of the element that 'invariably sells out to the highest bidder!' The vote was worthless to the Negro for 'it is not counted when cast' and the 'Yankees' who organized the whole idea of Negro suffrage had long since admitted to their mistakes. The ballot, Jones concluded, served only to debauch the Negro, prevent the democratic process, and preserve the saloon" (Crowe, 1968, p. 237).

The argument that blacks were a major impediment to the enactment of local and statewide prohibition measures served as a powerful rationale to strip them of their voting and political rights. Along with other tactics, pressure from the prohibitionists helped crush political participation, so that by the close of the century, black participation in the voting process was almost non-existent (Walton, 1969).

In addition to the focus on disfranchisement, prohibition campaigns were associated with a wave of virulent anti-black propaganda. Prohibition was urged in order to protect the white populace, particularly females, from the drunken debauchs of black men. The "negro" problem became a central issue in liquor reform. Sensational newspaper editorials were circulated condemning the liquor traffic as the cause of an alleged epidemic of sex crimes, as well as the massive wave of lynchings, riots, and other forms of racial violence that swept across the South. For example, Mr. Levy's "Nigger Gin" was specifically cited as the cause of rape and lynching in many Southern towns (Irwin, 1908):

> In every negro dive of the South, they sell certain brand names of gin, whose very names, for the most part, I can not mention here. Obscene titles, obscene labels, advertise by suggestion, by double meanings, that these compounds contain a drug to stimulate the low passions which have made the race problem such a dreadful thing in the South. These bottles do not contain what their labels imply; chemical analysis shows that the mixture is only cheap, blended gin, with a slight infusion, in some cases, of a sweetening which may be Benedictine. The viciousness lies in the double meanings, clear to every man who knows the Southern negro, in the pictures of naked white women on the labels, in ever greater obscenities. The suggestion that he do the nameless crime, to avenge which the Southern white lynches and burns, is before every negro consumer of gin.

> Is it plain now, the secret of many and many a lynching and burning in the South? The primitive negro field hand, a web of strong, sudden impulses, good and bad, comes into town or settlement on Saturday afternoon and pays his fifty cents for a pint of Mr. Levy's gin. He absorbs not only its toxic heat, but absorbs also the suggestion subtly conveyed, that it contains aphrodisiacs. He sits in the road or in the alley at the height of his debauch,

looking at that obscene picture of a white woman on the label, drinking in
the invitation which it carries. And then comes—opportunity. There follows
the hideous episode of the rope or the stake. (p. 10)

Ministers in Atlanta harped on the triple temptations of prostitution,
self-abuse, and negro women, along with the dooming consequences
of whiskey (Crowe, 1968). The renowned Reverend Wilbur Fisk Crafts
preached on the devil's girdle of liquor "which has changed so many
Negroes into sensual hyenas" and threatened the safety of the white
woman (Sinclair, 1964, p. 32).

The extreme racial tension and xenophobia of these passages led to
many blacks being the victims of terrorist attacks, murders, and lynchings. Conservative estimates suggest that about 1,200 blacks were lynched
between 1890 and 1900 (Aptheker, 1951).

THE AFTERMATH OF PROHIBITION: ALCOHOLIZATION OF BLACK LIFE

The racist propaganda associated with the prohibition movement
signaled the demise of a broad-scale anti-liquor movement in the black
community. Prohibition posed a painful dilemma for those blacks who
supported the temperance movement, but who were increasingly uncomfortable with its racist political agenda.

Statements on prohibition by black political figures were sparse and
ambiguous. The eminent black leader W. E. B. DuBois noted with dis-
appointment that prohibition was part of the "Promotion of Prejudice,"
which signaled a death blow to black progress. Even when blacks endorsed the movement, they failed to organize an active base of support
as they had in earlier periods. Other blacks registered a direct opposition
to prohibition legislation (*St. Paul Appeal*, August 22, 1891, pp. 17).

The disappearance of a major social movement for temperance among
blacks meant that temperance sentiments became extremely discretion-
ary and privatized. Thus, the temperance movement lost its broad appeal
as a movement that promoted social improvement and race advancement. Instead, it became a matter of individual preference few leaders
were willing to address publicly.

The decline of formal temperance activity paved the way for a new
cultural image of alcoholic beverages among blacks in the early decades
of this century. Liquor lost its power as a symbol of social oppression
and instead became associated with urbanity, sophistication, and freedom from oppressive Southern norms. These new values were rein-

forced by the increasing urbanization of the black population and the centrality of the nightclub culture in the Northern city.

Given their low social status and prominence in nightclub life, blacks became a prime market for whites peddling illegal alcohol (McKay, 1968) and, on a small scale, were drawn into the liquor distilling and trading business themselves.

Bootlegging gained in popularity among blacks in the South. Liquor manufacturing and trading flourished in areas in which farmland was poor and legal economic opportunities stifled. Tolerant folkways were another contributing factor (Winston & Butler, 1943). "The illicit liquor trade became almost respectable as well as profitable. A student put himself through a Southern theological seminary by selling bootleg liquor to congressmen" (Larkins, 1965, p. 26).

The massive shifts in black population that began after the turn of the century also intensified black alcohol consumption. Beginning around 1900, thousands of blacks began to leave the South in the "Great Migration" to Northern cities (Gwinnell, 1928). These blacks moved from farms, villages, and towns where social life was dominated by rural churches and family associations to areas where an extensive tavern and nightclub life and social scene flourished. In the urban setting, blacks quickly became a focus of the nightlife culture. As entertainers, musicians, and dancers, they were an indispensable element in this life-style. Blacks and whites from all strata of society mingled in the breakfast clubs, dance halls, and rounders activities in Harlem, Chicago, and San Francisco (Schoener, 1969; Daniels, 1980; Drake & Cayton, 1945).

It was probably during this period that a culture of heavy drinking, complete with its own folklore, began to flourish in some black circles. In the fiction and art works of this era, the nightclub and tavern emerge as major cultural institutions (Hughes, 1936, 1950; Mc Kay, 1968; Harrington, 1958). Langston Hughes' portrait of a Northern city "Juice Joint" (1936, p. 67) reveals how liquor became romanticized as an elixir of sensuality and gaiety. The poem describes a "gin mill on the avenue" (where "gin is sold in glasses finger-tall") with "singing black boys" and "gay dancing feet." This theme represented a major shift from the religious, temperance-oriented black literature of the nineteenth century. The image was partially created and directly supported by upper-middle-class whites who went "slumming" in the urban nightclubs and speakeasies to drink, to learn the Charleston, and to mingle with the exotic "jazz" crowd (Erenberg, 1981).

Although the nightclub culture did not necessarily pervade all aspects of black life, it made substantial inroads into many areas. The changes occurring during the prohibition years saw a decline in the traditional authority and influence of social institutions that restricted the use of alcohol, primarily the Protestant church, and a corresponding increase in activities that involved the use of beverage alcohol (cf. Frazier, 1966). In fact the church itself, long held in the vanguard of moral respectability—traditionally opposed to liquor, gambling, and other forms of "sinful" frolicking—relaxed its own moral code sufficiently to accommodate members and even their relatives who earned money through bootlegging and gambling. The churches were noticeably absent from efforts to abolish liquor in the community, as suggested by newspaper accounts of the period (Anderson, 1981).

The period also signaled the initiation of major alcohol-related problems in the black population. Statistics from this era indicate that there were disproportionate increases in death from alcoholism occurred among blacks (Carter, 1928), as well as in hospital admissions for alcoholic psychoses, particularly in the migrant black population (Malzberg, 1944). Cirrhosis mortality rates among blacks rose steeply after 1950, increasing substantially above white rates for the first time in the history of recorded vital statistics (Malin et al., 1980; Herd, 1985a).

IMPLICATIONS FOR CONTEMPORARY BLACK DRINKING PRACTICES

The juxtaposition of new urban drinking mores and the traditional temperance background of blacks appears to be reflected in the contemporary drinking patterns of blacks. As suggested earlier, the abrupt transition in the symbolic and social role of alcohol may be largely responsible for the current ambiguity in drinking norms and patterns within the black population. On the one hand, the persistence of traditional temperance values is evident in the prevalence of abstention and dry social norms among blacks. On the other hand, alcohol plays an important role as a social lubricant and luxury commodity in black life. The following section briefly examines the institutional matrices that reinforce both sets of values in black culture.

ABSTAINING AND "DRY" SOCIAL NORMS

The Protestant church, especially in its fundamentalist branches, is a major social force for abstinence in the black population. The church has had a long-standing role as the most prominent social, political, and cultural institution in black life. As described earlier, it was heavily steeped in temperance ideology throughout the nineteenth century. Anti-alcohol positions, as reflected in the doctrinal literature and church covenants and in the exclusion of alcohol from religious rituals, survive in many present-day churches frequented by blacks. The web of self-help and social activities connected with black fundamentalist churches is also virtually dry. The dry orientation even extends to secular groups that have social betterment as their goal (Borker *et al.*, 1980).

The importance of religious values in shaping black perceptions of alcohol use are illustrated by the data from a recent anthropological study of black drinking patterns among urban blacks (Herd, 1980). Respondents from fundamentalist backgrounds described what they believed were specific religious sanctions against beverage alcohol use. The following three excerpts from interviews probably typify the experiences of and beliefs held by many blacks raised in Baptist, Holiness, and other fundamentalist churches.

> We drink wine for the communion, but that's really grape juice. When I was growing up, in the church that I grew up in . . . I remember in the church covenant which we agreed on in communion, in unison, there was a line in which people were supposed to not drink at all. They said . . . reading the church covenant . . . "I will abstain from use and sale of intoxicating drink as a beverage." So that meant that you weren't supposed to drink if you were a Baptist—at least in that church. I don't get that from the church I belong to now.

> I know our churches' convenant says for us to abstain ourselves from all alcoholic beverages. We're not supposed to drink anything with alcohol in it at all.

> The church convenant, you know, it's an agreement that everybody's supposed to submit to (that) says you're supposed to abstain from drinking alcoholic beverages.

The influence of the church on family drinking patterns is suggested by the fact that many informants reported a pattern of non-drinking among parents and other relatives, especially females, with religious involvement often cited as the reason. The general pattern of alcohol use among parents, as reported by informants, revealed that alcoholic beverages were seldom kept at home or served with meals and were used only on holidays or for special events.

> We never kept liquor in the house. I don't keep it now. I drink, but I don't keep liquor. I don't; I drink outside. . . . I never remember having liquor in my house growing up. . . . My mother's family was like highly religious—my dad, my granddaddy, and my grandmother—you couldn't bring a drink of liquor in their house, period. . . . My daddy was from a little small Southern family. I don't think they were as religious as my mother's family, but . . . it

just wasn't a thing that you did. But on the other hand, my daddy's brother was a heavy, heavy drinker—right, but Dad's never touched a drop. And right now when my daddy takes a drink, I mean you would think that somebody had given him—what, raw tobacco or something—he just frowns and goes on so bad. But I don't ever recall liquor in my house, unless one of my uncles or someone came over and brought some with them for their consumption.

Another informant, a 25-year-old woman, describes a similar pattern in her household:

My mother doesn't drink at all and she thinks that it's wrong to drink. . . . Well, her religious concepts thinks it's wrong. . . . If someone brought some wine or something to the house, she wouldn't tell them not to drink it or—but she says, she doesn't and she wouldn't drink. . . . Sometimes on family occasions they [relatives] would drink a little but it wasn't a thing where there was always liquor around at family occasions either, 'cause my grandparents never had liquor in the house. They drink liquor very rarely, so it isn't just a thing where, they were having a dinner or a picnic or anything that liquor was bought for the occasion.

Blacks reared in this type of milieu typically express ambivalent views about alcohol. Even though they may now be drinkers, alcohol is regarded as a potent and dangerous substance, especially when consumed in excess. Many negative qualities are attributed to alcoholic beverages and are offered as reasons for drinking very little or not drinking at all. Alcohol is variously described as a depressant; a substance that will make you sick; or one that will make you lose physical or emotional control.

These are the themes described in the following set of responses by three different informants, who except for the first informant, are current drinkers:

Alcohol is really depressing. Physically you know, you lose your coordination, you don't feel things, you know—and your mind is dulled.

I don't like to drink if I'm depressed. . . . No, because drinking is a downer. It emotionally brings you down. . . . Drinking used to make me high, I'd much prefer smoking grass than drinking liquor because liquor does not give me the same kind of emotional high that smoking does. . . . I think with drinking you tend to be more shaky physically. Drinking makes me sick to the stomach. . . . If you drink a lot so that you really have a buzz on, or if I do so that I really am quite high on liquor, I'll probably be sick to the stomach before I get high. . . . It may be three glasses (of wine) or something that makes me start to feel nauseous.

I don't like to drink when at a really big party. . . . Cause I like to be in more control. When I drink, I think I'm under less control, and less in tune with

what's happening around me when I drink. Your senses are dulled, I think—
at least mine are when I drink.

Although anti-alcohol sentiments are sanctioned and reinforced in
many black social contexts, the focus on drinking establishments and
alcohol use associated with the prohibition era has also left a lasting
impression on black culture. Bars, taverns and nightclubs have retained
an important place in black society because they provide a place for
socializing, dancing, and listening to music. Alcohol is intrinsically associated with these establishments as it is in informal contexts—such as
house parties—which have the same focus. In these settings drinking
alcohol is regarded as an important symbol of sociability and pleasure
(Borker *et al.*, 1980, pp. 170–180).

The use of alcohol also emerges as an important way of displaying
wealth and enhancing one's personal prestige. Blacks are observed to
be conspicuous consumers of high-priced brands of liquor (Bauer, 1964).
Using expensive liquor as a way to communicate one's sophistication
and "class" seems to fit very well in a culture where a great deal of
emphasis is placed on enhanced self-presentation though elegant groom-
ing and luxury goods—for example, expensive clothing, cars, and jew-
elry.

As a holdover from prohibition, liquor also plays a key role in the
economy of black communities. Off-sale liquor establishments are regarded as one of the most viable forms of individual entreprenuership
available to blacks (Mosher & Mottl, 1981, p. 450). The liquor industry
views blacks as a primary market for distilled liquors, and this is very
visible in advertisements and promotional campaigns in local and national black publications.

In sum, the legacy of historical changes in the ideology of black
drinking is reflected in the coexistence of two different worlds of drinking
patterns in black life. The old temperance themes are reflected in the
vast network of church and quasi-religious self-help organizations in
which abstinence is the norm—a norm that is only ignored for ceremonial
occasions. In the other social world, as a holdover from the prohibition
era, alcohol is esteemed as a symbol of pleasure, status, and economic
worth.

DISCUSSION

The data presented in this study illustrate the importance of historical processes and societal changes on drinking patterns and alcohol-

related problems. The importance of social change, particularly at the political–economic level, has been given little attention in the current literature on blacks and alcohol. Most sociocultural literature focuses only on assumptions about internal cultural values or psychological factors as they now exist to explain black drinking patterns. Hence previous research has interpreted ambiguity in black drinking norms primarily as an outgrowth of status differentiation within black communities (Sterne, 1967, Sterne & Pittman, 1972) or as a correlate of poor self-esteem and self-disparagement among upwardly mobile blacks (Maddox & Allen, 1961).

Similarly alcohol problems and some forms of alcoholism among blacks are often interpreted in the light of cultural or individual level "stress" hypotheses. A persistent theme in the literature is that black drinking problems are a result of the "utilitarian" and "escape" drinking patterns learned during slavery to cope with poverty and oppression— patterns that are re-enacted under the stress and racism of contemporary ghetto life (Sterne & Pittman, 1972; Harper, 1976). This interpretation fails to explain the massive increase in black alcohol problems (e.g., cirrhosis of the liver—see Herd, 1985a) since slavery, despite major gains in social status among blacks.

As Makela (1975) notes in his critique of the sociocultural paradigms that are usually applied to alcohol problems, these theories focus on internal cultural attitudes while ignoring the role of alcohol control policies, alcohol consumption levels, and other social, economic, and political constraints on drinking behavior.

In contrast, the analysis presented here shows that attitudes toward alcoholic beverages among blacks have been profoundly affected by changes in the political–symbolic role of alcohol in the larger society and by major demographic shifts in the black population. These findings point to the need to focus on the interaction between ethnic specific norms and the larger sociopolitical structure in explaining the drinking pattern of blacks, and subcultural groups generally, in complex industrial societies.

REFERENCES

Aaron, P., & Musto, D. *Temperance in America: A historical overview.* Paper prepared for a Panel on Alternative Policies Affecting the Prevention of Alcohol Abuse and Alcoholism of the National Research Council, December 1979.

Anderson, J. New York Harlem, Part 3. *New Yorker.* July 13, 1981.

Aptheker, H. *A documentary history of the Negro people*. Vols. I and II. New York: The Citadel Press (1968, 4th Paperback Edition), 1951.

Bailey, M. B., Haberman, P. W., & Alksne, H. The epidemiology of alcoholism in an urban residential area. *Quarterly Journal of Studies on Alcohol*, 1965, 26, 19–40.

Bauer, R. A. The Negro revolution and the Negro market. *Public Opinion Quarterly*, 1964, 28, 647–648.

Blocker, J. S., Sr. *Retreat from reform: The prohibition movement in the United States 1890–1913*. Westport, CT: Greenwood Press, 1976.

Borker, R., Hembry, K., & Herd, D. *Black drinking practices study*. Report of Ethnographic Research to the Department of Alcohol and Drug Programs, State of California. Contract No. A-0259-8, 1980.

Cahalan, D., Cisin, I. W., & Crossley, H. M. *American drinking practices*, Monograph 6. New Brunswick, NJ: Rutgers Center of Alcohol Studies, 1969.

Carter, E. A. Prohibition and the Negro. *Opportunity*, 1928, 6, 359–360.

Cheagle, R. U. *The colored temperance movement*. Unpublished Master's thesis, Moorland-Springain Research Center, Howard University, 1969.

Cherrington, E. H. *Standard encyclopedia of the alcohol problem*, Vol. V. Westerville, OH: American Issue Publishing Company, 1929.

Clark, W., & Midanik, L. Alcohol use and alcohol problems among U. S. adults: Results of the 1979 national survey. In *Alcohol Consumption and Related Problems*, Alcohol and Health Monograph 1. DHHS Publication No. (ADM)82-1190. Washington, DC: U. S. Government Printing Office, 1982.

Crowe, C. Racial violence and social reform—origins of the Atlanta riot of 1906. *The Journal of Negro History*, 1968, 8 234–256.

Daniels, D. H. *Pioneer urbanites*. Philadelphia: Temple University Press, 1980.

Douglass, F. *Life and times of Frederick Douglass*. New York: Collier Books (1967 Ed.), 1892.

Drake, St. C., & Cayton, H. *Black metropolis: A study of Negro life in a northern city*, Vols. 1 & 2. New York and Evanston: Harper & Row (1962 expanded Ed.), 1945.

DuBois, W. E. B. *The Philadelphia Negro: A social study*. New York: Schocken Books (1967 Ed.), 1899.

Erenberg, L. A. *Steppin' out: New York night life and the transformation of American culture*. Westport, CT: Greenwood Press, 1981.

Frazier, E. F. *The Negro church in America*. New York: Schocken Books, 1966.

Genovese, E. D. *Roll Jordan, roll: The world the slaves made*. New York: Vintage Books, 1976.

George, C. V. R. *Segregated sabbaths*. New York: Oxford University Press, 1973.

Greene, L. J. *The Negro in colonial New England*. New York: Columbia University Press, 1942.

Gusfield, J. R. *Symbolic crusade*. Urbana: University of Illinois Press, 1966.

Gwinnell, W. B. Shifting populations in great northern cities. *Opportunity*, 1928, 6, 279.

Hannerz, U. What ghetto males are like. In N. E. Whitten & J. F. Szwed (Eds.), *Afro-American anthropology*. New York: The Free Press, 1970.

Harper, F. *Alcohol and blacks: An overview*. Alexandria, VA: Douglas, 1976.

Harrington, O. *Bootsie and others: A selection of cartoons by Ollie Harrington*. New York: Dodd, Mead, 1958.

Heath, D. A critical review of ethnographic studies of alcohol use. In R. J. Gibbins *et al.* (Eds.), *Research advances in alcohol and drug problems*, Vol. 2. New York: Wiley, 1975.

Herd, D. Religion and black drinking. In R. A. Borker (Ed.), *Black drinking practices study*. Report of Ethnographic Research to the Department of Alcohol and Drug Programs, State of California. Contract No. A-0259-8, 1980.

Herd, D. Prohibition, racism and class politics in the post-Reconstruction South. *Journal of Drug Issues,* 1983, *13,* 77–94.

Herd, D. Migration, cultural transformation and the rise of black cirrhosis. *British Journal of Addiction,* 1985a.

Herd, D. "We Cannot Stagger to Freedom": A history of blacks and alcohol in American politics. In L. Brill & C. Winick (Eds.), *Yearbook of substance use and abuse,* Vol. 3. New York: Human Sciences Press, 1985b.

Hughes, L. *One-way ticket.* New York: Knopf (1949 reprint), 1936.

Hughes, L. *Simple speaks his mind.* New York: Simon & Schuster, 1950.

Irwin, W. Who killed Margaret Lear? *Collier's Weekly,* 1908, *41,* 10.

Isaac, P. E. *Prohibition and politics in Tennessee 1885–1920.* Knoxville: University of Tennessee Press, 1965.

Johnson, G. G. *Ante-bellum North Carolina.* Chapel Hill: The University of North Carolina Press, 1937.

Kletzing, H. F., & Crogman, A. M. *Progress of a race.* New York: Negro Universities Press (1969 Ed.), 1897.

Koren, J. *Economic aspects of the liquor problem.* Boston, MA: Houghton Mifflin, 1899.

Larkins, J. R. *Alcohol and the Negro: Explosive issues.* Zebulon, NC: Record Publishing Company, 1965.

Levine, H. The discovery of addiction: Changing conceptions of habitual drunkenness in American history. *Journal of Studies on Alcohol,* 1978, *39,* 143–174.

Levine, H. Temperance and women in 19th century United States. In O. J. Kalant (Ed.), *Research advances in alcohol and drug problems, Vol. 5: Alcohol and drug problems in women.* New York and London: Plenum Press, 1980.

Lewis, H. *Blackways of Kent.* Chapel Hill: University of North Carolina Press, 1955.

Liebow, E. *Talley's corner.* Boston: Little, Brown, 1967.

McKay, C. *Harlem: Negro metropolis.* New York: Harcourt Brace Jovanovich, 1968.

Maddox, G. L., & Allen, B. A comparative study of social definitions of alcohol and its uses among selected male Negro and white undergraduates. *Quarterly Journal of Studies on Alcohol,* 1961, *22,* 418–427.

Makela, K. Consumption level and cultural drinking patterns as determinants of alcohol problems. *Journal of Drug Issues,* 1975, *5,* 344–357.

Malin, H., Kaelber, C., Sorenson, C., Dadovrian, C., & Munch, N. Trends of cirrhosis of liver mortality. *U. S. alcohol epidemiological data reference manual. Section I, Alcohol epidemiological data system.* Rockville, MD: National Institute on Alcohol Abuse and Alcoholism, 1980.

Malzberg, B. Statistics of alcoholic mental diseases. *Religious Education,* 1944, *39,* 22–30.

Mosher, J. F., & Mottl, J. R. The role of nonalcohol agencies in federal regulation of drinking behavior and consequences. In M. Moore & D. Gerstein (Eds.), *Alcohol and public policy: Beyond the shadow of prohibition. Report of a National Academy of Science's panel on alternative policies affecting the prevention of alcohol abuse and alcoholism.* Washington, DC: National Academy Press, 1981.

Netting, R. M. C. Beer as a focus of value among the West African Kofyar. In M. Marshall (Ed.), *Beliefs, behaviors, and alcoholic beverages.* Ann Arbor: The University of Michigan Press, 1979.

Pittman, D. *Alcoholism.* New York: Harper & Row, 1967.

Pottern, L. M., Morris, L. E., Blot, W. J., Ziegler, R. G., & Fraumeri, S. F. Esophageal cancer among black men in Washington, D. C. 1. Alcohol, tobacco and other risk factors. *Journal of the National Cancer Institute,* 1981, *67,* 777–783.

Quarles, B. *Black abolitionists.* New York: Oxford University Press, 1969.

Rogers, E., Goldkind, L., & Goldkind, S. Increasing frequency of esophageal cancer among Black Americans. *Cancer,* 1982, *49,* 610–617.

Rorabaugh, W. J. *The alcoholic republic.* New York: Oxford University Press, 1979.

Saint Paul Appeal. Meeting of anti-prohibition league of Iowa at Cedar Rapids. A big convention of Afro-Americans opposed to the present Iowa law. August 8, 1891, 1, 7.

Schoener, A. *Harlem on my mind.* New York: Random House, 1969.

Sinclair, A. *The era of excess: A social history of the prohibition movement.* New York: Harper & Row, 1964.

Smith, W. B. The persimmon tree and the beer dance. *Farmer's Register, VI.* (Shellbanks, VA April 1838). In Bruce Jackson (Ed.), *The Negro and his folklore in 19th century periodicals.* Austin: University of Texas Press (1967), 1838.

Stampp, K. *The peculiar institution.* New York: Knopf, 1961.

Sterne, M. W. Drinking patterns and alcoholism among American Negroes. In D. J. Pittman (Ed.), *Alcoholism.* New York: Harper & Row, 1967.

Sterne, M. W., & Pittman, D. *Drinking patterns in the ghetto.* St. Louis, MO: Social Science Institute, Washington University, 1972.

Umunna, I. The drinking culture of a Nigerian community. *Quarterly Journal of Studies on Alcohol,* 1967, *28,* 529–537.

U. S. Census Office. Tenth Census (1880), Vol. XII: Report on the Mortality and Vital Statistics of the United States (in two parts). Washington, DC: U. S. Government Printing Office, 1886.

Walton, H. *The Negro in third party politics.* Philadelphia: Dorance and Company, 1969.

Wharton, V. L. *The Negro in Mississippi, 1865–1890.* Chapel Hill: University of North Carolina Press, 1947.

Winston, S., & Butler, M. Negro bootleggers in eastern North Carolina. *American Sociological Review,* 1943, *8,* 692–697.

11

Alcohol
Cultural Conceptions and Social Behavior among Urban Blacks

ATWOOD D. GAINES

INTRODUCTION

A concern with beverage alcohol in black society has a long history. Frederick Douglass (1892) saw alcohol as a means masters used to control their slaves during holiday seasons, although Genovese, in his *Roll, Jordan, Roll* (1976) notes that slaves were also concerned about drunkenness in their masters. After emancipation, historians note, whiskey was often served at social events in order to manipulate the black vote in the South (Franklin, 1974). Alcohol seems also to have served as a device for keeping Blacks in their proverbial Southern place (Lee, 1944) and Southern prohibitionists tried to use the weak support for prohibition among Blacks as a cause for disenfranchisement (Walton, 1970; Wharton, 1947). Such attempts recall earlier Southern efforts to deny rights of state residency if an emancipated slave used alcohol (Guild, 1939). Some contemporary authors continue to suggest that alcohol, alcoholism, and arrests for alcohol-related behaviors are still some means

ATWOOD D. GAINES • Department of Anthropology and Department of Psychiatry, Case Western Reserve University and Medical School, Cleveland, Ohio 44106.

employed by a "White" establishment to control "Blacks" (see Staples, 1976).[1]

Through time, some views have changed about alcohol and "Blacks," but much has remained the same. The folk theory of the disinhibiting effects of alcohol is still with us. And there is still the propensity to lump together people with more or less known African, usually West African, ancestry as a homogeneous group which by unexamined, implicit processes, has remained unitary (although the members of this "group" did not originally form a single cultural or biological unit). It is assumed that in ideas and behavior, group members exhibit a unity of belief and behavior such that anything attributed to one member person is attributed to all; or, conversely, the "group" has no culture, only homogeneous responses to homogeneous negative social and economic experiences.

This chapter has three goals: (1) to analyze alcohol research professionals' conceptualizations, assumptions, folk theories, and consequent problems related to "Blacks" as a research population (see footnote 1); (2) to present some new ethnographic data and material on black cultural theories related to alcohol; and (3) to suggest a methodology for the study of alcohol and other drugs in their cultural context in an attempt to solve the intensive-extensive and subjective-objective dilemmas in social research.

In this chapter, I can but summarize the material from three research projects in which I was involved. In these projects, variation in beliefs and practices, rather than consistency in the putative group, has been the principal finding. Some of the sources of this variation will be discussed below. It will be shown that all is not variation. Some of these beliefs and practices differ in source and form and thus are not variants of other beliefs and practices. The confounding of the two is caused by the American system of social classification.

This chapter will necessarily be concerned with two levels of folk theories, theories derived from cultural or subcultural knowledge marshaled to explain particular phenomena in the actor's social field. The

---

[1] The terms *White* and *Anglo* are not used interchangeably here. The former term has the same problems associated with its use and its putative referent as does the term *Black*, which is a focus of the present paper. I would use *Anglo* to exclude many people often labeled White, for example, Nicaraguan-, Italian-, Hungarian-, Russian-, French-, Lithuanian-, Spanish-, or Czechoslovakian-Americans. *White* appears where it is used by informants or in the literature. "Black" appears in the text to show that the literature or informants seem to be assuming a unity of all people of African descent. Black (unmarked) should be used only to apply to people who are both of African descent and who bear black culture. The nominative forms are herein capitalized, the adjectival form is not. In part, I am here merely updating the insights of Boas (1940), the founder of American anthropology and an early antagonist of racist thinking in science and society.

first level of folk theory is that of the bureaucratic, social scientific, and medical professionals, including alcohol funding agents, traditional alcohol researchers, and providers. I will focus on their system of social classification, a folk conceptual or cognitive system that uses salient criteria to classify people and to explain social behavior.

The second level is composed of the folk theories and explanatory models of people classified as belonging to a single group by the folk theories in the first level. This chapter, then, will consider how folk theory, in both science and society, contributes to the ethnography of alcohol and the anthropology of science (see Gaines & Hahn, 1982; Hahn & Gaines, 1985 for other research of biomedical scientific theory and practice).

SOME RESEARCH ON ALCOHOL IN BLACK CULTURE

The data presented here derive from three different research projects. The results of the research will be used in the discussion of conceptions of social behavior related to alcohol. All the research projects discussed were conducted in Northern California, in San Francisco, Oakland, Richmond, and other Bay Area cities, as well as the city of Stockton northeast of the Bay Area.

THREE PROJECTS

In the first project (1976), I worked with my anthropological colleagues in interviewing providers involved in all facets of alcohol treatment, including medical and social detox programs, residential programs, community counselors (publicly paid), autonomous, unaffiliated community counselors, prison alcohol program officials, community groups (Black, Hispanic, Native American), and individuals in various communities (see Frankel, Gaines, Reingold, & Robles, 1978). We attempted to gain an understanding of the providers' conceptions of alcohol use, abuse, problem behavior, and treatment in their service.

In the second research project (1977), I directed the interviewing of "Black" and "White" adolescents in the East Bay. I interviewed the "Blacks," some 80 male and female adolescents. This group included a group of middle- to upper-middle-class male and female high school students from Oakland who were participants in a college-bound program at an East Bay university. These would have been called the "black bourgeoisie" by Frazier (1931) or members of the "respectable upper class" by Drake and Cayton (1945). However, as I will show below, perhaps they should be considered not as a subdivision of black culture, but rather as belonging to a different cultural tradition. Most of the 80 adolescents interviewed, however, were lower- to lower-middle-class residents of the city of Oakland.

The third research project was a long-term, major effort concerning the role of alcohol in two black communities in San Francisco. This research, conducted 1979–1980, was funded by the State of California. The proposed research employed a cultural, that is, a symbolic, interpretive approach (see Geertz, 1973; Rabinow & Sullivan, 1981), wherein the intersubjective world of the actors was the main focus of the study. I was interested in the way in which alcohol was conceptualized, the ways in which different beverages and drinking and non-drinking contexts might be understood and categorized, and in the explanations for and definitions of problematic and non-problematic drinking. Close attention to the definitions of terms was essential for even the most obvious terms such as "beer," "wine," or "non-drinker," and to social categories, and conceptual categories, and discourse about alcohol. In short, my interest was in the semantics of alcohol and the frameworks of experience that guide human action (Geertz, 1973) in these communities.

AN INTERPRETIVE APPROACH TO ALCOHOL

For this last project my proposal called for a two-stage research effort. The first stage was to be long-term ethnographic research in two selected black communities in San Francisco. I intended to conduct this stage of the research as anthropological fieldwork, that is, as ethnography through participant observation.[2] The ethnographic research was to take a semantic, symbolic approach in order to understand the place and conceptualization of beverage alcohol among members of these two communities. These data would then be used to construct a survey instrument to obtain quantitative data. The construction and adminis-

[2] In order to ensure a fresh perspective, I conceived of hiring females to conduct the ethnographic research, since it was evident that most other research on alcohol tended to show not only a middle-class Anglo bias, but also a male bias. I expected that more could be learned about such topics as women and alcohol, as well as others, through the use of women researchers. My desire to hire women (Black) researchers (and a woman research coordinator) was opposed by the funding agent, the research firm for which I did the research, and the sociological colleagues selected by the firm to give research assistance. Fortunately I prevailed and in the end the two women researchers assisted the research coordinator in writing the final report (see Borker, Herd, & Hembry, 1980) as I left the study before its completion.

tration of the instrument constituted the second stage of the proposed research.

An interpretive perspective was advocated in order to develop appropriate domains and means of questioning (e.g., a culturally appropriate lexicon, interrogative forms), and to avoid what seem to be major problems with the traditional survey approach to the study of alcohol. Traditional sociological approaches generally assume that the social and cultural worlds of the research target population(s) are known and that one need only construct a questionnaire to tap known patterns of behavior, which can then be explained *post hoc*. Such approaches cannot take into account potentially very different conceptually organized worlds. Without prior knowledge, it is not possible to tap such forms of folk understandings as folk theories and explanatory models that guide behavior and interpret experience. In the sociological tradition, researchers tacitly assume that they know enough of the cultural ideas and categories, if they are so recognized, to ask relevant questions framed in appropriate syntactical and lexical as well as conceptual forms.

This relatively simple proposed research format appears to be a means of approaching the common extensive–intensive and subjective–objective dilemmas encountered in social scientific research. Cultural data can be gathered during the ethnographic stage and employed in the instrument that generates the quantitative data gathered in the second, survey stage. However, until other similar projects are completed, the problem of comparability looms large. I did not concern myself with this question for to do so would have reduced the potentially innovative aspects of the research project. In order to produce comparable data, one must pose the same, often inappropriate, questions used in other, often culturally limited, if not insensitive, surveys. An interpretive, as opposed to an empiricist, ethnographic research orientation allows the systematic exploration of the subjective understanding of alcohol and alcohol-related behaviors. From such a data base, a survey instrument can be constructed to tap cultural understandings in culturally appropriate ways.[3]

Because the research considered folk conceptions and theories, it was appropriate that the conceptions of the funding agency, the collaborating/advising social scientists, and the providers involved in the re-

[3] The length of time to conduct such research seems altogether reasonable for an anthropologist. However the State of California in this instance did not really understand the reason for the ethnographic research. The overseers of social research in most State agencies are familiar with sociological survey research and as a consequence thought of the ethnographic research as mere window dressing.

search also be considered. Their assumptions, explanatory models, and folk theories were thus also objects of study and as such serve as foils for the present analyses. I have suggested that such a dual focus might be adopted formally in any contract social anthropological research (Gaines, 1981).

PROFESSIONAL ASSUMPTIONS AND FOLK THEORIES

In the first project, my anthropological colleagues and I were interested in uncovering various notions about alcohol use, abuse, problems, and treatment among providers working in the alcohol field. We recognized that we were dealing with ideologies, culturally constituted systems of classification, assumptions, and conceptualizations rather than social or biological reality. An immediate source of friction between us and the research group for which we worked arose from the group's assumption that the conceptual framework of caregivers constituted or mirrored empirical reality. Our recognition of the range of beliefs about alcohol and the folk bases of these ideas influenced the development of our cognizance of the importance of investigating the folk theories of the co-researchers, funding agents, and providers, along with those of the research population; but we had little time to explore the former systematically.

The second research project, conducted for a non-profit research and education firm, was intended to test the relation between strength of adolescent ethnic identification and problem drinking. The hypothesis was that "Blacks" who identified strongly with black history and culture, as it was defined, would be less likely to have problems with alcohol. It was assumed that conscious knowledge about the achievements of Blacks could be measured and that it constituted an index of ethnic identification, that is, identification with the ethnic group. Ironically the research seemed to show that those who were most embedded in black culture actually knew the least about black achievements and, further, that these Blacks were more likely either to have problems with alcohol or to be total abstainers. The research formulators believed that identity could be equated with knowledge and the latter with "race." Variation in the specified group was seen as a function of "ethnic identification." Cultural differences among members of the (putative) group were wholly unanticipated.

The "Request for Proposal" from the State of California, which led to the development of the third and largest project noted above, specified that the State wished to evaluate research proposals that could provide information upon which new policies and programs might be based. Specifically, the State called for research that would "improve the ability of the Department of Alcohol and Drug Abuse, the counties, and other appropriate agencies and groups to plan and implement programs to more effectively and efficiently decrease alcohol related problems among Blacks in California." The state's primary goal, stated clearly at times, was to derive statistics on the number of "black alcoholics." With these data, it was thought, the number of alcohol programs and other ancillary activities that might be needed could be established. That is, the resources necessary to deal with a problem of a defined size among "Blacks" could be ascertained by the number of people with alcohol problems, both the former and the latter as defined by the funding agent.

The research request had several difficulties embedded in its formulation which the research I designed attempted to overcome.[4] These included how the State and its sociological consultants defined "problem" and the social group "Black" in which these problems were believed to occur. Both of these difficulties will be discussed below. Other problems associated with this and other contract social anthropological research have been analyzed elsewhere (see Gaines, 1981).

One major difficulty with such research involves the folk system of social classification used by professionals to classify individuals having African ancestry, a problem shared with lay persons in America. For the State and for sociological colleagues, this was not a problem. But among students of culture, it is well known that anyone, properly socialized, can acquire any given cultural ideology. As Boas (1940), the founder of American anthropology, long ago demonstrated, there exist no human "races," a fact amply confirmed by contemporary research (Montagu, 1963; Hiernaux, 1975; Watts, 1981). He further noted that cultures, including such things as language and religion and behavior in general, are demonstrably unrelated to the putative races (Boas, 1940). However, my bureaucratic and sociological colleagues (and some anthropological ones) believed, and doubtless still believe, that if black culture existed only "Blacks" could share it and, furthermore, it must be distinct from anything non-blacks manifested. They reasoned, from folk racial assumptions, that if there existed a distinctive black culture, then all those so classified should partake of it.

Another folk theory was to be found among research colleagues and the staff of the funding agencies. This theory held that no cultural

---

[4] My departure from the project was occasioned by some of the problems associated with contract interdisciplinary research with personnel problems mentioned elsewhere (see Gaines, 1981).

differences would be found among "black people" related to alcohol because alcohol problems have universal causes. This perspective was one that was widely shared, especially among biomedical professional and sociological advisors, many of whom had had alcohol problems of their own at one time or another. (For studies of the folk cultural basis of Biomedicine and biomedical practice, see Gaines & Hahn, 1982; Hahn & Gaines, 1985.) In their view, explanations in terms of cultural differences were merely another of the many evasions or rationalizations seen with alcoholism, a problem that could be defined *a priori*.

Thus, two major folk theories were found to be held by both the funding agency and the project research group: a racial, biological determinist theory and a quasi-biomedical theory. The latter conceives of a disease or condition unaffected by culture (an entity about which the proponents were unclear). A related or sometimes related notion was also found among the professionals. In this view, "oppression and discrimination" may be, or have been "triggering" events or conditions for the development of a "pathological" condition, defined without regard for cultural context or meanings in "empirical" terms and usually constituted by measures of quantity and frequency of consumption and functional problems.

In contrast, I construed the object of study to be a cultural ideology, not a race of people. It could not be otherwise, since the group specified is a group only in terms of a folk conceptual system. Individuals so classified have very heterogeneous social, economic, and genetic heritages. Most, if not all, of the people in this category have substantial European as well as African ancestry [not to mention Native American (see Hallowell, 1976)]. The social category "Black," in lay society is thus analogous to the old scientific typological approach in biology compatible with racial stereotyping (Watts, 1981). Individuality is ignored as individuals are seen only as manifestations of a putative standard type (Watts, 1981; Montagu, 1963); striking variation is thus ignored. "Races, like all units of taxonomic classification, are arbitrary categories that exist only in the minds of classifiers, not in nature itself" (Watts, 1981, p. 10).

It should be noted also that one might even add the qualification of "West" to "African," since populations of Africa vary considerably in both genotype and phenotype and the ancestors of American Blacks are primarily from West Africa. African populations do not form a single group on any morphological, genetic, linguistic, or cultural basis (Alland, 1973; Franklin, Jacobs, & Bertrand, 1981; Hiernaux, 1975). In fact, about 1,000 biologically distinct populations have been identified in sub-Saharan Africa (Franklin et al., 1981, p. 498). People classified as "Black" in America do not form a single behavioral, social, genetic, or biological unit (Boas, 1940) due to differences in origin, social experience, and a differential, massive genetic exchange with Europeans and Indians over the centuries.

We may observe here that "racially specific" diseases are highly problematic. For example, the African ancestors of those called Black in America largely came from river areas of West Africa, with lesser numbers coming from the Congo (now Zaire). It is in only these two areas of Africa that the severe form of sickle cell anemia is found. Were these people from East or South Africa, the problem of sickle cell anemia would not have appeared in America, for the trait is rare to non-existent among populations in those areas. We note that the trait is found at high levels in many "White" populations around the Mediterranean (Italians, Greeks, Tunisians, Kuwaitis, Algerians, etc.), as well as among people in India, Indonesia, and the Philippines (Livingstone, 1958). The condition is thus unrelated to "race." Rather, it is related to disease environments in which hyperendemic or holoendemic malaria has occasioned an adaptive biological response, sickling, in the *local* populations (Livingstone, 1958).

The problematic assumptions are, then, as follows: It is assumed that there are a number of people who, because they are known to have African ancestry, constitute a single, unique social group or community; this specificity suggests that this community has specific needs and problems; it is also assumed that service needs may be ascertained by collecting statistical information about "alcohol problems" and that alcohol problems constitute a single domain of discourse in which problems may be universally defined and are eluctable through the employment of statistical measures. Although other assumptions are implicit, these will suffice to frame the argument. The following considers the cultural constructions of alcohol for both the informants and the research professionals involved in the projects described above.

CULTURAL CLASSIFICATIONS OF BEVERAGE ALCOHOL

Explanatory Models

In analyzing cultural classificatory systems of beverage alcohol, I use an important formulation in medical anthropology, that of the explanatory model (EM). The concept of the EM, advanced by Kleinman (1980), suggests that individuals and healers alike develop cognitive models of particular sickness episodes. Explanatory models are held by the individuals who constitute the "local health care system" (LHCS) (Kleinman, 1980). The LHCS has three sectors: (1) the popular sector, which includes individuals and families; (2) the folk sector, which includes folk healers; and (3) the professional sector, which includes professional healers. Individuals in each of the sectors hold various models of sickness and apply them to specific sickness episodes in themselves or significant others. The theories of alcohol service and research professionals mentioned above are kinds of professional EMs. A given EM is seen as involving one or more of the following conceptualizations: the etiology of the sickness, its course, the sickness' processes and progress, its outcome, and appropriate therapeutic intervention(s) for the sickness.

Also of importance and considered below are cultural conceptions of bodily processes and structures (ethnophysiology), actions of drugs and other substances (ethnopharmacology), and conceptions of person seen as underlying and giving coherence to EMs (Gaines, 1985). All of these may be seen as topics to consider in an effort to gain entry into the intersubjective world of the actors, the world of their lives and their actions.

CULTURAL CONCEPTIONS

It is known that in some cultures alcoholic beverages are considered a part of the cultural domain of food. This is the case in the Mediterranean culture area, but it is not the case in Northern Europe. Black culture distinguishes alcohol from food, but at the same time it recognizes certain analogous qualities. Thus, many informants stated that alcohol is unsuitable for consumption with meals. Alcoholic beverages are seen as appropriate to consume at other times without food. One does not begin a meal with a cocktail or apéritif, drink wine with dinner, or conclude a meal with a liqueur. The reason given is usually that of "taste," that is, the taste of alcoholic beverages makes them unsuitable for consumption with foods. There are special occasions, however, when exceptions to the general rule are made. The exceptions involve celebrations such as those held for defined holidays, especially Thanksgiving, July 4th, and Christmas, and family celebrations (weddings, birthdays). At funerals, which are large gatherings, it is expected that family members and close friends of the deceased will accompany the deceased's family back to the family home after the funeral services and there, all will be served a very large meal (generally buffet style). Alcohol is made available at such gatherings.

Research indicates that there are many "Blacks" who consume alcoholic beverages with meals and who see no gustatory conflict between alcohol and food. Such people may have been reared this way or have learned the custom through travel or from Anglo friends (Borker, Herd, & Hembry, 1980). Individuals of African ancestry who regularly socialize with (or as) Anglos or others such as Italian-, French-, or Mexican-Americans seem not to use the classificatory system that excludes beverage alcohol at mealtime.

Many informants, adult and adolescent, stated that only on some festive occasion would there be alcohol present in the home. This was true even for individuals who are abstainers themselves and whose parents did not drink. Many of the adolescents interviewed reported that family gatherings provided them with the their first opportunity to sample an alcoholic beverage. Generally, the alcohol seems to be provided by a relative and given in a happy celebratory mood for the purpose of adult entertainment.

The absence of alcohol may be related to another cultural conception, according to which alcohol is seen as a lubricant for social interactions, but is inappropriate for such instrumental activities as business gatherings. Since alcohol is excluded at mealtime and is used on festive occasions only, it may be associated in people's minds with good times and hence is used when "times are bad" or for the "blues" as a (psychological) remedy. Its exclusion at mealtime also may create a notion of a special or exceptional nature of alcoholic beverages and reduce the experience or familiarity of culture members with alcohol, unlike the French or the Italians, thus making the development of problematic (to an outsider) drinking perhaps more likely. Research indicates that "Blacks" learn to drink at an earlier age than other groups, but drinking patterns are likely to change in later years; that is, individuals may reduce the amount of their drinking or stop altogether in later life (see Klatsky, Friedman, Siegleaub, and Gerard, 1977; Maddox, 1970; Higgins, *et al.*, 1977).

A DUAL SYMBOLIC CLASSIFICATORY SYSTEM

The ethnographic data point to the existence of a dual symbolic classificatory system in black culture which provides a framework for the interpretation of experience and a model for social action (see Geertz, 1973) related to beverage alcohol. Dual symbolic classificatory systems have appeared in the anthropological literature for some time (Hertz, 1969; Fox, 1975; Needham, 1973). In such systems, elements are linked to one another through culture-specific associations of particular perceived properties. These elements than can be contrasted with another category, the elements thus forming binary oppositions. In urban black culture, such a system of classification of alcoholic beverages appears.

Contrasting terms include "light" and "dark" beverages, which are also characterized as "soft" and "hard," respectively. In this system, wine and beer are "soft" relative to distilled spirits. The system must be understood in context, since the terms are used in a relative, not an absolute sense.

For instance, although wines are generally considered soft *vis-à-vis* hard liquor (distilled spirits), some wines may be considered hard or "rough." Such wines are those considered "cheap" and/or "nasty." They are also called "rotgut" and are considered hard compared to other wines. Darker beverages, such as whiskey, bourbon, and the like, are thought to be harder than lighter spirits, such as gin, vodka, or rum. In this instance, the term soft also implies that the beverage has a less deleterious effect on the body. This classification relates to and constitutes portions of ethnopharmacological and ethnophysiological systems that should be further analyzed.

The informants suggested that consumption of the lighter distilled spirits has less of an effect on the body regardless of its proof. Thus, an informant could consider that he had "cut down on drinking," to be acting conscientiously with regard to his health, because he had stopped consuming darker beverages and had switched to lighter ones, although the proof remained virtually unchanged. It appears that the dimensions of soft and hard also encode the dichotomy of feminine versus masculine, which in turn encodes "refined" and "unrefined." The system outlined here provides some insight into drinking patterns and strategies encountered in the field. For example, males tend to drink those drinks associated with maleness—hard, dark beverages, including red wines and cheap wines or rotgut depending on their status. But when a male wishes to appear less rough, a bit refined, he may choose those beverages associated with the refined, feminine side of the dichotomous scheme and may drink, or say he drinks, "mellow" (good) scotch or even lighter drinks such as vodka, gin, or rum. Or one might choose red wine over distilled spirits, thus again using the system to select an appropriate beverage for the particular social event.

Using (unconsciously) the dual classificatory system, adolescent females could explain in detail the strategies they would use, or had used, in managing an impression of themselves with male friends. As an example, a young woman (17 years old) normally preferred to drink vodka. But when she dated someone new, she would manage the impression that she was not hard, rough, or unrefined by concealing her normal preference and pretending to prefer a wine, a sweet red wine most often, or an even "lighter" white wine. The use of feminine elements to signal "refinement" in general is widespread in black culture and explains the use of haircurlers, nail polish, makeup, and fur coats among the most masculine of ghetto figures, pimps and players (see Milner & Milner, 1972).

It is noteworthy, too, that there are different expectations for males and females regarding alcohol consumption in black culture. For example, it was considered totally unacceptable for a female to appear drunk in public. Such behavior appears as the *sine qua non* of moral bankruptcy. This is part of an elaborate dichotomy involving differential expectations for men and women in black culture (Aschenbrenner, 1975; Stack, 1974) analogous to the tradition of the American South (the font of black culture) and the Mediterranean culture area (Davis, 1975; Gaines, 1982a; Peristiany, 1975).

Interestingly, other "Blacks" participating in the college-bound program and some adolescents interviewed in ethnically diverse communities saw no reason for different expectations for males and females. Nor did they know of or understand the principal social dichotomy in black culture between domestic respectability and (public) life in the streets. They further assumed that proof, not color, was the important factor in determining potency. The members of this group manifested an ignorance of the logic of street corner life and drinking (see Liebow, 1967; Anderson, 1978) and they held an egalitarian view of male–female drinking: what was an approved beverage for a male was appropriate for a female and if drunken behavior was unacceptable for females, it was also unacceptable for males. These "Blacks" thus indicated very different conceptualizations of alcohol than did the others.

Good and Bad Alcohol

In addition to the classification of light and dark and light and heavy, with their connotations of lesser and greater potential effect on the body, there is in black culture the notion that there are "good" and "bad" types of alcoholic beverages. Bad beverages are those considered "cheap" (that is, disvalued, *not* inexpensive), such as rotgut wines (Thunderbird Ripple or Twister). Not only are such beverages considered to be without merit, they are also believed to have a more negative impact on body function than other beverages. It is believed that cheap wine will cause bodily harm that will not be caused by the consumption of approved wines, although the percentage of alcohol may be even more in the latter. Wines of quality and distilled spirits of quality are believed to have not only a "smoother" or "mellower" taste, but are also believed to have an analogous salutory, "mellowing" effect on the body. Hence in this ethnopharmacological view, the alcohol's potentially negative impact on the body is mitigated when quality beverages are used (see Borker *et al.*, 1980).

Off Brands

An interesting category of beverage alcohol is that of "off brands." This appellation is applied to beverages considered beyond the pale. It is believed that such beverages are "nasty" in taste and worthless, although they may often be expensive. ("Nasty" is a generic term for anything considered to have an offensive taste whether or not the actual taste is known; the term is also applied to anything that is disliked. In Anglo culture, the term has taken on the connotation of "really great," "beautiful," "fun," as in the lyric from the 1984 Randy Newman song "I Love LA," which states, "I'm drivin' around . . . with a big, nasty redhead by my side . . .").

However off brands are often labels well respected by Anglos. "Off brand," as an appelation, only appears to serve as another term for cheap labels. In fact it refers to *unknown* marks or labels. The vast majority of wines considered to be of high quality from California's 1970 wave of new vineyards (e.g., Mirassou, Stag's Leap, J.P. Lohr, Firestone, Moët) would be considered off brands because they are unfamiliar. It was a source of amusement for the adolescents in black culture that Anglos are consumers of off brands because, as these adolescents stated, it shows that "White people are really gullible"; "[they're] silly enough to try just any old thing." The term can thus serve as a marker of ethnic boundaries as well as an expression of racism.

BEER, ALE, LIQUOR, AND ALCOHOL

A discussion of beer in black culture brings us to some distinct terminological problems. First, the term does not in fact refer to beer. Rather, it refers to malt liquor and/or to ale. The term did refer to beer among the college-bound group. These adolescents and other interviewed who lived in various plural communities around the Bay Area also knew of beer kegs and keg parties. Interestingly, not a single one of the youths from (all) "black neighborhoods," some 60 or so, had *ever* heard of the term *keg* or knew to what it referred.

The term *alcohol* when used by bearers of black culture often refers only to distilled spirits. Thus, young and old people will say that they "don't drink alcohol" for a variety of reasons including "taste," strength, weight problems, loss of control, or problems with significant others (Borker *et al.*, 1980). However, they are referring not to beer or wine, which they may in fact regularly consume, but only to distilled or hard liquor. The term *liquor* itself is similarly frequently employed to refer to distilled spirits, not to beers, ales, or wines.

Certain brands of malt liquor are considered to be much more potent than others, although again, the alcohol content is the same. For example, Old English 800, is called "crazy eight" because of its "powerful" effect on the drinker. Malt liquors also are seen as being rough, strong, dark forms of beverage and are thus the choice of those who consider themselves, like their drinks, to be "bad." Of course to be "bad" in black culture is also to be good (see Milner & Milner, 1972; Brown, 1972).

Beer, properly speaking, is viewed very differently. Beer is considered a very mild, weak beverage more closely related to soft drinks than to alcohol. It is believed to be virtually without negative impact on the body. When beer is consumed among "Blacks," it is those who are members of the "respectable" domain culture and not street corner habitués or those who do not belong to black culture. Among the respectables, beer consumption is acceptable for such out-of-doors drinking as sporting events, whereas the consumption of alcohol (in other forms) is not. Thus, at sporting events, beer may be consumed as a thirst quencher and no stigma accrues to the drinker. But the public consumption of hard liquor is considered inappropriate and to mark those community members lacking respectability. However, sometimes respectables may "hang on the corner" as well (see Anderson, 1978; Liebow, 1967).

CHARACTER AND CONSUMPTION

An interesting element, implicit above and seen over and over in black culture, is the association of specific kinds of beverages or consumption patterns and personal character. Judgments of personal character and moral worth are frequently made on the basis of certain "indicators" such as type of alcohol consumption and place of consumption. Thus, people who drink off brands are unworthy of respect. Similarly, those who drink rotgut are highly suspect. We see in this construction how character and consumption are related. Those who consume "bad" drinks in public places are thus stigmatized not only in terms of place and kind of alcohol consumed, but more importantly in terms of their moral being. This construction is of considerable importance in problem definition and help-seeking and for research on alcohol use among bearers of black culture.

Sociological and biomedical approaches focus on the ubiquitous measures of quantity and frequency (see Maddox, 1970; Stern & Pittman, 1972; Klatsky *et al.*, 1977) as indicators of alcohol problems. However it appears that drinking in black culture becomes problematic depending on the *quality* of beverage and *place of consumption*. And the conclusion reached is a moral rather than a biomedical (including psychiatric) one. In the biomedical EM, it is assumed that alcoholism is a treatable problem, which is an attribute of persons. In black culture, the person *is* a "bad actor," "no good," without respectability, and therefore beyond the pale; he or she is not a person *with* a problem. Such a person is in need not of treatment, but rather of ostracism.

PROBLEM IDENTIFICATION AND THERAPY

A CULTURAL CLASSIFICATION OF DRINKERS

Respectable Drinking

As indicated, just as types of beverage alcohol are classified, so too are people in terms of their consumption of particular beverages. In general there seem to be several "respectable" relationships with alcoholic beverages (also see Lewis, 1955). Abstainers are considered respectable, and this is one not infrequent relationship to alcohol in black culture (Harper, 1976a; Stern & Pittman, 1972). Individuals who "sip," or drink lightly are called "sippers." Sippers are also worthy of respect (also see Bourne, 1976). So too are weekend drinkers, even if they appear to an outsider to have weekend "binges." "Heavy" drinking does not automatically invite the concern of significant others and seems not to be seen as problematic or significant to self or others (see Bourne, 1976).

Problem Drinking

The notion of "problem" surfaces only if other significant elements of a symbolic complex are also present, such as solitary or public drinking. These latter behaviors indicate aspects of non-respectable involvement with alcohol. Those of the street, the non-respectables, consume alcohol in public places, although not all street corner habitués are non-respectable types (see Anderson, 1980; Liebow, 1967). Non-respectables include gang members (Rohrer & Edmonson, 1960), although insiders may feel that public consumption of alcoholic beverages is an expression of group solidarity within a gang (see Kaiser, 1969).

A central element in this classification is that public consumption of alcoholic beverages (not including beer) is disvalued by respectable people, those of the domestic and church domains, although valued in the life of the streets. It is clear that normative drinking in black culture is preeminently social (see Harper, 1976a; Stern & Pittman, 1976; Rohrer & Edmonson, 1960). Drinking is meant to enhance conviviality and the presentation of self, but it appears that the social context should be private and involve only an audience of known individuals, including family and friends. We see then that notions about quality of beverages and context of consumption are central to the black culture's evaluation of drinking behaviors. As the evaluative criteria include quality and place/context of consumption, wineheads are seen as the archtypical problem drinkers, although the term does not necessarily imply Skid Row alcoholism. Because of this perspective, definitions in individuals' EMs about problem drinking differ from biomedical or social service personnel. As noted, "heavy" drinkers are not necessarily problem drinkers. The assumption of "problem" seems to depend more on such indicators as family relationships, job maintenance, and especially the type and context of beverage consumption. As a black alcohol counselor explained, "a (black) man who has a job and is still with his family wouldn't believe he could have an alcohol problem, no matter how much he drank."

"Social drinkers" are those who drink only in the company of others and only on "social occasions." Regardless of how much is consumed by such drinkers, it is unlikely that a diagnosis of a drinking problem would be made by self or others. As indicated above, the beverage one consumes can also play a large part in determining if a problem is present. That is, one who consumes "quality" beverages is much less likely to note a problem with alcohol than is a person who drinks less respectable beverages, the consumption of which may be defined in and of itself as a problem. For example, one black woman stated that even though she was in a recovery home, she did not have an alcohol problem. Her explanation for this was that she drank only "good brandy" (Korbel) and not "that ole cheap stuff them others be drinkin'; that ole wine and stuff."

ALCOHOL AND THE PERSON

Another issue of importance concerns problem identification. This is the cultural conception of the nature of the person. Researchers have become increasingly aware of the importance of cultural conceptions of the person and their relationship to cognitive process and social actions related to health and illness (Conner, 1982; Gaines, 1979, 1982a, 1985; Hahn, 1985; Kapferer, 1979; Shweder & Bourne, 1982). Of importance here is the centrality of the notion of person in the formation of etiological hypotheses and therapeutic modalities (the latter are based upon and made sensible by the former) (Fox, 1964; Gaines, 1979, 1982a, 1985) and the success of such therapies (Kapferer, 1979; Gaines, 1985).

Informants' discourse indicates that persons and their alcohol problems are seen as unity. That is, person conceptions are such that behaviors attributed to alcohol by others, including the professional folk theories and EMs, are perceived as expressions of the inherent characteristics of the person. As such, behavioral complexes are not separable, problematic entities. They are aspects of personhood. It is difficult therefore to see problems as separate from persons, to see them as distinct entities or to see them as amenable to some form of treatment. This same view is held by professional and lay Chicano informants as well.

This person conception is further complicated by the fact that persons are seen as immutable from birth, except by divine stroke or some other miraculous transformation. This view is much like that of the French and other Mediterranean conceptions of person and that of the American South (see Gaines, 1982a). Thus if a problem is perceived, individuals may conclude that little or nothing can be done by a person him- or herself. The problem is only a "natural" part of the person or the person is only "being him- (or her)self." If a problem is finally noted, and it is felt that something can be done, recourse to helping agencies is organized by cultural ideas relating to indigenous EMs, especially the etiological theories contained therein, and also to conceptions of appropriate caregivers and the nature of social reality.

ETIOLOGY AND HELP-SEEKING

First, we should note that black culture is familistic (but does not exhibit a single, unique family form, see Aschenbrenner, 1975). Individuals are enmeshed in and rely upon wide family networks. These networks include people outsiders would refer to as friends, but who are conceptualized as relatives in the cultural kinship system (Aschenbrenner, 1975; Stack, 1974; Hannerz, 1969). This kinship idiom is found also in Southern America and in the Mediterranean (Davis, 1975; Gaines, 1982a; Gilmore, 1982; Peristiany, 1975). As a consequence, it is difficult, if not impossible, for individuals to seek treatment from "strangers," such as those in alcohol treatment centers. It is assumed that unrelated others are dangerous and without concern for nonfamilial others. Several informants and alcohol counselors mentioned that "Blacks" often did not trust "Whites" who might counsel them. They did not want, as one put it, "white folks messin' with they head."

Given the importance of religion (especially fundamentalist Protestantism in the lives of people of the American South, the tradition
from and in which black culture developed, and still thrives), it should
be noted that the definition of problem may be in a domain wholly
distinct from that of the caregivers. Drinking problems may be seen in
a religious context, as products of sin, or the Devil's work. The appropriate source for relief would not be a secular state or community agency,
but rather a trusted religious figure or church (see Manning, 1977). Some
of the caregivers seen as appropriate may be actually certified professionals in the health field, such as the Christian psychiatrists I have
studied in the American South (see Gaines, 1982b, 1985).

Secular etiological theories are important to consider as well. Such
folk theories, aspects of EMs, are similar in some respects to religious
etiological theories; that is, they assume an external agency is responsible
for a problem. Informants frequently attributed alcohol problems (by
which is generally meant the "problem" of people who drink in public,
such as winos, see, for example, Harper, 1976b) to racism and related
issues of oppression, discrimination, and economic difficulties. The
problem was always seen as residing *outside* the individual. With such
a conceptualization, it makes little sense for one to pursue individual-
centered therapies, but makes group and divinely inspired therapies
plausible. Several directors of residential programs felt that religious
conversion was one of the most successful means of treating alcohol
problems in "Blacks." One residential home was in fact upstairs from a
church (fundamentalist) and a number of the residents were in regular
attendance.

Other folk theories place the blame for drinking problems on an
individual's microsocial world: that of relatives, lovers, suitors, and work.
These are seen as "unavoidably" fraught with hazards and problems.
Drinking is thus seen as a sensible, understandable response to frustrations in the macro- or microsocial world or and for the "blues," although
biomedicine tells us that alcohol actually exacerbates dysphoric affect.
As regards this folk theory, a look at an insider's evaluation is instructive.

In my interviews with a county prison alcohol program director, it
was clear that the director was well aware of the folk theories mentioned
above. He also recognized that they made the type of treatment he could
offer very problematic in terms of efficacy and the faith his charges could
invest in his program. When queried about his knowledge of these folk
theories, he indicated his awareness of them although he had some
difficulty accepting their validity. He explained the reasons for his skep-
ticism. He said many of the people he knew and worked with in city
and county government were "White," and therefore were not subject

to the negative impact of economic hardships, racism, and oppression claimed as etiologically significant by "Blacks and other minorities." He said, "They [Whites] have nice homes, nice jobs, nice families, cars and income and so forth; but *they're* mostly alcoholics [most of them are alcoholics]. So, I don't know about whether it's really racism or economic problems that make them [Blacks and Chicanos] drink."

Another dimension of the treatment problem was raised by several informants and concerned confidentiality of those in treatment. First, it should be noted that a central form of social control in black culture, and Southern culture in general, is gossip [and this is true also of Mediterranean cultures (see Bailey, 1974)]. People very much fear that others will learn of their "coming and goings," their "business" and thereby do them harm. Some informants suggested that people might use treatment centers if they could be sure that their "business" would not be "spread all over town" by caregivers. One solution they suggested would be to have centers for people in one city located in another, at some remove. Thus, individuals in City A would have centers for them in City B, whereas those in City B would have centers for them in City A. Informants felt this would remove the fear of gossip as a barrier to seeking help, as clients would be less likely to encounter people known to them.

Discussions with health-care providers, including clergy, about their relationship with formal providers indicate that such individuals as ministers referred problems to formal providers only if they were personally known to them. A notion of familiarity thus pervades referrals and seems to result in referrals being made to individuals who are not necessarily equipped to handle alcohol problems, but who are known to help with other problems and known to the referrer.

Informants who seemed not to be members of black culture held views much like those of the service providers and other professionals in that alcoholism was seen as a condition in need of treatment, including medical treatment. The problem was further located within individuals, thus making individual treatment strategies meaningful. Again, we see variation among the members of a putative group. The precise meaning of such variation is considered below.

CULTURAL VARIATION AND DIFFERENCE

VARIATION

An individual's experience with alcohol, including personal involvement or that of a close relative or friend, seemed to produce differential knowledge among informants. A strong home life precludes a life in the street, the latter variation in black culture being seen as negative and intimately alcohol-involved. Because the consumption of alcoholic beverages is closely tied to family gatherings for holidays or family fêtes, the size of one's kin network can influence one's experiences with alcohol. Similarly, such events as divorce, moving, and other social changes affect one's kin group and therefore social experience with family gatherings, not to mention the potential trauma of such changes. Interaction with other ethnic groups similarly influences patterns.

Certainly alcohol in black culture is seen as enhancing the sociability of gatherings, including talking or "rapping" (Brown, 1972), and projecting a particular image of self (Stern & Pittman, 1976; Borker *et al.*, 1980) and thus has a very important and positive role in black culture. The many possible variations of individual experience lead to variations in belief and action relating to alcohol in black culture, including the important cultural conceptualizations that categorize positive and negative life-styles within the culture.

It is important to consider, and even stress, the variation within black culture that is usually overlooked by researchers (but see Hannerz, 1969; Borker *et al.*, 1980). But it is perhaps even more important to recognize that much of what appears to be variation among "Blacks" is not variation, but rather cultural difference.

DIFFERENCE

Americans with African ancestry do not now, nor have they for some time, experienced homogeneous genetic, social, or economic forces and histories. As people of African descent never formed autonomous cultural units in America (see Hallowell, 1976), individuals had and have widely varying experiences that America's concern for, and the symbolic system relating to, color consistently ignores, although it is clear that recent changes indicate shifts in the racial classificatory system (see Blu, 1980).

Hallowell, one of the deans of psychological anthropology, in his early studies of Native Americans, for instance, was surprised to find that many did not speak their aboriginal language (but rather French) and that some Indians, including at least one chief of the last century, were biologically "White," and that others were "Black" (Hallowell, 1976). Similarly it is because of past and present differential social experience that we find considerable variation in beliefs, behavior, and biology (Boas, 1940) among people of African ancestry (most of whom happen to have about as much or more European ancestry).

Several points may be stressed here. First black cultural theories directly and indirectly related to alcohol affect behavior and cognitive processes related to the use and abuse of beverage alcohol. We can but note here that black culture seems not to be unique, but rather is a variant of what might be termed Southern culture which is shared with Southern "Whites." As examples, one might cite a number of common-alities of Southerners, whether "White" or "Black," including dialects, authoritarian and enthusiastic religious traditions, the coincidence of political and religious leadership, matrifocality, machismo, familism, cuisine, gossip as social control, "campanilismo," codes of honor and shame (see Gaines, 1982a for more on this relationship), rootwork, and even such culture-specific folk illnesses as hexing and falling out (Weidman, 1979).

Second, the seemingly obvious should be stressed: Many people who have more or less African ancestry do not share black culture. Many such people from other shores are becoming increasingly evident on the American scene and include, for example, Cubans, Jamaicans, Haitians, Ethiopians, Nigerians, and Puerto Ricans. To call all these people "Black" is misleading and inappropriate.

More often overlooked are the American "Blacks" who do not carry black culture. One might mention, for example, 1984's Miss America, Vanessa Williams, film star Jennifer Beals, singer/actress Irene Cara (who is actually Cuban and Puerto Rican), Pro Football's Franco Harris, or Olympic ice skater Tai Babalonia. All these people have at least some known African ancestry but are not in fact generally considered "Black" by the general public. The same is true for a great many athletes, es-pecially baseball players, from the Caribbean. These perceptions are indicative of the shift in the classificatory system in America (see Blu, 1980).

There are now numerous communities with significant numbers of such people; for example New York and its northern suburbs; the various "heights" communities surrounding Cleveland; areas of Oakland, San Francisco, and Los Angeles, California; the Chicago suburbs; and Washington, D. C. and its outlying suburbs.

The sciences, especially, should not continue to adhere to traditional American folk biological theories of social classification. Researchers should adopt a "cultural" theory of human difference wherein identity or group-ness is based upon behavior, as one finds in contemporary France (Gaines, unpublished), and not putative biology. Such a system would not *imply* the concept of race as do the terms *White* and *Black*, even when they are called "ethnic" terms. The racial implication of the terms is the reason that the former Black Muslims, now the Muslim Community in America, is thusly called and why they refuse to use the terms *Black* and *White*.

The continued use of such terms implying a unity of people with more or less African ancestry suggests a (no longer very subtle) form of racism. This is evident in certain academic, and especially political-economic, circles in which it is frequently said that "though it is recognized that racial labels are merely social convention, they have the same consequences for all people of African ancestry" (Fox, personal communication, 1982). In point of fact, the homogeneity is entirely fictional and tends to be used to prop up a moribund theory.

In other empiricist circles, such as sociobiology, we find a similar manifestation of folk theory in the guise of science. Van den Berghe (1981) has tried to combine Marxism and sociobiology in analyzing ethnicity in what is doubtless a nineteenth-century empiricist's dream. Although it is an insightful, well written study, the synthesis is bound by the shackles of a nineteenth-century empiricism which is unable to grasp the "fact" of the cultural bases of social categories. Color is not *intrinsically* significant as is evident in France (Gaines, unpublished) and as Malcolm X learned from his travels in North Africa and the Middle East.

No extant bases exist, except implicit or explicit racism, for the continued conceptual lumping together of people who have widely divergent backgrounds. New, nonbiological (and non-"lumpen" economic) social classifications should be employed in the sciences, which in this seem to be lagging behind changes in the popular sector. Such terms should refer to the actual significant cultural differences and not to implicit, putative "biological" differences. It is hoped titles such as that of the present chapter, which only seem to focus on a real, empirical social group, will become increasingly rare, if not extinct.

SUMMARY

This chapter has focused on several aspects of alcohol in science and society. Ethnographic research on alcohol in black culture indicates that variations occur within that cultural tradition and that many people assumed to be a part of black culture by virtue of their more or less African ancestry actually exhibit aspects of another cultural tradition. The material presented shows certain aspects of the semantic reality of alcohol in black culture, including cultural classifications of beverage alcohol, problem identification, therapy, and help-seeking. A consideration of the folk theoretical basis of research professionals' social categories highlighted for us the problematics of racial categories employed to describe research populations. This study suggests that research should focus on cultural communities and ideologies and not on implicit or explicit racial groupings, as such groupings are things of the mind rather than of the natural world.

Variation and difference among people assigned the social classi-fication "Black" were discerned through a symbolic, interpretive ap-proach. This approach seems to have provided some new insights and data, more of which might have been obtained had the methodology outlined above been completed.[5] The approach may help obviate the sometimes ethnocentric, narrow empiricism of traditional sociological research. Beliefs and behaviors discussed here may be seen as elements of cultural systems. They are not based upon biological or economic homogeneity.[6] I have also briefly suggested the regional affiliation and basis of black culture. With these understandings we may, for example, suggest that, contrary to members of the category itself, there is no unitary "black experience" (or "family") common and unique to all peo-ple in America with African ancestry. Such claims often equate biology and behavior as does the wider society.

Similarly, political economic, biological, and sociobiological theorists who continue to stress and/or assert behavioral or ideological common-alities of people with African ancestry manifest a resistance to thinking in other than racial terms, terms that are perhaps more integral to the theories they expound than has been generally recognized.

REFERENCES

Alland, A., Human biological diversity in Africa. In E. Skinner (Ed.), *Peoples and cultures in Africa*. Garden City, NY: Doubleday, 1973.

Anderson, E., *A place on the corner*. Chicago: University of Chicago Press, 1978.

Aschenbrenner, J., *Lifelines*. New York: Holt, Rinehart and Winston, 1975.

Bailey, F. G. (Ed.). *Gifts and poisons*. New York: Shocken Books, 1974.

Blu, K., *The Lumbee problem*. Cambridge, England: Cambridge University Press, 1980.

Boas, F., *Race, language and culture*. New York: The Free Press, 1940 (reissued 1966).

Borker, R., Herd, D., & Hembry, K. *Ethnographic report for the black drinking practices study*. 3 Vols. Berkeley, CA: Source, Inc., 1980.

---

[5] Because of my departure and that of the other anthropologists after the completion of the ethnographic research and write-up, the survey was not informed by the ethnographic research data. The material presented here thus summarizes material derived from the ethnographic research portion of that project and other research I conducted through formal and informal interviews done from an interpretive perspective.

[6] The best extant bibliography on alcohol and "Blacks" is that by Daniel N. Maltz (1980), compiled as a part of the third project described herein. It is available through the Applied Anthropology Documentation Collection, University of Kentucky, Lexington, KY 40506.

Bourne, P., Alcoholism in the urban black population. In F. Harper (Ed.), *Alcohol abuse and black America*. Alexandria, VA: Douglass Publishing Company, 1976.

Brown, H. R., Street talk. In T. Kochman (Ed.), *Rappin' and stylin' out: Communication in urban black America*. Urbana: University of Illinois Press, 1972.

Conner, L., The unbounded self. In A. Marsella & G. White (Eds.), *Cultural conceptions of mental health and therapy*. Dordrecht, Holland: D. Reidel, 1982.

Davis, J. *People of the Mediterranean*. London: Routledge & Kegan Paul, 1975.

Douglass, F. *Life and times of Frederick Douglass*. New York: Collier McMillan, 1892.

Drake, St. C., & Cayton, H. *Black metropolis*. 2 Vols. New York: Harper & Row, 1945.

Fox, J., On binary categories and primary symbols. In R. Willis (Ed.), *The interpretation of symbolism*. New York: Halsted Press (J Wiley), 1975.

Fox, J. R., Witchcraft and clanship in cochiti. In A. Kiev (Ed.), *Magic, faith and healing*. New York: The Free Press, 1964.

Frankel, S., Gaines, A. D., Reingold, J., & Robles, N. *Conceptions of alcohol use, abuse and problem behavior in three California counties*. Monograph of the Social Research Group and the School of Public Health, University of California at Berkeley. Berkeley: Social Research Group, 1978.

Franklin, J. H., *From slavery to freedom*. New York: Knopf, 1974.

Franklin, R. R., Jacobs, C. F., & Bertrand, W. E. Illness in black Africans. In H. Rothschild (Ed.), *Biocultural aspects of disease*. New York: Academic Press, 1981.

Frazier, E. F. *The Negro family in the United States*. Chicago: University of Chicago Press, 1931.

Gaines, A. D. Definitions and diagnoses. *Culture, Medicine and Psychiatry*, 1979, 3, 381–418.

Gaines, A. D. Hard contract: Issues and problems in contract social anthropological research. *Kroeber Anthropological Society Papers*, 1981, 59–60, 82–91.

Gaines, A. D. Cultural definitions, behavior and the person in American psychiatry. In A. Marsella & G. White (Eds.), *Cultural conceptions of mental health and therapy*. Dordrecht, Holland: D. Reidel, 1982a.

Gaines, A. D. The twice-born: Christian psychiatrists and Christian psychiatry. In A. D. Gaines & R. A. Hahn (Eds.), *Physians of Western medicine: Five cultural studies*, Special Issue, *Culture, Medicine and Psychiatry*, 1982b, 6(3).

Gaines, A. D. The once- and the twice-born: Self and practice among psychiatrists and Christian psychiatrists. In R. A. Hahn & A. D. Gaines (Eds.), *Physicians of Western medicine: Anthropological approaches to theory and practice*. Dordrecht, Holland: D. Reidel Publishing Company, 1985.

Gaines, A. D. *Ethnicity as a cultural system: "Race" and culture in Strasbourg*. (Unpublished).

Gaines, A. D., & R. A. Hahn, (Eds.) *Physicians of Western medicine: Five cultural studies*. Special Issue, *Culture, Medicine and Psychiatry*, 1982, 6(3).

Geertz, C. *The Interpretation of cultures*. New York: Basic Books, 1973.

Genovese, E. *Roll, Jordan, roll: The world the slaves made*. New York: Random House (Vintage Books), 1976.

Gilmore, D. Anthropology of the Mediterranean area. In A. Beals, B. Siegel, & S. Tyler (Eds.), *Annual reviews in anthropology*. Palo Alto, CA: Annual Reviews, Inc., 1982.

Guild, J. *Black laws of Virginia*. Richmond, VA: Whittet and Shepperson, 1939.

Hahn, R. A. A world of internal medicine. In R. A. Hahn & A. D. Gaines, (Eds.), *Physicians of Western medicine: Anthropological approaches to theory and practice*. Dordrecht, Holland: D. Reidel Publishing Company, 1985.

Hahn, R. A., & Gaines, A. D., (Eds.) *Physicians of Western medicine: Anthropological approaches to theory and practice*. Dordrecht, Holland: D. Reidel Publishing Company, 1985.

Hallowell, A. I. American Indians, white and black. In A. I. Hallowell, *Contributions to anthropology: Selected papers of A. Irving Hallowell.* Chicago: University of Chicago Press, 1976.

Hannerz, U. *Soulside.* New York: Columbia University Press, 1969.

Harper, F. (Ed.) *Alcohol abuse and black America.* Alexandria, VA: Douglass Publishing Company, 1976a.

Harper, F., Summary, issues, and recommendations. In F. Harper (Ed.), *Alcohol abuse and black America.* Alexandria, VA: Douglass Publishing company, 1976b.

Hertz, R. *Death and the right hand.* Glencoe, IL: Free Press, 1960.

Hiernaux, J. *The peoples of Africa.* New York: Scribner's, 1975.

Higgins, P., Albrecht, G., & Albrecht, M. Black–White adolescent drinking. *Social Problems,* 1977 *25,* 215–224.

Kaiser, L. *The vice lords: Warriors of the streets.* New York: Holt, Rinehart and Winston, 1969.

Kapferer, B., Mind, self and other in demonic illness. *American Ethnologist,* 1979, *6,* 110–133.

Klatsky, A., Friedman, G., Siegleaub, A., & Gerard, M. Alcohol consumption among white, black or oriental men and women. *American Journal of Epidemiology,* 1977, *105,* 311–323.

Kleinman, A. *Patients and healers in the context of culture.* Berkeley: University of California Press, 1980.

Lee, A. Techniques of social reform. *American Sociological Review,* 1944, *9,* 65–77.

Lewis, H. *Blackways of Kent.* Chapel Hill: University of North Carolina Press, 1955.

Liebow, E. *Tally's corner.* Boston: Little, Brown, 1967.

Livingstone, F. Anthropological implications of sickle-cell gene distribution in West Africa. *American Anthropologist,* 1958, *60,* 533–562.

Maddox, G. Drinking and abstinence. In G. Maddox (Ed.), *The domesticated drug: Drinking among collegians.* New Haven, CT: College and University Press, 1970.

Maltz, D. *Black drinking practices study archival research report to the Department of Alcohol and Drug Abuse.* Berkeley, CA: Source, Inc., 1980.

Manning, F. The salvation of a drunk: Play and reality in Pentecostal ritual. *American Ethnologist,* 1977, *4,* 397–412.

Milner, C., & Milner, R. *Black players.* New York: Bantam Books, 1972.

Montagu, A., The concept of race in the human species in the light of genetics. In A. Montagu (Ed.), *Race, science and humanity.* Princeton, NJ: Van Nostrand, 1963.

Needham, R. (Ed.). *Right and left: Essays on dual symbolic classification.* Chicago: University of Chicago Press, 1973.

Peristiany, J. G. (Ed.). *Honour and shame: The values of Mediterranean society.* Chicago: Midway Reprints, 1975.

Rabinow, P., & Sullivan, J. (Eds.), *Interpretative social science.* Berkeley: University of California Press, 1981.

Rohrer, J., & Edmonson, M. (Eds.). *The eighth generation grows up.* New York: Harper & Row, 1960.

Shweder, R., & Bourne, E. Does the conception of person vary cross-culturally? In A. Marsella & G. White (Eds.), *Cultural conceptions of mental health and therapy.* Dordrecht, Holland: D. Reidel, 1982.

Stack, C. B. *All our kin.* New York: Harper & Row, 1974.

Staples, R. *Introduction to black sociology.* New York: McGraw-Hill, 1976.

Stern, M. Drinking patterns and alcoholism among American Negroes. In D. Pittman (Ed.), *Alcoholism.* New York: Harper & Row, 1967.

Stern, M., & Pittman, D. The significance of drinking among housing project youth. In

M. Stern & D. Pittman (Eds.), *Drinking patterns in the ghetto.* Vol. 1. St. Louis, MO: Washington University Social Science Institute, 1972.

Stern, M. & Pittman, D. Alcohol abuse and the black family. In Frederick Harper (Ed.), *Alcohol abuse and black America.* Alexandria, VA: Douglass Publishers, Inc., 1976.

Van den Berghe, P. *The ethnic phenomenon.* New York: Elsevier, 1981.

Walton, H. Jr. Another force for disenfranchisement: Blacks and the prohibitionists in Tennessee. *Journal of Human Relations,* 1970, *18,* 728–738.

Watts, E. The biological race concept and diseases of modern man. In H. Rothschild (Ed.), *Biocultural aspects of disease.* New York: Academic Press, 1981.

Wharton, V. L. *The Negro in Mississippi, 1865–1890.* Chapel Hill: University of North Carolina Press, 1947.

Weidman, H. Falling-Out: A diagnostic and treatment problem viewed from a transcultural perspective. *Social Science and Medicine,* 1979, *13B,* 95–112.

IV

Case Studies
American Indians

12

Indians, Ethnicity, and Alcohol
Contrasting Perceptions of the Ethnic Self and Alcohol Use

JOAN WEIBEL-ORLANDO

INTRODUCTION

On May 15, 1804, Lewis and Clark set out from St. Louis, Missouri on what was to be a two-year scientific expedition to and personal adventure in the as yet uncharted Great American Desert. The goals of their mission included the initiation of peaceful relations between their fledgling nation and the tribal peoples encountered in their travels along the Missouri River and its tributaries to the Pacific Ocean. As representatives of the U. S. government, Lewis and Clark were well supplied with gift items by which peaceful relations with tribal groups were to be cemented. These gifts included medallions minted by the U. S. Treasury to be given specifically to potential tribal allies, foods, beads, European-style clothes, blankets, and kegs of whiskey (Biddle, 1962).

The entire stock of whiskey was traded less than halfway through the expedition. Among some of the Indians who came to powwow and trade with the explorers, the demand for whiskey was so great that the expedition leaders had to ration its distribution; other tribal groups re-

JOAN WEIBEL-ORLANDO • Neuropsychiatric Institute, University of California, Los Angeles, California 90024.

fused to trade for it, however. Liquor, as a standard trade exchange item, had been established by the first western European fur traders or *courier du bois* as early as the mid-1700s. By 1804, it was a well entrenched cultural and economic currency among the Northern Plains Indians Lewis and Clark were to encounter (Dailey, 1968).

Important to this book, however, is the range of beliefs about alcohol and its meaning for Indian people among the tribal groups the Lewis and Clark expedition encountered 180 years ago. The Sioux thought of alcoholic beverages as *mni wankan* or "sacred water," a magical substance that, they believed, allowed them to communicate more easily or directly with the spirit world (Mohatt, 1972). Its power to effect an altered state of consciousness was so desired that the one recorded event of potential hostility between the expedition members and an Indian village was prompted by Lewis and Clark's refusal to ply each member of the Sioux entourage with more than one-quarter of a glass of whiskey.

> Tuesday, September 5, 1804—The chiefs and warriors from the camp two miles up the river, met us, about fifty or sixty in number, and after smoking delivered them a speech. We then invited the chiefs on board, and showed them the boat, the airgun, and such curiosities as we thought might amuse them: in this we succeeded too well; for after giving them a quarter of a glass of whiskey, which they seemed to like very much, and sucked the bottle, it was with much difficulty that we could get rid of them. They at last accompanied captain Clark on shore in a periogue with five men: but it seems they had formed a design to stop us; for no sooner had the party landed than three of the Indians seized the cable of the periogue, and one of the soldiers of the chief put his arms around the mast; the second chief who affected intoxication then said, that we should not go on, that they had not received presents enough from us. (Biddle, p. 52)

The Arikara, on the other hand, roundly refused the whiskey offered them and, in no uncertain terms, told the expedition leaders why:

> Wednesday, October 10, 1804— . . . as we were desirous of assembling the whole nation at once, we [also] invited the chiefs of the two upper villages to a conference. They all assembled at one o'clock, . . . we then made or acknowledged three chiefs, one for each of the three villages, giving to each a flag, a medal, a red coat, a cocked hat and feather, also some goods, paint and tobacco, which they divided among themselves. On our side we were equally gratified at discovering that these Ricaras made use of no spiritious liquors of any kind, the examples of the traders who bring it to them so far from tempting, having in fact disgusted them. Supposing that it was as agreeable to them as to the other Indians, we had at first offered them whiskey; but they refused it with this sensible remark, that they were surprised that their father should present to them a liquor which would make them fools. On another occasion they observed to Mr. Tabeau, that no man could be their friend who tried to lead them into such follies. (Biddle, p. 63)

These historical accounts illustrate the problems inherent in attempts to formulate a global Indian point of view or belief system with regard to alcohol consumption. The expanding range of beliefs and practices pertaining to alcohol consumption among contemporary American Indian tribal groups exemplifies Benedict's (1934) notion that societies choose their particular cultural configurations from an arc of possible behaviors. Tribal differences with regard to the function of alcohol in their societies are further cross-cut by such intra-tribal influences as sex, age, religious affiliation, and family drinking histories (Weisner, Weibel-Orlando, & Long, 1984).

Given the reality of regional, tribal and intral-tribal variation, how does one begin to talk in holistic terms about American Indian drinking practices and beliefs in a book with the organizing principle of variation in experience with alcohol among major ethnic American groups? To begin, non-tribal Americans view Indians as another global ethnic group. The critical thematic thread that unites disparate tribal groups into a larger social entity is the imposition of the label *Indian*, an essentially political act. The process by which the ethnic category Indian was formulated has been and still is, two-pronged. Although non-tribal Americans originated, perpetuated, and cherished certain Indian stereotypes, Tribal Americans also developed and exploited their own ethnic stances *vis-à-vis* non-Indians. Ablon (1964) was among the first to acknowledge just such a process. How the ethnic category Indian is defined and redefined by both non-Indians and the labeled tribal people and the influence of alcohol in shaping the ethnic group stereotype had cultural and political ramifications that reverberate to this day.

A review of the Indian and alcohol literature, particularly the subset of political and/or historical works (Churchill & Larson, 1983; Dailey, 1966, 1968; Lurie, 1971; Mail & McDonald, 1981; Mosher, 1975; Officer, 1971), underscores the important role the distribution and regulation of alcohol beverages played in the evolving political and economic relations between North American tribal peoples and the emerging "Americans." It is also important to acknowledge Indians as an ethnic category in any discussion of Indian alcoholism treatment programs, for it is at this level of specificity that most alcoholism interventions are currently addressed. It is just such homogenizing, social categorizing, and unfounded generalizing about Indians, based on imposed stereotypes not of their own making, that have been the bane of most Indian alcoholism treatment programs.

The us/them dichotomy, which epitomizes the structure of ethnic relations, is exemplified by the social distance between American Indians and other ethnic American groups (Fiske, 1977; Fiske & Weibel, 1980). Not only are obvious demarcations differentiating red man from white man perpetuated, but also Indians clearly enunciate the differences in their historical-cultural backgrounds from all other racial group *imigrés* to North America. Afro-Americans and Mexican-Americans were at a loss to understand the reluctance of American Indians to empathize with, much less, join their civil rights activities in the 1960s and 1970s (Wax, 1971). However, most Amerian Indians are certain that their grievances against the colonial power that has denied or systematically divested them of their rights since the days of the earliest settlement are fundamentally different from those grievances held by the descendants of imported slave laborers or impoverished European *imigrés* (Steiner, 1968). If American Indians identify with any other ethnic groups, they are the equally colonialized and disenfranchised third-world *indigene's*, with whom Indians feel an affinity (Churchill & Larson, 1983).

To the 1.2 million people who claim Indian heritage in the United States (U. S. Census, 1980) and consciously express their ethnic group membership in a variety of public arenas, their unique social category in our mosaic society is a valid and salient status to be sustained, enhanced, enacted, and articulated. Ethnic group membership and ethnic stance is maintained in a number of ways. This chapter describes two mechanisms already well entrenched in Indian interactional style *vis-à-vis* other Indians and non-Indians, as well as an evolving third ethnic posture. These presentations of the ethnic self are intrinsically influenced by strong and conflicting beliefs about alcohol use and its meaning for Indian people.

Sacred separation is the first ethnic stance to be explored. The notion that in-group/out-group membership can be defined by the dichotomous criterion of spiritual purity versus secular profanity is not a new concept to anthropology (Levi-Strauss, 1963; Van Gennep, 1909). This sociostructural mechanism allows certain Tribal Americans to enact a popular American Indian stereotype, the noble spiritualist. Inherent in this ethnic persona is the belief Indians and non-Indians alike share about a golden age of unadulterated Indianness, unspoiled by western European-inspired dissipations. In a restorative vein, all those things eschewed as *not* Indian are profane and to be discarded. Alcohol, the "white man's disease,"[1] heads the list of rejected profanities. All those things traditionally Indian are revered, raised to totemic import, and adopted as one's birthright. The most obvious expressive forms of sacred community are ceremonial involvement and maintenance, resurrection, adop-

---

[1] All terms in quotation marks are used indigenously.

tion, or reinvention of traditional forms of Indian spiritual thought and practice.

Fifth Sunday Sings, a day-long convocation of Indian Christian churches, exemplifies the demonstration of sacred separation through ceremonial performance.[2] The Sun Dance complex, a public demonstration of atonement, faith, courage, and ability to endure hardship for the spiritual benefit of self and community is a second example of the ritualization of sacred separation. Prohibitions on drinking alcoholic beverages or ingestion of drugs are strongly enforced at both the Fifth Sunday Sings and the Sun Dances.

Profane separation or deviant solidarity is a second ethnic stance some Indians choose to present within their ethnic group as well as to non-members. The demonstration of this world view involves the flagrant misuse of alcoholic beverages in a collective and public context. Antithetical to sacred separation, the notion of deviant solidarity again is not novel. Lurie (1971) early on talked about the "longest on-going protest movement" in referring to American Indian drinking behavior. Until 1951, the sale of alcoholic beverages on Indian reservations was prohibited by federal law (Officer, 1971). On Sioux and Navajo reservations, it is still prohibited by tribal law. Laws governing the liquor trade with Indians have a long and bitter history in the United States (Dailey, 1968; Mosher, 1975). Written most usually at the behest of concerned tribal leaders or empathetic missionaries, the original intent of the protective, if paternalistic prohibitive laws have long been lost on Indians interested either in alcohol consumption or self-determination. In a perverse interpretive twist, prohibition has come to represent one further example of Anglo domination of Indian life. The illicit drinking party, drinking houses, making a run to a border town and stocking up, and the hazards of making it back to appointed drinking spots, are risks elevated to recreation, that is, "hellraising," "having fun," "putting one over on the law and order" (Everett, 1972, 1973; Hill, 1978). White (1970) calls the Indian drinking party a "culture of excitement." These acts of defiance, that is, sharing rides, pooling money, and passing the bottle among the "bros," create opportunities for shared experience. The practice of these activities eventuate in routinized, expected patterns of behavior—cultural knowledge that demonstrates the "insider's view" and thereby validates ethnic group membership.

More recently, a third ethnic stance has emerged that can be characterized as an enactment of the belief in the Indian as moderate and/or

---

[2] These Protestant Indian communities are not to be confused with Native American Church membership, yet another ceremonial mechanism by which sacred separation is enacted.

rational man. The notion of self-monitoring one's drinking behavior, known indigenously as "maintaining," and the development of social control mechanisms by which public group behavior is monitored and excessive consumption of alcoholic beverages is mitigated are examples of this third perspective on alcohol consumption and its function in contemporary Indian life.

The powwow, a secular performance of traditional tribal music and dance, is an Indian ceremonial event incorporating elements of all three ethnic stances. The powwow is essentially a social event in which ethnic identity is enacted through a dramatic public presentation of symbolically ethnic dress, food, music, and dance. Reports of early twentieth-century powwows indicate that heavy drinking and the expected complement of antisocial behavior (fights, accidents, arrests, etc.) were common occurrences during the week-long celebrations (Mails, 1978). More recently, drinking at powwows has been heavily sanctioned, monitored, and constrained (Weibel & Weisner, 1980). Currently, public inebriation has all but been eliminated at powwows through the skillful use of several social control mechanisms invoking the notions of separation of sacred and profane (secular) space. Associated beliefs maintain that drinking alcohol beverages is behavior relegated to profane (secular) space and that areas designated as sacred are not to be profaned by alcohol consumption. This world view represents a recent cultural invention, a manipulation of traditional symbols and a sanctification of traditionally secular activities in order to control and mitigate certain undesirable (un-Indian) behaviors such as alcohol misuse.

METHODS

SAMPLE POPULATION

The models of ethnic identity and the function of alcohol consumption in the demonstration of these ethnic personas presented in this chapter are based on ethnographic observations of Indian drinking behavior from 1978 to 1982. Observations were carried out in California and Oklahoma urban centers and rural communities and on the Sioux and Navajo reservations as well as the towns that border them (Weibel, 1981, 1982).[3]

[3] This study was funded by grants #0A-0256-7 A-1-5 from the California State Department of Alcohol and Drug Programs and #RO1AA04817-1-3 from the National Institute of Alcoholism and Alcohol Abuse, and was administered through the Alcohol Research Center at UCLA.

The 1980 census indicates that the American Indian population is almost evenly divided between rural and urban locations (U. S. Census, 1980). This is in sharp contrast to the essentially rural cast of the Indian residence pattern in 1960 (U. S. Census, 1960). Since 1953, significant numbers of American Indians have migrated from rural and reservation settings to major industrial centers, often with the encouragement and financial assistance of a federal program that came to be known as Relocation (Ablon, 1964; Garbarino, 1971; Graves, 1970; Weibel, 1977). Consequently our study of Indian drinking practices began in 1978 with 155 Navajos, Sioux, members of the Five Civilized Tribes from eastern Oklahoma, and indigenous California Indians in the urban Los Angeles area. These tribal groups were chosen from among the more than 100 tribal groups represented in Los Angeles because of their large numbers, both in Los Angeles and in the nation as a whole, and their widely divergent views concerning the use and function of alcoholic beverages in their communities (Price, 1978; Stratton, Zeiner, & Paredes, 1978). In ensuing years, our sample populations swelled to include 248 rural, indigenous California Indians and Navajos, Sioux, and Cherokees living in their respective reservations and territories.

DATA COLLECTION

Two data collection methods were employed. First, an intensive life history interview was used to elicit self-reports of early exposure to alcohol and other substances, levels of traditionalism, lifetime drinking cycles, and basic demographic data, as well as measures of psychological adjustment to urban life, indicators of stress, and medical sequelae of substance abuse and individual strategies developed to self-monitor drinking.

Second, our field staff became participant–observers in a wide range of drinking and non-drinking settings frequented by Indians.[4] The staff visited urban Indian bars, powwows, community meetings, Indian Centers, clinics, churches, and after-hours gathering places and were invited to private house parties. Over 400 hours of structured and unstructured observations were made. Each type of setting was visited at least four times, so that generalized, interactional, drinking, or alternative behav-

---

[4] The field staff was made up of Native American research assistants Bernadine (Bunny) Lindquist (Seneca), Eva Northrup (Hopi/Cherokee), Gene Herrod (Creek), Homer Stevens (Kickapoo), Shirley Bates (Karok), David Hostler (Hoopa), Betty Owen (Wiyot), Ernest Marshall (Hoopa/Karok/Yurok), Logan Slagle (Cherokee), Janie Jones (Cherokee), Charlotte Ortiz (Sioux), and John Eagleshield (Sioux). Their "insider" perspectives added significantly to the focus and reliability of the field work data.

ior patterns could be established for each setting. Ethnographic data provide not only a validation of the self-reported drinking behaviors elicited through interviews, but also a description of the kinds of situational and individual mechanisms that regulate the drinking characteristics of the various drinking settings.

THEORETICAL PERSPECTIVES

The social settings in which Indians gather on a regular basis and the habituated behaviors displayed in these settings constitute institutions, in that institutions are groups of people organized according to conventionalized rules or norms in which the facilities at hand are used to carry out activities that have a function (meet a need). They also constitute institutions, in that an institution is the relationship or behavioral pattern of importance in the life of a community or society (Malinowski, 1944). Rather than formally chartered institutions, however, the social drinking and non-drinking milieus of Indians constitute nonformal, but nonetheless regulated and codified institutions in which beliefs about and rules of correct comportment are often implicit rather than explicit. Further, the rules of correct comportment are the functions of environmental and sociocultural features of each setting.

Finally, the institutions to be described constitute a range of social arenas in which certain ethnic stances, or personas, can be explored, enacted, validated, and intensified. These ethnic stances are not static, but rather evolving; they can be tribe specific and are influenced by such environmental features as urban/rural location, ethnic mix of the audience, and most importantly for this book, personal drinking preferences.

FINDINGS

SACRED SEPARATION

The Fifth Sunday Sing

Christianity was introduced to Native North Americans as early as the seventeenth century by western European missionaries (Dailey, 1968). Christianity is so deeply entrenched in certain Indian societies that its original colonialistic premise goes unexamined by contemporary Indian adherents (Gardner, 1969; Stratton *et al.*, 1978; Weibel & Weisner, 1980).

Although all the observed Fifth Sunday Sings are held in Indian Protestant churches in Los Angeles, the ceremonial institution originated among rural, eastern Oklahoma Christian Indians (Foreman, 1934; Stratton et al., 1978). The so-called Five Civilized Tribes (Cherokee, Choctaw, Chickasaw, Seminole, and Creek) of eastern Oklahoma are generally split between fundamentalist Christian and pre-Western contact belief systems (Weibel & Slagle, 1983). Although some Cherokee Keetoowahs and Creek Stomp Dance devotees have migrated to urban centers, by and large they practice their traditional religious institutions "back home," in tribal rather than urban settings. This is not true of Christian Indians in urban settings. Even though congregations may number 40 or less, at least six Indian Christian churches flourish in Los Angeles to this day. Involvement in these religious communities constitutes large segments of the members' social lives, much as it does in the rural Oklahoma communities from which these urban Indian churchgoers emigrated (Gardner, 1969; Weibel, 1977; Weibel & Weisner, 1980).

Fifth Sunday Sings are held in a host church on the last Sunday of every month that has five Sundays (four times a year). They are attended to a greater or lesser degree by members of all six Indian churches in Los Angeles.[5] The Sing, a mixed social and sacred event strongly influenced by fundamentalist Christian doctrines and ritual structure, is a day-long affair for most participants. Informal singing of old-time Gospel favorites begins at nine in the morning and continues long into the evening. Food preparation begins days beforehand and is the responsibility of the female members of the host church. A nightly potluck feast of ethnic favorites, including *benaha* (boiled cornmeal mush balls wrapped in corn husks) and *pashobi* (hominy corn stew), provides a midday break. Socializing is the order of the time of feasting.

Alcohol consumption is minimal to non-existent at a Sing. In fact only the most clandestine drinking occurs. The no-drinking sanction is so ingrained that the subject is not even raised in the public announcements that punctuate the day's activities. The few Fifth Sunday Sing regulars who have drinking problems abstain during this event or drink covertly at some distance from the church property. Overt drinking would be highly censured.

The most powerful dimensions in determining drinking behavior in this setting are beliefs about the sacredness of the event and ethnic group identity. A strong prohibitionist tradition prevails in the fundamentalist Indian churches of eastern Oklahoma (Weibel, 1977). The peo-

---

[5] Two church groups in Los Angeles are made up largely of Southwest Indian groups. By and large their members rarely attend the Fifth Sunday Sings.

ple who attend Sings have been enculturated, since childhood, in the tradition of regular attendance at all-day church meetings, week-long revivals, summer church campground conferences, and negative attitudes about alcohol consumption.

Abstinence is a way of life for many Fifth Sunday Sing regulars, particularly the women. Those of the churchgoing community who do drink heavily are considered deviants and as "having a problem." Rather than being ostracized, they are lovingly and tolerantly welcomed into the fold in the hope that one day they will "see the light," end their "backsliding," and stop drinking. They are expected, however, to refrain from drinking during church events, and most do. The unspoken prohibition against drinking in that setting is so pervasive that one regular, a favorite singer and pianist, informed me,

> I stayed away last Fifth Sunday Sing because I had been drinking heavily for a while and couldn't face my friends and family. I knew they were unhappy about my "backsliding." It would make them sad to see me this way. Nobody would have a good time, so I stayed away. (Choctaw man, 62)

Another Fifth Sunday Sing regular explained:

> I never had a drink in my life. And to my knowledge none of my five kids ever have. You see, we're Christians. (Choctaw woman, 67)

Out of curiosity, she attended a few powwows when she first came to Los Angeles. Her antipathy to drinking is so complete that she refused to attend future powwows because of the drinking she observed in those settings (minimal though it was). As far as she was concerned, powwows are "not what my people do," "heathen," and "sinful." For her, ethnic identity is intrinsically tied to church membership, sobriety, and the distinction between "civilized Indian" and "heathen Indian" ceremonial participation.

The Sun Dance

The great early Sioux Sun Dances had their origins in vows made by individuals during times of extreme crises. Desperate people declared that, in return for the Great Spirit's help, they would seek a vision, undergo solemn purification ceremonies, and do the Sun Dance—making, at that time, whatever physical sacrifices were necessary to the fulfillment of the vow (Mails, 1978). With the encroachment of white settlers and armed forces, the Sun Dance theme shifted from associations with the hunt and war complex to survival of the culture within severe confines. It remained for some Indian nations the vital core of their answer to life's problems and needs.

With the advent of the reservation system, sundancing was greatly curtailed and, ultimately, outlawed in 1904. Its practices, however, continued, albeit *sub rosa*, for 70 years (Mails, 1978). Since passage of the Religious Freedom Act in 1978, sundancing has had a dramatic resurgence. There are so many hamlets or influential families on the Sioux reservations who now sponsor annual Sun Dances that at least one Sun Dance a week is held throughout the summer.

The majority of contemporary vows or pledges are in response to either personal addiction to alcohol or the devastation of alcohol abuse that plagues the Sioux reservation generally (Mails, 1978). It is, in the view of the medicine men and tribal leaders, a curse of incredible proportions. Drinking at or coming drunk to a Sun Dance is anathema; those who do so are severely reprimanded and removed from the premises. Posters advertising a Sun Dance emphasize the spiritual and traditional nature of the dances and boldly warn that no alcohol or drugs will be permitted on the site. Cameras and tape recorders, further trappings of white society, are also excluded. Importantly, white spectators are not excluded. The boundaries of ethnic identity are more keenly felt and delineated when the "others" are present.

Usually Sioux Sun Dances are held in remote places (out of society) where no commercial (polluting) facilities are available. The dance area or mystery (hoop) is formed by erecting a shade arbor around the circumference of a circle left open to the sun. The exact center of the open circle is marked by the Sacred Tree or sun pole. The Mystery Hoop or Circle are terms synonymous with the Hoop of the Nation or primary ethnic classification (tribe). Special reverence is shown in this area of sacred separation. One may enter only at prescribed times and through one narrow access route (Densmore, 1918; Lowie, 1954; Powers, 1969).

The pledgers begin the ceremony by separating themselves from the community and entering the sweatlodge. There they participate in the *Inipi,* or rite of purification. After initial prayers and ritual passings of a sacred pipe, the flaps of the sweatlodge are secured, making it airtight and dark inside. Completely shut off from the outside world, the darkness is said to represent the sinfulness of the soul and ignorance, from which the participants must now purify themselves before they dance.

The Hoop is a sphere of limitless energy or power. To be within it during a Sun Dance is to be exposed to divine power. The Sun Dancer is transformed during the four days of dancing into a special channel or tube through which God bestows blessings on all the nation. All *Inipi* participants are cleansed from sin and ignorance in order to become pure tubes through which God will be willing to work to dispense his power to nourish and enlighten all things. Pure tubes of power are, of necessity, alcohol-free. Therefore, a highly valued critical role (Sun Dancer) provides the mechanisms by which an Indian demonstrates at once independence from external pollutants, ethnic group membership, and a nexus with sacred power.

Temporary abstinence, marshaled by the tenets of allowable participation in high status ritual performance, provides the model of ideal cultural behavior and attitude as concerns the use of alcoholic beverages among the Sioux. Ideally abstinence, a sacred separation from external pollutants as demonstrated in ritual performance, is generalized to daily life. To those for whom abstinence is a way of life, intrinsically associated with ethnic identity, alcohol misuse is "the white man's disease" for which Indians have no indigenous cure. One can only keep oneself free from its alien and all-consuming evil power through purification and sanctification in their many cultural forms.

PROFANE SEPARATION

Indian Bars

As a result of legislation restricting acquisition of alcoholic beverages by American Indians, the well-documented clandestine Indian drinking party has taken on the attributes of an expected adult recreational activity among those rural Indians who live in so-called "dry communities" and who wish to drink and "party" (Lurie, 1971; Leland, 1978; May, 1975; Kemnitzer, 1972). The terms "to party" or "partying" describe the flamboyant drinking behavior typical of Indian bars (Hill, 1978; Leland, 1981; Weibel, 1981). Aggressiveness associated with heavy public drinking is documented at length in the Indian and alcohol literature (Everett, 1978; Hill, 1978; White, 1970). The "partying" drinking style is also athletic and competitive. Bar drinking and "partying" is not considered "social drinking." This term is considered applicable only to "white man's drinking" and is said to have "nothing to do with the way we drink." In this sense "to party" differentiates in-group from outgroup drinking behavior (Leland, 1981; Weisner *et al.*, 1984).

Contrary to Price's (1978) opinion about the diminishing role of urban Indian bars as social service institutions, we found seven bars in Los Angeles that have large, relatively stable, Indian clienteles. Each of these bars can be categorized by one of four distinct types of customers who patronize them: Skid Row inhabitants; younger, newly arrived, and working-class people; upwardly mobile and athletic club cliques; and community leaders. Each bar type appears to have evolved its own set of rules about acceptable drinking comportment. These rules eventuate in distinctive drinking styles among the clienteles (Weibel, 1981; Weibel & Weisner, 1980). Some participants in our study frequented all four types of Indian bars, and their drinking and interactional styles appear to shift with the setting. However, most individuals frequent, with greatest regularity, the bar setting that serves a clientele closely paralleling their own life-style, community status, and preferred drinking style (Weibel, 1981).

"Partying"

The peripatetic weekend drinking party is a well-documented phenomenon of Indian drinking (Kemnitzer, 1972; Kuttner & Lorincz, 1967; Levy & Kunitz, 1974; Littman, 1964; MacAndrew & Edgerton, 1969). In the urban setting it takes on the further adaptation of bar hopping or "making the rounds," that is, traveling from one Indian drinking setting to another, usually ending up at an established, open air, after-hours spot. Relatively unrestricted drinking, socializing, "49ing,"[6] and occasional acts of verbal and physical violence occur at this public area in downtown Los Angeles every weekend from bar closing time (2 A.M.) to sunrise.

Crucial to an understanding of the significance of "partying" to Indians who do so with regularity are the following environmental and social characteristics of the phenomenon. The Indian drinking party is composed almost exclusively of Indians; it takes place within the context of a predominantly non-Indian social milieu; and it is embedded in a potentially hostile environment. Its peripatetic nature necessitates breaking certain laws of the dominant society, that is, driving while under the influence of alcoholic beverages, public drunkenness, etc. Its communal nature provides the setting for displays of ethnic solidarity, critical mass, cooperation, and premeditated acts of deviance. Its public quality allows for the demonstration of the "Indian way of drinking," which contrasts sharply with white middle-class standards of acceptable public drinking behavior.

The Indian drinking party identifies its members as a people willing to take calculated personal risks in their opposition to a prohibiting external social force. Violation of laws not of their own making and

---

[6] "49ing" is a secular song and dance convention that follows most powwows and many social events. Singers and dancers follow the drummers to the perimeter of the powwow grounds. With linked arms the group sings protest, popular, and occasionally scatological songs as they circle the drum.

believed to be unfair, demeaning, and paternalistic is elevated to political protest and translated into a demonstration of personal integrity and freedom of action. Unwilling to accept the inferior status of ward to the federal government's guardianship, drinking Indians view alcohol consumption as symbolic action—a group demonstration of a deeply held personal belief in their right to total self-determination.

Seemingly conflictual beliefs in, and demonstration of, personal independence and the propensity of individuals to form strong affective bonds or common interest groups to age set groups are traditional elements of tribal life, particularly among the Plains Indians (Hoebel, 1977; Erikson, 1950; Powers, 1969; Lowie, 1954). The well-documented young men's drinking party (Graves, 1971; Waddell, 1971) has cultural prescriptions in such close-knit social and recreational organizations as the Plains raiding and hunting party (Lowie, 1954). Thematic elements of the raiding party have symbolic and behavioral parallels in the contemporary drinking party ethos. Performing acts of bravery and daring, taking physical risks, flaunting one's supremacy in the face of a hostile enemy in order to shame him, deliberately breaking the peace, ritualizing theft and abduction, and inflicting bodily harm upon the enemy were and are actions taken to demonstrate one's exuberant manhood and enhance one's status in the peer group and the community.

The exclusivity of the Indian drinking party, its patterned, ritualistic format and displays of private, in-group membership behavior in public settings exemplify *profane separation*. In attempts to establish a "turf for the bros," the urban Indian drinking party clearly defines for the hapless non-Indian spectator/interloper his outsider status relative to the "partying bros."

PEOPLE WHO CAN HOLD THEIR LIQUOR—AN EMERGING
THIRD ETHNIC STANCE

Maintaining

Drinking in urban Indian bars is essentially the social lubricant for the more important social–interactional aspects of the settings—the sociability, comfortableness, and well-being one feels in "hanging out with one's people" (Waddell, 1971) and favorite "drinking buddies" (Graves, 1970). Sociability wanes when people overimbibe. For the good of the "party," most people in Indian bars attempt to monitor their drinking in such a way that they don't "get loaded and spoil the fun."

Because Indian bars are the setting for potential trouble (fights, police raids, group arrests), it is important to minimize behavior that would make it necessary for the management to call the police. To achieve this goal, most Indian bar "regulars" adhere to the principle of "maintaining." Maintaining is not an Indian-specific concept and behavioral pattern. Rather, it is a cross-ethnic, subgroup bond. To maintain, one drinks "to get a buzz on" and to "feel good." One drinks continually and steadily. However, maintainers monitor their physical and cognitive states in such a way that their behavior continues to be socially acceptable. People who drink until they are "sloppy drunk," "a pest," "not able to carry on a decent conversation," or "passed out" are subject to the negative sanctions of their barroom peers. They are made the butt of practical jokes and verbal slurs, are fair game to the "jack roller," and are usually publicly censured by friends or mates.

However, people who maintain are assets to the drinking party. These heavy drinkers are able to keep up the flow of non-serious conversation and "Indian humor" in a steady stream of repartees and teasing insults. They are still able to dance, shoot pool, buy the next round, take care of themselves or friends in case of a physical attack, and drive home after the bars close. Maintainers contribute positively to the ongoing social interactions of the bar. They do not become a burden to their drinking companions, start fights, or muscle-in on another person's "party." In other words, they monitor their drinking and comportment in such a way that their drinking behavior constitutes what Gusfield (1981) has described as drinking competence.

For those who say they maintain, the behavior is an adaptation to the urban Indian drinking milieu (Weibel, 1981). Acutely aware of the negative stereotype of Indians as not being able to handle liquor (MacAndrew & Edgerton, 1969) and the arbitrary way the police enforce public inebriation restrictions, particularly in the areas in which Indians congregate to drink (Graves, 1970), increasing numbers of Native Americans have adopted maintaining as a precautionary measure. It is the control mechanism by which one avoids police involvement and community censure. On a personal and social–interactional level, it is the control mechanism by which one participates and augments the ongoing "partyness" of the bar setting.

The Powwow: Microcosmic Ordering of Sacred and Secular Domains

Although participation is pan-tribal, the powwow is an embodiment of the Plains Indians' dance traditions (Nurge, 1970). Though modified, and spectacularized, its structure remains essentially that of the Omaha Grass Dance, a social dance complex that is said to have been formulated in the early 1800s (Densmore, 1918). It is practiced in both rural and urban settings. Originally a social dance rather than religious ceremony, the contemporary powwow, particularly in urban areas in which continuity of ethnic identity is felt to be under greatest stress, has taken on an aura of religiosity uncharacteristic of its rural counterpart (Weibel & Weisner, 1980).

The Urban Model.   A powwow is held every weekend somewhere in the Los Angeles area throughout the fall, winter, and spring months. At powwows, Plains chants are sung to the beat of sacred drums. Men and women, teenagers, and small children all wear elaborate Northern and Southern Plains dance regalia and perform the social and honoring dances of their traditions.

In urban areas powwows are held usually indoors in a municipally funded recreation center. They are attended predominantly by Indians, although there is usually a smattering of non-Indian spectators. The tribal makeup of powwow participants is primarily Plains Indians, with a few Southwestern and Eastern Woodlands tribes represented. The Plains tribes, as a culture, maintain weaker prohibitions against drinking than do the eastern Oklahoma tribes (Child, Barry, & Bacon, 1965; Hurt & Brown, 1965; Kemnitzer, 1972; Lemert, 1954, Stratton *et al.*, 1978). Consequently, many people who attend powwows have life histories of regular alcohol use.

Our ethnographic observations indicate that some drinking occurs at urban powwows. However, strong sanctions are placed on drinking in urban powwow settings. The sacred nature of the drum, chants, and some dances is stressed in the powwow emcee's occasional public admonishments about drinking in that setting. Drinking is prohibited within the dance circle—a sacred place. However, the more pragmatic concern of the hosting powwow clubs—that is, to maintain public facilities as powwow sites—exerts an even stronger negative sanction on drinking at urban powwows. Participants are repeatedly warned of the possibility of losing the use of public auditoriums if drinking gets out of hand. Another mechanism used to ensure minimal drinking is an internal system of security guards made up of male members of the host powwow clubs.

If an individual desires to drink at a Saturday night powwow, he usually does so discreetly in the parking lot or in a car parked some distance down the street and away from the recreational facilities. Most participants refrain from drinking during the event. Participants who drink heavily do so covertly or after the powwow's end in an Indian bar or at home.

The Rural Model.   Rural powwows are attended by urban as well as rural Indians. They are held outdoors, over an extended time period, and are almost exclusively Indian in makeup. The settings provide considerably more privacy and protection from public scrutiny than do urban Saturday night powwow settings. These annual events usually begin on a Friday night and extend over a holiday weekend.

The unboundedness of time and the rural, private, and predominantly Indian dimensions of the setting all allow for relatively unrestricted drinking among those powwow participants who view the event as essentially social and who characteristically drink heavily. People who drink moderately or abstain continue to do so in this environment. However, people who drink heavily, but who would refrain from doing so in the urban-time-bound powwow environment are not expected to refrain from drinking in this setting. At rural powwows, license is given to chronic heavy or binge drinkers to "kick back," "raise a little hell" (Hill, 1978), and to engage in "some serious drinking."

The sacred aspect of the chanting and drumming restrains drinking at powwows. Conceptually, the dance circle or arena is not one designated spiritual place, but rather concentric rings of diminished spatial sacredness. The drum is the heart of the dance, the center of its energy. There are strong prohibitions around the drum and its immediate area, the most sacred place of the powwow setting.

The next division of space is the ring around the drummers in which the participants dance. Although this space is less sacred than the drum circle, the dancers have more intercourse with the spiritual center of the dance than do the spectators. For this privilege, the dancers observe certain prohibitions. As a demonstration of reverence, the women must wear at least a dance shawl over their shoulders as they move around the drum. The elaborately costumed male dancers take great care in the assembly of their dance dress. Certain feathers, flutes, fans, and medicine pouches, handed down from one generation to the next, are prayed over or are especially prepared and arranged by the family for the dances. A strong drinking prohibition also applies to this segment of the dance arena.

Non-dancing spectators encircle the dance arena in a third concentric spatial ring. In this transitional space sacred and secular elements of the event converge. The assembly is intertribal, pan-ethnic, and multigenerational. Some drinking occurs in this transitional space.

The heaviest drinkers restrict their ingestion to a fourth space located in the outer region of the encampment beyond the circle of spectators. This highly secular space includes food and handicraft concessions and camping and parking areas.

The spatial separation of sacred and secular powwow activities parallels the Levi-Strauss (1963) model of sacred and profane space, the symbolic separation of places inhabited by men (society) and spirits (nature). In fact the analogy of the Levi-Strauss model of sacred and profane space can be expanded one step farther. Open fields or wooded areas often lie beyond the most secular areas of the rural powwow. This area, in effect, symbolizes the separation of the man-ruled world (society) and the domain of the bush (a place in which manmade rules no longer apply). Much of the alcohol-induced, antisocial behavior—the sequelae of three days of continuous drinking (fighting, passing out, seducing)—occurs in this space.

Recently Indians, aware of the possible medical, social, political, and legal implications of three or four days of unrestricted drinking, have developed certain indigenous precautions and proscriptions about unchecked drinking at rural powwows. At a three-day powwow given in celebration of the end of the Longest Walk,[7] stringent precautions were imposed to limit substance abuse among participants. Signs that forbade alcoholic beverages, drugs, and weapons were displayed everywhere. Scores of security guards, alert to signs of discord or agitation, patrolled the parkgrounds with walkie-talkies. Onlookers suspected of being intoxicated or troublemakers were stopped, searched, and if intoxicated, escorted away from the campgrounds. People who did drink were spatially separated from the non-drinking majority—in the parking lot, in campers, away from the activity itself. The sacredness of the pipe ceremony at sunrise was stressed.

Summary and Discussion

Even for settings in which heavy drinking is expected behavior, indigenous drinking control mechanisms have been developed. Such strategies as spatial segregation and the manipulation of time and symbol mitigate the more devastating effects of three or four days of binge drinking by the heavy-drinking cliques who attend rural powwows and Indian bars specifically for that recreational activity.

A third ethnic stance seems to be emerging. Consistent with themes of Red Power and self-determination, and beyond flaunting displays of perverse ethnic pride, increasing numbers of American Indians are integrating their unique historical-cultural heritage with twentieth-century

[7] The Longest Walk occurred in 1978. Pan-tribal protest group members and non-Indian sympathizers walked from San Francisco to Washington, D. C. in order to foster mass media coverage of the plight of contemporary American Indians.

experience. Associated with integration and/or actualization of the ethnic self are pragmatic decisions about drinking in rational ways. It is no longer necessary to drink defiantly and with a self-destructive vengeance or to reject totally "demon rum" to "act Indian." Rather, some Indians believe that alcoholic beverages can be consumed in ways that demonstrate Indians do know how to "hold their liquor."

IMPLICATIONS FOR TREATMENT

Although indigenous control mechanisms and attitudes about moderate use of alcohol are evolving, alcohol abuse continues to be the foremost medical and social problem in contemporary American Indian populations (Mail & McDonald, 1981; Price, 1975; Snake, Hawkins, & LaBoueff, 1977). Evaluation studies point to the disappointingly low success rates of existing alcohol rehabilitation programs in Indian communities (Kline & Roberts, 1973; Lang, 1974; Snake et al., 1977; Towle, 1975; Weibel, 1981). Increasingly, evaluators, treatment personnel, and potential clients deplore the cultural bias of existing alcoholism intervention programs and call for the integration of more traditional forms of healing practices in programs with large numbers of Indian clients (Albaugh, 1973; Bergman, 1971; Garcia, 1973; Kahn, Williams, Galvez, Lejero, Conrad, & Goldstein, 1975; Shore, 1972; Weibel, 1981).

"Winos" and Alcoholics: A Problem of Definitions

Definitional differences impede the development of effective alcoholism interventions in Indian communities. In more than 400 interviews with both rural and urban American Indians over the past five years, we asked participants to define alcoholism. Because of such influences as level of acculturation, tribal affiliation, and individual attitudes about and experience with drinking, the range of answers is too wide for us to talk about an Indian consensus on the definition of alcoholism. Among the more acculturated participants, answers tend to have a textbook quality: "a psychologial and physical dependency on the drug, alcohol." For the former or recovering alcoholic, the definition takes on dogmatic qualities: "Anyone who feels he needs a drink in a social situation is an alcoholic." People who have experienced the effects of alcohol abuse in their family or community talk about it in more psychosocial terms: "An alcoholic is the person who doesn't know when to stop," "has a problem," "is hurting his family with the drinking," "makes a fool of himself in public," "doesn't act like a Cherokee."

In fact, Indians do not use the label *alcoholic* very much to define abuse drinkers. Rather, they talk about people who, most professionals in the field would call alcoholics as "winos," "bums," or simply "guys who drink too much." Considerable time is spent in Indian alcoholism treatment centers teaching usually a coerced client, first, the concept of alcoholism as an illness, second, convincing him that he is indeed an alcoholic and ultimately that he needs to do something about it. More often than not, the client of an Indian alcoholism treatment facility is in treatment as the lesser of two evils. Usually arrested for public drunkenness, vagrancy, or at the insistence of his family, his choice to place himself in treatment rather than to go to jail introduces him to a resocialization process by which his propensity "to party," "hang out with the winos," "raise a little hell," and "get loaded" (positively perceived activities) is relabeled alcoholism (a negatively perceived label). Resistance to this labeling process greatly reduces the effectiveness of treatment.

Treatment Modality Types

The high recidivism rates deplored by many alcoholism treatment personnel are due in part to the lack of fit between the client's world view and life experience and the program treatment modality. The Indian and alcohol literature is rife with observations of diversity of acculturation levels, attitudes, and tribal philosophies, as well as sociohistorical influences that shape drinking behavior and levels (Berreman, 1956; Boatman, 1968; Codere, 1955; Curley, 1967; Hamer, 1965; Levy & Kunitz, 1974; MacAndrew & Edgerton, 1969; Mail & McDonald, 1981; Maynard, 1969; Price, 1975; Stratton *et al.*, 1978; Topper, 1974).

Over the past five years, we visited and observed over 40 Indian alcoholism recovery homes as well as traditional healers in California, South Dakota, New Mexico, Arizona, and Oklahoma. From the range of treatment approaches we encountered, a continuum model of treatment typologies was developed. The alcoholism treatment programs we visited can be categorized across a range of six basic types. The range spans programs that advocate the medical model approach to alcoholism intervention, with no accommodation to cultural differences among their Indian and non-Indian clients, to a traditionalist approach by which medicine is administered by an indigenous Indian healer with no recognition or employment of Western medical intervention.

Several Indian alcoholism treatment programs advocate the disease theory of alcoholism. They stress that treatment is first and foremost addressed to the immediate symptoms of the disease alcoholism and only secondarily if at all to the social and psychological malaise of the client. Further, since alcoholism is a disease, it requires specific interventions that transcend the vagaries of age, sex, socioeconomic level, or cultural experience. These programs tend to be strongly influenced by the tenets of Alcoholics Anonymous (AA) and tend to parallel the prevailing medical model of alcohol treatment.

A second approach to Indian alcoholism treatment is characterized by its socio-psychoanalytic orientation. Although clients and staff deal initially with the maintenance of sobriety and a return to physical health, greater emphasis is placed on the identification and resolution of psychosocial problems of which abusive drinking is seen as a symptom. The goal of this approach is to develop more positive coping strategies. Program personnel in this category tend to believe that interpersonal and intra-psychic disturbances, the antecedents of abusive drinking, are pan-cultural phenomenon to which certain psychotherapeutic techniques can be applied successfully, regardless of cultural experience. An eclectic mix of Western psychoanalytical treatment models is imposed. Again this model is not new, but analogous to widely practiced psychosocial models of alcoholism treatment.

A third type of alcoholism intervention program is Indian-cliented, Indian-staffed, and usually Indian-run. These programs use both AA and psychoanalytical and/or psychotherapeutic techniques as counseling strategies. There is a general consensus among staff members of this type of program that Indians treating other Indians makes for optimal client–counselor rapport. And if the Indian counselor is also a recovering problem drinker, his function as a positive role model for the novice abstainers is a further asset. Although these programs are demographically Indian, their treatment approaches are essentially Western medical or psychosocial in orientation. We have labeled this category the assimilative model.

A fourth intervention type attempts some modification of treatment modalities in recognition of the patient's specific cultural background. This type of program might have one or more Indian counselors. Posters with Indian themes might be hung about the living and meeting areas. Group therapy sessions include discussions of the Indian historical and cultural influences that pertain to the use of alcoholic beverages among American Indians. Non-Indian counselors are sensitized to basic elements of Indian life and social–interactional style. The *culture-sensitive model* is our rubric for this category.

A fifth type of treatment program incorporates Indian values and ceremonial curing practices into standard alcoholism intervention strategies. Attempts are made to express commonly held knowledge about alcoholism intervention in terms compatible with Indian thought and cultural experience. Clients are encouraged to involve themselves in spiritual quests based on tribal lore. A return to or intensification of the practice of traditional Indian skills and/or social activity is also encouraged, as well as membership in AA, group therapy, and other types of social systems supportive of sobriety. We have labeled this type of alcoholism intervention the *syncretic model*.

The final category in our continuum of treatment typologies consists of a dependence upon the assistance of traditional healers or spiritual leaders to combat the craving for alcohol to the exclusion of other forms of therapy. For obvious reasons we call this treatment type the *traditional model*.

Each alcoholism intervention type was exemplified by at least one program we visited. However, the majority of the programs are best described as culture-sensitive or assimilative. Although most treatment personnel voice a certain empathy with traditional forms of curing, the practicalities of administering publically funded intervention programs subject to federal regulation restrict attempts at syncretism to not much more than arm's-length experimentation or facilitation initiated only at the insistence of a client. The discrepancies between treatment approaches and prevailing healing practices further impede alcoholism intervention in Indian communities.

Sacred Separation and Sobriety

Finally, the lack of effectiveness of Indian alcoholism intervention programs can be attributed to an incomplete understanding of changes over time in attitudes about and function of intoxicants for Native Americans. For instance, ethnographic accounts of the indigenous use of fermented or hallucinogenic drugs among Native North Americans point to the highly ritualized nature of their ingestion and the spiritualistic connotations of the altered states of consciousness produced by their use (Bergman, 1971; Pascarosa & Futterman, 1976). Clearly before the introduction of the "white man's disease," intoxication was associated, with a few exceptions, with sacred endeavors and enhancement or facilitation of the sacred encounter. However, early drinking models strongly at variance with the original use of intoxicants among Indians (the carousings of the north woods trappers; celebrational use of alcohol by explorers, cowboys, loggers, and fishermen; and the prohibitionist stance of certain Christian sects) have greatly altered that position. Today for most Indians, alcohol use is associated with the secular and recreational aspects of community life. Conversely, abstinence from alcohol beverages is associated with the sacred elements of community life.

Indian alcoholism intervention programs with the highest rates of sustained client sobriety are those that integrate a variety of spiritual elements and activities into their treatment strategies. As an earlier chapter by Rodin suggests, AA membership is an embodiment of sacred separation. Membership in such religious groups as the Native American Church or any number of fundamentalist Christian denominations or the assumption of the role of pipe carrier, Sun Dancer, or medicine man have necessitated and perpetuated abstinence from alcoholic beverages for many Indian former alcohol abusers. The American Indian Movement, initially a strictly political force, now defends itself as a religious movement and strongly advocates abstinence from all intoxicants for its members. Involving the concept of sacred separation as a viable ethnic stance and abstinence as one of its demonstrable forms may be a culturally appropriate intervention strategy and the effective first step toward sustained sobriety for contemporary American Indians who they, their family, friends, and community feel "drink too much," "have a problem with alcohol," or, what is even worse, suffer from "the white man's disease."

REFERENCES

Ablon, J. Relocated American Indians in the San Francisco Bay area: Social interaction and Indian identity. *Human Organization,* 1964, *23,* 296–304.

Albaugh, B. J. *Ethnic therapy with American Indian alcoholics as an antidote to anomie.* Paper presented at the 8th Joint Meeting of the Professional Association of the U. S. Public Health Service, 1973.

Benedict, R. *Patterns of culture.* Boston: Houghton Mifflin, 1934.

Bergman, R. L. Navajo peyote use: Its apparent safety. *American Journal of Psychiatry,* 1971, *128,* 695–699.

Berreman, G. D. Drinking patterns of the Aleuts. *Quarterly Journal of Studies on Alcohol,* 1956, *17,* 503–514.

Biddle, N. *The journals of the expedition under the command of Lewis and Clark.* Vols. I and II. Norwalk, CT: The Easton Press, 1962.

Boatman, J. F. *Drinking among Indian teenagers.* Unpublished master's thesis, University of Wisconsin, Milwaukee, 1968.

Child, I. L., Barry, H., & Bacon, M. K. Cross cultural study of drinking: Sex differences. *Quarterly Journal of Studies on Alcohol,* 1965, *3,* 49–61.

Churchill, W., & Larson, P. An anti-colonialist perspective on Native American substance abuse. *Talking Leaf,* 1983, *48,* 4–8.

Codere, H. Review of alcohol and the northwest coast Indians. *American Anthropologist,* 1955, *57,* 1303–1305.

Curley, R. T. Drinking patterns of the Mescalero Apache. *Quarterly Journal of Studies on Alcohol,* 1967, *28,* 116–131.

Dailey, R. C. *Alcohol and the North American Indians: Implication for the management of problems.* Paper presented to the 17th Annual Meeting of the North American Association of Alcoholism Programs, Addiction Research Foundation, 1966.

Dailey, R. C. The role of alcohol among North American Indian tribes as reported in the Jesuit relations. *Anthropologica*, 1968, X, 45–57.

Densmore, F. *Teton Sioux music.* Washington, DC: U. S. Government Printing Office, 1918.

Erikson, E. *Childhood and society.* New York: Norton, 1950.

Everett, M. W. *Verbal conflict and physical violence: The role of alcohol in Apache problem solving strategies.* Lexington: University of Kentucky Monograph, 1972.

Everett, M. W. *"Drinking and trouble": The Apachean experience.* Paper presented to the Society for Applied Anthropology, Tucson, Arizona, 1973.

Fiske, S. J. Rules of address: Navajo women in Los Angeles. *Journal of Anthropological Research*, 1978, 34, 73–91.

Fiske, S. J. Inter-tribal perceptions: Navajo and pan-Indianism. *Ethnos*, 1977, 5, 358–375.

Fiske, S. J., & Weibel, J. Navajo social interactions in an urban environment: An investigation of cognition and behavior. *Bulletin of Southern California Academy of Sciences*, 1980, 79, 19–37.

Foreman, G. *The Five Civilized Tribes.* Norman: The University of Oklahoma Press, 1934.

Garbarino, M. S. Life in the city: Chicago. In J. O. Waddell & M. Watson (Eds.), *The American Indian in urban society.* Boston: Little, Brown, 1971.

Gardner, R. E. The role of a pan-Indian church in urban Indian life. *Anthropology UCLA*, 1969, 1, 14–26.

Garcia, M. F. *Analysis of incidence of alcohol intake by an Indian population in one state of the U.S.A. (Montana), in relation to admissions to a psychiatric hospital for treatment and hospitalization.* Paper presented at the IXth International Congress of Anthropological and Ethnological Sciences, 1973.

Graves, T. D. The personal adjustment of Navajo Indian migrants to Denver, Colorado. *American Anthropologist*, 1970, 72, 32–54.

Graves, T. D. Drinking and drunkenness among urban Indians. In J. O. Waddell & M. Watson (Eds.), *The American Indian in Society.* Boston: Little, Brown, 1971.

Gusfield, J. Managing competence: An ethnographic study of drinking, driving and the context of bars. In *Social Drinking Contexts*, Research Monograph #7, Washington, D.C.: NIAAA Clearing House, 1981.

Hamer, J. H. Acculturation stress and the functions of alcohol among the Forest Potawatomi. *Quarterly Journal of Studies on Alcohol*, 1965, 26, 285–302.

Hill, T. W. Drunken comportment of urban Indians: 'Time-out" behavior? *Journal of Anthropological Research*, 1978, 34, 442–467.

Hoebel, E. A. *The Plains Indians: A critical bibliography.* Bloomington Indiana Unversity, 1977.

Hurt, W. R. Brown, R. M. Social drinking patterns of the Yankton Sioux. *Human Organization*, 1965, 24, 222–230.

Jorgenson, J. G. *The Sun Dance religion: Powers to the powerless.* Chicago: University of Chicago Press, 1972.

Kahn, M. W., Williams, C., Galvez, E., Lejero, L., Conrad, R. D., & Goldstein, G. The Papago psychology service: A community mental health program at an American Indian reservation. *American Journal of Community Psychology*, 1975, 3, 81–97.

Kemnitzer, L. A. The structure of country drinking parties on the Pine Ridge Reservation, South Dakota. *Plains Anthropologist*, 1972, 17, 134–142.

Kline, J. A., & Roberts, A. C. A residential alcoholism treatment program for American Indians. *Quarterly Journal of Studies on Alcohol*, 1973, 34, 860–868.

Kuttner, R. E., & Lorincz, A. B. Alcoholism and addiction in urbanized Sioux Indians. *Mental Hygiene*, 1967, 51, 530–542.

Lang, G. M. C. Adaptive strategies of urban Indian drinkers. Doctoral dissertation, University of Missouri-Columbia, 1974. (U.M. Order #75-20132.)

Leland, J. Women and alcohol in an Indian settlement. *Medical Anthropology*, 1978, 2, 85–119.

Leland, J. The context of Native American drinking: What we know so far. In Research Monograph #7 *Social Drinking Contexts*. Washington, DC; NIAAA, 1981.

Lamert, E. M. Alcohol and the northwest coast Indians. *University of California Publications in Culture and Society*, 1954, 2, 304–406.

Levi-Strauss, C. The structural study of myth. In *Structural Anthropology*. New York: Basic Books, 1963.

Levy, J., & Kunitz, S. *Indian drinking: Navajo practices and Anglo-American theories*. New York: Wiley (Interscience), 1974.

Littman, G. *Some observations on drinking among American Indians in Chicago.* Proceedings of the 27th International Congress on Alcohol and Alcoholism. Vol. 1 Frankfurt-am-Main, Germany, 1964.

Lowie, R. *Indians of the Plains*. New York: McGraw-Hill, 1954.

Lurie, N. O. The world's oldest ongoing protest demonstration: North American Indian drinking patterns. *Pacific Historical Review*, 1971, 40, 311–332.

MacAndrew, C., & Edgerton, R. B. Indians can't hold their liquor. In C. MacAndrew & R. B. Edgerton (Eds.), *Drunken comportment*. Chicago: Aldine Publishing Company, 1969.

Mail, P. D. *American Indian alcoholism: What is not being done?* 110th Annual Meeting of the American Public Health Association, Montreal, Canada, 1982.

Mail, P. D., & McDonald, D. R. *Tulapai to Tokay: A bibliography of alcohol use and abuse among Native Americans of North America*. New Haven: Human Research Area Files Press, 1981.

Mails, T. E. *Sundancing at Rosebud and Pine Ridge*. Lake Mills, IA: Graphic Publishing Company, 1978.

Malinowski, B. *A scientific theory of culture and other essays*. Chapel Hill: University of North Carolina Press, 1944.

May, P. A. Arrests, alcohol and alcohol legalization among an American Indian tribe. *Plains Anthropologist*, 1975, 20, 129–134.

Maynard, E. Drinking as part of an adjustment syndrome among the Oglala Sioux. *Pine Ridge Reservation Bulletin*, 1969, 9, 35–51.

Mohatt, G. The sacred water: The quest for personal power through drinking among the Teton Sioux. In D. C. McClelland, W. N. Davis, R. Kalin, & E. Wanner (Eds.), *The drinking man*. New York: The Free Press, 1972.

Mosher, J. F. *Liquor legislation and Native Americans: History and perspective*. Berkeley: University of California, Bolt School of Law, 1975.

Nurge, E. *The modern Sioux: Social systems and reservation culture*. Lincoln: University of Nebraska Press, 1970.

Officer, J. E. The American Indian and federal policy. In J. O. Waddell & O. M. Watson (Eds.), *The American Indian in urban society*. Boston: Little, Brown, 1971.

Pascarosa, P., & Futterman, S. Ethnopsychedelic therapy for alcoholics: Observations of the peyote ritual of the Native American Church. *Journal of Psychedelic Drugs*, 1976, 8, 215–221.

Powers, W. *Indians of the northern plains*. New York: Putnam, 1969.

Price, J. Applied analysis of North American Indian drinking patterns. *Human Organization*, 1975, 34, 17–26.

Price, J. Urban ethnic institutions. In J. A. Price (Ed.), *Native studies: American and Canadian Indians.* Toronto: McGraw-Hill Ryerson, Ltd., 1978.

Shore, J. H., & von Fumetti, B. Three alcohol programs for American Indians. *American Journal of Psychiatry,* 1972, *128,* 1450–1454.

Snake, R., Hawkins, G., & LaBoueff, S. *Report on Alcohol and Drug Abuse Task Force Eleven: Alcohol and drug abuse.* Final report to the American Indian Policy Review Commission, Washington, D.C., 1977.

Steiner, S. *The new Indians.* New York: Harper & Row, 1968.

Stratton, R. Variations in alcohol problems within the Oklahoma Indian population. *Alcohol Technical Report Oklahoma City,* 1977, *6,* 5–17.

Stratton, R., Zeiner, A., & Paredes, A. Tribal affiliation and prevalence of alcohol problems. *Quarterly Journal of Studies on Alcohol,* 1978, *39,* 1166–1177.

Topper, M. C. Drinking patterns, culture change, sociability and Navajo adolescents. *Addictive Diseases,* 1974, *1,* 97–116.

Towle, L. H. Alcoholism treatment outcomes in different populations. In *Research treatment and prevention: Proceedings of the Fourth Annual Alcohol Conference on the National Institute on Alcohol Abuse and Alcoholism,* Washington, DC (SUDOCS #HE 20 8314.974), 1975.

U. S. Census. U. S. Census Population: 1960 Subject Reports. Nonwhite Population by Race. Final Report PC(2):-1C. Table 10, p. 12. Washington, DC: U. S. Government Printing Office, 1960.

U. S. Census. Census of the Population and Housing, Final Population and Housing Count, U. S. Summary, Series PHC 80-V, Table 1, p. 4. Washington, DC: U. S. Government Printing Office, 1980.

Van Gennep, A. *The rites of passage.* London: Routledge & Kegan Paul, 1909 (Reprinted 1960.)

Waddell, J. O. "Drink friend": Social context of convivial drinking and drunkenness among Papagos in an urban setting. In E. Chafetz (Ed.), *Proceedings of the First Annual Alcoholism Conference of NIAAA,* Washington, DC: U. S. Printing Office, 1971.

Wax, M. L. *Indian Americans: Unity and diversity.* Englewood Cliffs, NJ: Prentice-Hall, 1971.

Weibel, J. *Native Americans in Los Angeles: A cross-cultural comparison of assistance patterns in an urban environment.* Unpublished doctoral dissertation, University of California, Los Angeles, 1977.

Weibel, J. There's a place for everything and everything in its place: Environmental influences on urban Indian drinking patterns. In *Social Drinking Contexts* (Research Monograph No. 7). Washington, DC: NIAAA Clearing House, 1981.

Weibel, J. American Indians, urbanization and alcohol: A developing urban Indian drinking ethos. In *Special population issues.* Washington, DC: National Institute on Alcohol Abuse and Alcoholism, 1982.

Weibel, J., & Slagle, L. Unpublished Progress Report of Native Healers in Alcohol Rehabilitation Project to NIAAA. Washington, DC, 1983.

Weibel, J., & Weisner, T. *An ethnography of urban Indian drinking patterns in California.* Report presented to the California State Department of Alcohol and Drug Programs, 1980.

Weisner, T., Weibel-Orlando, J., & Long, J. Serious drinking, white man's drinking and teetotaling: Predictors of drinking level differences in an urban Indian population. *Journal of Alcohol Studies,* 1984, *45*(3), 237–250.

White, R. A. *The lower class "culture of excitement" among contemporary Sioux.* Lincoln: University of Nebraska Press, 1970.

13

Navajo "Alcoholism"

Drinking, Alcohol Abuse, and Treatment in a Changing Cultural Environment

MARTIN D. TOPPER

This chapter views Navajo "alcoholism" within the framework of abusive drinking, and not from any one conception of the "disease" of alcoholism. Abusive drinking is any use of alcoholic beverages that contributes to the development of a physical illness, a mental or emotional disorder, or the destabilization of an economic or social relationship that would not have otherwise been destabilized. This broad definition of alcohol abuse is employed because no single definition of "alcoholism" can adequately describe the variety of behaviors and clinical symptoms encountered when one studies or treats Navajos whose lives have been changed by frequent or poorly timed intoxication. When providing services to Navajos who suffer from alcohol abuse, this broad definition of alcohol abuse provides more insight into the manner in which intoxication contributes to the patient's maladaptation than any of the more strictly defined constructs, be they medical, psychological, or traditional

This chapter reflects the opinions of the author and does not reflect the opinions of any other individual or any governmental organization.

MARTIN D. TOPPER • Winslow, Arizona 86047.

Navajo in origin. It is precisely for this reason that I have developed the multidimensional approach discussed here.

INTRODUCTION

Navajo culture is changing rapidly. It is no longer possible to speak of Navajo drinking without specifically referring to a cohort of drinkers and their level of acculturation. The work of Levy and Kunitz (1974) and Topper (1970, 1974, 1980) indicates that drinking among the Navajo cannot be viewed as a single set of culturally patterned behaviors from which there are only minor departures.

Contemporary Navajo culture is becoming increasingly complex. The 8,000 Navajos who returned from Fort Sumner in 1868 were a relatively homogeneous group. However, a population increase of 2,000%, the expansion of reservation lands by over threefold, and the development of various federal and tribal education, health, and employment programs on the reservation since the end of World War II has led to the evolution of at least five Navajo subcultures. They are the very traditional rural elderly, the semi-acculturated rural Navajo, the wage-work families of the agency towns, the wage-work families of the border towns, and the acculturated Navajos of the large cities. Recent improvements in phone service, broadcasting, and reservation roads and the widespread ownership of pickup trucks have provided the Navajo a number of methods of transmitting cultural changes. This makes it still possible to view Navajo culture as a largely unified cultural entity. However, when one speaks of drinking and the cultural perception of appropriate and deviant drinking behavior, there is no single set of cultural values and behaviors that can be said to represent all, or even a clear majority, of Navajos.

The complex, multilinear evolution (see Steward, 1955) of Navajo culture has created a situation in which it is difficult to describe and define Navajo drinking or develop a unimodal treatment system for reducing alcohol abuse by Navajos and alleviating some of its impacts. Given this complex situation, this chapter will attempt to define the scope and nature of Navajo drinking and "alcoholism" from both the traditional Navajo perspective and from the perspective of the acculturative forces experienced by contemporary Navajos. From this broad viewpoint, it will then be possible to explore the principles of diagnosis and treatment of abusive drinking among the several and diversifying Navajo subcultures.

SETTING

The Navajo Tribe is composed of approximately 154,000 people who reside in a 30,000 square mile area of Arizona, New Mexico, and Utah. About 24,000 square miles of this is reservation. The rest is federal, state, and private land. Residence is primarily rural; however, there are a number of "agency towns" on the reservation and "border towns" in adjacent communities.

The Navajo speak a Southern Athabascan language. Archaeological evidence indicates that their ancestors most likely migrated to the Southwest United States from southwestern Canada between A.D. 1200 and A.D. 1500. After arriving, the Navajo gradually evolved from a hunting and gathering people who lived in small bands to a people who lived in dispersed clan segments in extended family residence groups. Subsistence was by a combination of horticulture, sheepherding, and raiding. It was raiding that led to the "round-up" of almost 8,000 Navajos by Colonel Kit Carson in 1863 and the "Long Walk" to a five-year imprisonment at Fort Sumner, New Mexico.

In 1868, the Navajos who survived the ordeal of captivity returned to a newly formed reservation located in the Chuska, Lukachukai, and Carrizo mountains on the Arizona–New Mexico border. The Navajo population rapidly increased and the reservation grew to its present size by 1934. About the same time, it was determined that the reservation was badly overgrazed. In 1933, the Bureau of Indian Affairs decided to impose a program of phased livestock reduction. By 1939, this program had cut the number of livestock in half. Combined with the increasing Navajo population, this cut meant that the Navajos could no longer be self-sufficient on their traditional subsistence economy.

This change had a major impact on the evolution of Navajo culture. As White (1959) has theorized and Aberle (1963) demonstrated, major economic changes frequently lead to major changes in the social and philosophical subsystems of a culture. Livestock reduction forced many Navajos who had marginal livestock holdings to seek employment in the wage-work economy that was evolving on and off the reservation. As the economy of the reservation grew with increasing federal and tribal programs and with the development of a limited private sector, a separate subculture of Navajo wage workers began to appear (see J. Levy, 1960; Levy & Kunitz, 1974; Topper, 1974).

These wage workers became the vanguard of cultural change on the reservation. Aberle (1963) has described the important role they played in the establishment of the Navajo Native American Church. Kunitz and Levy (1981) and Topper (1981) have described how wage workers began residing near job sites in such agency towns as Window Rock and Tuba City. And I have discussed the development of new patterns of drinking in agency towns (Topper, 1974, 1980). The emergence of Western Stomp Dance, Cat Dance, and finally Disco Dance drinking during the 1960s, 1970s, and 1980s made it clear that the agency town was a focal point for the development of new drinking patterns among adolescent and young adult Navajos.

The early 1980s brought a slowing of reservation economic development. However, Broudy and May (1982) indicate that the Navajo population is still growing at a rate that is four times the national average. In 1978, the average Navajo was 16.4 years old, and today there are 75,000 Navajos under the age of 17. As a result of the limitations on grazing, this young population wil largely enter the wage-work market. However, the unemployment rate on the Navajo Reservation is now estimated at 56%. Given current federal and tribal program reductions and the limited size of the reservation's private sector, it is clear that the stage is being set for new cultural changes among the Navajo that will be equally as dramatic as those of the 1930s.

There can be little doubt that these changes will have an impact on drinking behavior. Broudy and May (1982) have already described an increase in what they call "behaviorally related deaths" among the Navajo from 1965 to 1978. At present, these authors (1982, p. 7) claim that approximately 40% of all Navajo deaths are due to suicide, homicide, accidents, and mental disorders (including alcoholism). Given this, Broudy and May (1982, pp. 15–17) see increasing alcohol abuse and behaviorally related illnesses among the Navajo as an evolutionary process leading in the direction of increased morbidity and mortality. This process is part of a broader evolutionary change in Navajo culture that has been brought on by a lagging economic growth and rapid increases in population. These conditions have generated considerable emotional stress and have significantly contributed to recent increases in medical, social, and psychological pathology. This is borne out by the over 130% increase in the rate of criminal homicide, suicide, and cirrhosis of the liver since 1965 and by the fact that the alcohol-related death rate for the Navajo is 20 times the national average.

Much of this morbidity and mortality is found among teenagers and young adults. However, alcohol-related illness can be found among Navajos of all ages. Therefore, it is important to conceptualize treatment by first looking at traditional Navajo drinking and then seeing how it has been changed by the process of acculturation. This view provides a broad perspective both for assisting Navajos who remain traditional and for helping those who have adopted some of the values and drinking patterns of Anglo-Americans. Such a broad approach is important because, although the increase in socioeconomic stress may now be falling heaviest on those Navajos who depend upon wage work for their living, the strong Navajo extended-family system mandates that the economic, social, psychological, and medical burdens of change and change-related abusive drinking will be more or less distributed throughout the reservation population.

TRADITIONAL NAVAJO DRINKING

Traditionally the Navajo viewed drinking as one of a number of behaviors that were classified as "bad life" or *doo yá'át'ééhgo iina'da'*. Alcoholic beverages were not native to the Navajo and were first acquired through trade with Europeans. Even though they eventually learned to make a home brew called *tołbáhá*, drinking never quite became a socially approved activity.

A duality in attitude toward drinking gradually evolved among the Navajo. This is reflected both in Navajo mythology and the Navajo language. According to Navajo mythology, drinking was created some time after the "good life" or Navajo way of life (*yá'át'ééhgo iina'*) was "given" to the people by the gods and Holy People. Therefore drinking is not a part of the life that "was intended" for the Navajo. Instead, it is the product of the evil intentions of beings like Coyote, who interfered with the plans of the creators. Even so, drinking is a rather mild form of bad life. It is considered to be *tsi'na'ada'* or "wild, reckless" behavior, which is not as dangerous for the Navajo People as mental illness (*bini'bihoodit'į'*) or witchcraft (*adigash*).

Linguistically, Navajos divide drinkers into two categories: the "alcoholics" (*yéego da'adláanii*) and those who simply drink (*da'adláanii*). The drinking of alcoholics is socially unacceptable because they drink in the following manner: They primarily drink alone; they drink frequently; they drink with strangers and other non-kinsmen; they "drink up" scarce family and extended family resources; and worst of all, they try to "solve their problems" through drinking rather than through the help offered by medicine men. Those who were simply drinkers used alcohol in a more socially acceptable, although not condoned manner. Their drinking generally followed the following rules: It was infrequent; it was not to the point of serious intoxication; fights were not allowed; financial or other economic resources were not wasted, either through the expenditure of money or through the destruction of personal or family property; the alcoholic beverage was shared; and the drinking took place primarily among kinsmen. If drinking adhered to these principles, then it often provided an environment of conviviality in which family relationships were strengthened instead of damaged.

In addition, drinking also provided a viable means of escape from the pressures of living in a subsistence-level culture in which economic resources were scarce and continual anxiety and stress were generated by poverty, economic uncertainty, and the concentration of scarce resources in the hands of a few rather powerful families that had large herds. It is clear that the "escape" or narcotizing psychodynamic function of alcohol (see R. Levy, 1958) was important in traditional Navajo drinking as a means of temporary relief from these psychological pressures. Furthermore, the work of Levy and Kunitz on the Kaibeto Plateau (1974, p. 77) demonstrated the importance of the sharing of alcohol in symbolically reaffirming the economic bond and principle of economic redistribution between relatively wealthy individuals and those who depended upon them. Therefore, even though alcohol may have been considered "bad life" in traditional Navajo society, it had important social, psychological, and economic functions when used in a controlled manner.

DRINKING PATTERNS

Four basic patterns of drinking evolved on the Navajo Reservation[1] during the last half of the nineteenth and the first half of the twentieth centuries. These were largely compatible with other aspects of Navajo culture and, except for the drinking of "alcoholics," functioned to support the social and economic subsystems of Navajo culture. These drinking patterns were the house party, the drinking of older men in groups, the drinking of young men in groups, and the drinking of alcoholics.

These four patterns of drinking evolved slowly and were well integrated into the traditional Navajo way of life. Spicer (1972, p. 60) has called this type of cultural change "incorporative integration." Basically this form of acculturation takes elements of foreign cultures and allows them to be "accepted into a culture in such a way that they enhance the existing organization of that culture." Alcohol was (and still is) prohibited on the Navajo Reservation, but this prohibition has never been strictly enforced. Given this, and the relative isolation of the Reservation from the American mainstream, an environment existed in which drinking patterns from the "dominant" culture could be assimilated by the

---

[1] This statement refers only to on-Reservation drinking patterns. Levy and Kunitz (1974) have clearly shown that Navajo drinking patterns off the Reservation did not conform to traditional Navajo drinking patterns as early as the first part of the twentieth century.

Navajo and given new meanings that were largely supportive of the Navajo way of life.

The house party was the only drinking environment in which Navajo women had a socially acceptable drinking role. These parties were often held during events that had greater than normal economic, social, and/or religious significance. They included major healing ceremonies, marriages, female puberty rites, hogan building, dam building, Spring sheep shearing, harvesting, the butchering of such large animals as cattle or horses, and other events that required Navajos to temporarily interrupt their isolated existance and come together for their mutual benefit.

House-party drinking normally took place in the evenings when other activities were either finished or had temporarily come to a halt. At these times, people had usually gathered indoors and were talking about matters of personal interest. Drinking would often start when one of the older men would produce a bottle of wine, offer the first drop to Mother Earth, take a drink, and pass the bottle around. All adults who felt like it would then take drinks and pass the bottle on. Sometimes, adults would allow children to take small drinks. As one bottle was finished, the same man or another man might produce another bottle. This sharing of alcoholic beverages (usually wine or home brew) would continue until most of the adults in attendance were mildly intoxicated. Serious intoxication was usually prevented by strong cultural values concerning sharing. Few dared to take too big a drink from the bottle. In addition, the drinking was usually brought to a close by an older male who, seeing that people were beginning to get intoxicated, would take the last drink from the bottle, pour the last drop on the ground for Mother Earth, and exclaim *bíigha'*, which means "enough," but which also connotes satisfaction. Such an exclamation was normally sufficient to end the drinking, and those who wanted more knew that it was time to take their liquor and go into the desert and drink.

A second traditional Navajo drinking pattern in which a great deal of social control was evidenced was the group drinking of older men. This drinking often occurred during traditional ceremonies. At times it would also take place after a house party. This form of drinking was frequently found at Enemy Way ceremonies during the summer. It usually began in the early evening as participants, well-wishers, and onlookers were gathering for one of the three nights of outdoor dancing that conclude the ceremony. Before the outdoor singing and dancing began, men who were related to each other by kinship or by marriage would often decide to share a few bottles of wine. At this time, the men would "pitch in" and one of them would go and make a purchase from a bootlegger. After acquiring several pints of strong red wine, the members of the group would then go into the desert where they would not be spotted either by the police or their wives. Normally the group was composed of *hastoi* or "elder men," who were from their mid-30s on up in age. Sometimes a few younger kinsmen or in-laws would be invited along to socialize and to "learn what they could" from the older men.

The actual drinking took place in the desert about one-quarter to one-half mile from the site of the ceremony. The men usually chose a place behind a hill where the sand was soft and where they would not be detected easily. They would sit in a circle, begin to talk, and open the first bottle. The first few drops would be offered to Mother Earth and then the man who opened the bottle would drink from it and then pass it "sunwise" around the circle to prevent the occurrence of any untoward supernatural incidents. If someone had cigarettes, these might also be lit one at a time and passed around the circle.

As they shared the wine, the men would talk about their work, their herds, and the ceremony. Often, as a treat to younger men and as a means of showing their competence and position in life, the older men would share some of their ceremonial knowledge. These discussions could become quite competitive if two medicine men were present and if they had a little too much to drink. At such times a considerable amount of knowledge about ceremonies and their myths of origin might be disclosed. In this way drinking served as a vehicle for cultural transmission.

It is important to note that although the older men would often brag about their abilities, accomplishments, and knowledge, there were few discussions of disputes between drinkers, their domestic relations, and other controversial topics. The men were gathered to share a drink and to talk about things that were of mutual interest. The function of this gathering was to reinforce family relationships and to recognize and accept each individual's status as a respected member within the family unit. The former was accomplished by sharing the financial resources to buy the wine, the latter by the demonstration of competence through what one had to say.

In addition, this drinking environment allowed enough individual competition so that interpersonal animosities could find ventilation through bragging. If the bragging became too competitive, the antagonists could be separated by their kinsmen who had a strong stake in keeping behavior under control. However, the result of such competitive behavior could be therapeutic in that an individual who wanted to reestablish a relationship could enter into a mutual recognition of the opponent's special talents. In this manner, the ground could be laid for the settling of these undiscussed disputes at a later time.

The third pattern of traditional drinking was that of younger men in groups. This drinking conformed to the same general pattern as that of the older men. The topics of conversation, however, differed. The young men often talked about their prowess in athletics or of their abilities to attract and hold the attention of young women. The drinking behavior of the younger men served the same social and psychological functions as the drinking behavior of older men.

The fourth traditional drinking pattern was that of the alcoholics (*yéego da'adláanii*). This type of drinking involved more than one drinking pattern because it was possible for both men and women to drink in an unacceptable manner and each sex had its own style of disapproved drinking. The traditional form of alcoholic drinking seen among the men on the reservation was that of the isolated drinker. This was an individual who went off to the bootlegger by himself and either consumed his alcohol alone or consumed it in a clandestine manner in a public place, such as the local trading post. The reason that his drinking was so heavily stigmatized was that it took the individual away from the economic tasks that he or she was obligated to perform and it did not involve any sharing of "drinks" among kinsmen. This sharing of alcoholic beverages symbolized other reciprocities and redistributions that were central themes in the Navajo economy and in Navajo culture in general. My work on traditional Navajo daily life (Topper, 1972) has indicated the degree of interdependence members of traditional Navajo households have upon each other and the necessity of pooling labor to operate the Navajo subsistence economy successfully. Kluckhohn and Leighton (1958, p. 110) and Witherspoon (1975, p. 86) have argued this as well. Without cooperation between households, illness, injuries, and local economic setbacks could lead to serious hardship and, in some cases, starvation.

Given the above, the man who spent his money selfishly on his own drinking was symbolically denying his social and economic obligations. At first, kinsmen and affines would try to help by counseling him and/or by holding ceremonies. But those around him would ultimately begin to view his continuing drinking as a drain on precious resources that could be better spent on those who could and would reciprocate. The prognosis for such a man in Navajo society was poor.

The traditional female alcoholic was a person who drank in the company of men when they drank in groups in the desert or who hung around the bootlegger's house or the trading post and traded sexual favors for liquor. Such behavior was a threat to the stability of traditional Navajo households because of its economic and sexual implications. In a young woman, such behavior could lead to divorce and increase the strain on her extended family who would then have to help support her

children. An older woman drinking in this manner could become a
burden on her married children. Again kinsmen and affines would view
such drinking as a form of selfishness because the drinker spent time
drinking instead of helping out. This lack of reciprocity would stigmatize
her as much as her sexual activities would. The degree to which a woman
was ostracized would depend upon several factors. However, those who
spent much of their time drinking or who abandoned young children
had futures that were not much brighter than those of their male counterparts.

CHANGES IN NAVAJO DRINKING PATTERNS DUE TO ACCULTURATION

Navajo culture has changed significantly since the livestock reduction of the 1930s and the induction of many young men into the armed
forces during World War II. With this change has come a change in
drinking patterns. Traditional drinking patterns are still evident in many
areas of the reservation. However, today they exist side by side with
patterns that have been greatly influenced by off-reservation drinking.
These newer patterns differ significantly from the older ones in that
drinking no longer occurs more or less exclusively among kinsmen and
affines. The drinking cohort often forms more or less spontaneously at
various events and places. For example, drinking now occurs at agency
town community centers, in bars located in or near the bordertowns, at
powwows both on and off the reservation, at reservation high school
games, at local agency town fairs, and in vacant lots and homes in the
bordertowns. In many of these newer drinking environments, the drinker
has little or no ability to control whom he or she drinks with.

These newer drinking environments place the Navajo drinker in
the company of strangers. This seriously disrupts the socialization function of drinking. The disruption occurs because the drinker is no longer
drinking in the company of relatives and affines with whom there are
clearly defined rules for social interaction while drinking and with whom
there are clear economic relationships. There is little that binds most
drinkers in these new environments. In addition, Navajos not related
by kinship can be quite suspicious of each other. In the early 1800s,
unrelated bands of Navajos frequently raided each other's family residence groups. Even into the 1940s, travel by traditional Navajos, especially women and children, was quite limited, and many Navajos had
little contact with strangers. Also the fear of being witched by other
Navajos who were "jealous" was, and still is, prevalent. Given this,

most social, economic, and emotional ties have traditionally been limited to kinsmen and affines. Strangers of any culture have not been easily accepted.

This traditional attitude has formed the basis of the attitudes of the current generation. One might think that the increasing exposure to strangers through schools and wage-work employment would have desensitized young Navajos to the stresses of interacting with strangers. However, the opposite seems to have occurred. Many of the problems that young Navajos have in socializing with strangers are the result of their boarding school and wage-work experience.[2] Boarding school relationships are, for the most part, temporary. The children who enter the boarding schools find an environment in which the dormitory aids, teachers, school administrators, and even other students change from year to year. This turnover discourages the development of close relationships. Instead, the students learn to relate to categories of others. Relationships with adults become depersonalized, formalized, and restricted to such specific domains and areas of interaction as the classroom, the dormitory, the counseling office, and the principal's office. This constriction of relationships prevents many young children from developing close emotional ties to adults. In addition, the shortage of boarding school staff complicates the relationship of the children to each other. The ratio of children to dormitory aids can approach 100:1.[3] The ratio of students to teachers is frequently more than 25:1. These ratios mean that the children are always in competition for the attention and affection of adults. Because competitiveness is frowned upon in traditional Navajo culture, many dormitory children withdraw from adults and develop a society of their own.

A society the dormitory children develop shields them from what they perceive as a lack of warmth and caring on the part of the over-

---

[2] Although it is my observation from both ethnographic research and clinical experience among the Navajo that employment and boarding school experiences are major contributors to the development of the problems many young Navajos have in relating to strangers and to the recent steady increase in the levels of socio- and psychopathology among young Navajos in general, it is important to note that in any specific case, one must account for such complicating factors as the stability of a child's household, the quality of parenting, the availability of family resources, and the genetic history of the patient's blood relatives in attempting to understand how a specific individual became an abuser of alcohol. However, when viewing contemporary Navajo culture in general, it is difficult to uncover any other factors in the institutional structure of that culture external to family life that have as great an impact on the development of psychopathology and maladaptive behavior as employment (including unemployment) and boarding school experiences.

[3] Ratios this high are often found in the larger dormitories during the night shift.

worked staff. However, this children's society is in itself a highly competitive environment in which children frequently do not develop lasting relationships. The older children in each dormitory become leaders. In return for their attention, they often demand loyalty and favors from the younger children. In this situation, many children feel that acceptance is based primarily upon what they have and what they can provide. This mercenary quality of acceptance leads to a situation in which dormitory children do not often develop firm, trusting relationships with their peers or with older children.

Employment for wages on or off of the reservation reinforces these attitudes that approval is contractual in nature and that one must compete for the approval of one's leaders. Wage employment is by its very nature contractual. Approval for on-the-job performance is provided largely in the form of monetary compensation. This aspect of the wage-work contract is well understood by the average boarding school student. However, personal acceptance, emotional support, and protection are not often parts of a contract between an employer and an employee. It is easy for the former dormitory student to fail to understand this. If this happens, then problems in his or her relationship with the employer can develop, especially if the employer is non-Navajo and never went to boarding school. Such difficulties can develop when a Navajo employee requests that an Anglo- or Spanish-American employer become involved in his or her personal affairs. Such requests are often rejected, and this can generate feelings of loss and anger that may reinforce feelings of worthlessness and non-acceptance the young person may already have (see Topper, 1983). Instead of satisfying unmet needs, the employment situation may only add to the suspicion of strangers and feelings of worthlessness. Such frustrations may only provide further impetus for seeking escape in the narcotizing effects of alcohol.

The net impact of acculturation appears to be that only the escape or narcotizing function of alcohol remains for many young Navajos who drink in non-traditional environments. Given the fact that these people are an ever-increasing segment of the Navajo population, a major trend toward a new and dangerous form of drinking is underway. Those who drink for escape in non-traditional environments find themselves using a disinhibiting, depressant drug among strangers about whom they feel ambivalent. Furthermore, they drink in environments in which traditional Navajo rules for social control of drinking do not apply, and for which, there has not been the development of non-traditional social controls. Finally they frequently bring with them considerable anger and frustration concerning their economic and perhaps social condition. Given these factors, the increasingly high rate of alcohol-related morbidity and mortality among young Navajos is explainable. Many of these young people are neither culturally nor emotionally prepared either to tolerate the stresses of non-traditional drinking environments or to experience the emotional release or satisfaction that such drinking provides for people of other cultures. Clearly, then, these newer forms of drinking are not as therapeutic as the more traditional ones.

THE PROBLEMS OF TREATMENT

Treating alcohol abusers who belong to a large minority population with several layers of acculturation is at best difficult. The Navajo situation is further complicated by considerable underemployment and unemployment that create a high level of means–goals disjunction in the population. When faced with such a complex cross-cultural situation, it is usually best to begin to look at the problems of treatment in general as a prelude to the development of treatment designs. By first examining the basics of cross-cultural treatment, it will then be easier to discuss the principles of diagnosis and treatment upon which a Navajo-specific treatment system may be viewed. This broad approach allows maximum flexibility.

In order to diagnose and care for patients in a cross-cultural context, I have found it valuable to conceive of diagnosis and treatment as occurring in four basic dimensions. They are the medical dimension, the psychological dimension, the socioeconomic dimension, and the cultural–historical dimension. Since evaluation and treatment are a two-stage process, it is best to view each stage separately in order to see how the four dimensions operate in each stage of the process.

MULTIDIMENSIONAL EVALUATION

The medical dimension of evaluation includes taking the patient's medical history and establishing current condition and needs. In addition to the standard physical examination and routine laboratory testing, the evaluation of the alcoholic patient focuses on the discovery of medical conditions that may affect the patient's ability to benefit from or tolerate further treatment. For example, a 60-year-old patient with Korsakoff's psychosis will have different treatment needs and capacities for recovery than a 20-year-old patient with no medical complications. This is not to say that the medical evaluation dominates the choice of treatment modality; in a number of situations, however, it does set limits within which treatment choices are made.

The second dimension of evaluation is the psychological. It involves taking the patient's psychological history and determining current mental status. It also has a major impact upon the choice of treatment. For example, a manic-depressive patient who self-medicates with alcohol may appear to be an uncomplicated binge-drinking alcoholic. However, the nature and course of manic-depressive illness is quite different from that of simple alcoholism, and a successful treatment outcome cannot be expected without syndrome-specific chemo- and psychotherapy. Therefore, psychological evaluation and the treatment of the patient depend upon the culturally sensitive application of accepted techniques of taking a psychological history, recording the patient's complaint, and determining current mental status.

Evaluation along the medical and psychological dimensions provides a picture of the patient's ability to become physically well and to experience psychological growth. These first two dimensions of evaluation also provide a core of information for determining the degree of risk the patient is facing at the time of intake. If the degree of risk is high, then an emergency admission may be required to conserve the life or mental health of the patient or to protect his or her loved ones.

However, many alcohol abusers do not require emergency care at the time of intake and the socioeconomic and cultural-historical dimensions are clearly relevant to the initial evaluation. These dimensions differ from the medical and psychological in that they involve an evaluation of both the patient and the cultural conditions that form the patient's milieu. The socioeconomic and cultural-historical dimensions view the patient's current functioning and the way that the patient is required to function in the community. In short, the therapist determines if the patient is doing well and if the patient's community is doing well, given the problems that both must resolve if successful adaptations are to occur.

The socioeconomic dimension focuses on the specific position patients occupy in the society and economic system of their local community and on the degree to which the community can meet patients' social and economic needs. With respect to this dimension one asks what the patient does for a living; if the patient resides in a nucleated or extended family household; if the patient is married and has children; if all members of the patient's family are in good health; if the patient has relatives or in-laws living nearby; if the patient is financially solvent and has adequate housing; if there is adequate housing and employment in the local community; if the local community has a high crime rate or other major legal or social problems; if the patient is active in the government or political system of the local community; if there are adequate governmental, charitable, and treatment resources available locally; and a number of other questions that may be necessary to place the individual patient within the local socioeconomic matrix. These questions provide vital background information for determining how well the patient is functioning; how well the patient's household is functioning; how well the local community is functioning; and how useful the local community can be in offering supportive services to a specific patient and his or her household. In this way, one can determine if the patient has realistically adapted to the local environment and if the local environment, or the patient's family, contributes to his or her psychopathology.

The fourth dimension is the cultural-historical. This dimension views the patient from the perspective of long-term economic, social, and philosophical practices and beliefs held by the cultural group to which the patient belongs. The use of this perspective in evaluation and treatment requires that the diagnostician/therapist take the time to become familiar with the history of the culture or cultures from which the local patient population is derived. Such familiarity is essential if the care provider is to understand what it means for the patient to participate in and experience the activities of daily life that constitute the socioeconomic dimension of the patient's existence. The contributions that are made by viewing this dimension include the ability to determine the patient's level of acculturation and his or her ability to profit from Western psychotherapies; the ability to view the patient's family structure and its functioning in light of the norms of patient's culture; the ability to understand how the patient's condition would be viewed by members of his or her own culture; the appreciation of the traditional psychotherapeutic resources available in the patient's culture; the estimation of how capable the patient might be of benefiting from these traditional therapies; the identification of traditional social and economic resources in the patient's culture that could be employed to assist the patient; the evaluation of the patient in light of the patterns of psychological development that are normative in the patient's culture; and the appreciation of the prescribed and proscribed ("taboo") behaviors that are an important part of the patient's culture.

In all, the cultural-historical dimension provides an integrating framework for viewing the other three dimensions of evaluation and treatment. It allows one to see the patient's medical, psychological, and socioeconomic status in light of a larger system of belief and behavior. When evaluating a patient and when providing treatment, this overview can be very useful in determining a patient's level of psychopathlogy, in viewing how this form and level of psychopathology fit or do not fit within the patient's cultural milieu, in setting goals for treatment, and in evaluating treatment outcomes.

MULTIDIMENSIONAL TREATMENT

When the evaluation is completed, the process of treatment begins. In working with American Indians who have suffered from abusive drinking and other psychological and behavioral disorders, I have found it useful to think of treatment as having three basic elements. They are medical intervention, psychotherapy, and environmental intervention. Each of these three elements of care is, like the process of evaluation, multidimensional in nature. Even though each element is grounded in one or two of the basic four dimensions, it cannot be successful without being integrated with all the others.

The first element, medical intervention, includes treatments that stabilize the basic medical problems of the patient. These include such somatic interventions as psychotropic drugs and electroconvulsive therapy, treatment of major medical disorders of significant others in the patient's life, and the recognition of and allowance for the use of traditional native psychotropic medications and physiotherapies when providing care. This first element obviously is focused within the confines of medicine. This is especially true for emergencies. However, when one enters into the non-emergency and other, less restrictive areas of care, the roles of the other dimensions become more important. The psychological dimension looms large in any medical intervention in which psychotropic medication is used. The socioeconomic dimension is important when choosing between similar treatments of differing expense. And, the cultural-historical dimension is important in presenting treatment in a manner that is compatable with the native culture.

The second element of treatment is psychotherapy. This form of treatment is primarily aimed at using verbal interaction to help the patient experience psychological growth by gaining a better understanding of himself or herself and by resolving psychological conflicts and reintegrating split-off aspects of the ego. However, psychological treatments are not without their medical, socioeconomic, and cultural-historical impacts. The medical aspects and impacts of psychotherapy include the remission of vegetative signs of depression; the motivation of the patient to seek treatment for medical disorders that complicate his or her emotional state; and the termination or alleviation of substance abuse.

The socioeconomic aspects and impacts of psychotherapy are many. These include improvements in marital, family, and extended family relationships; improvements in social relationships on the job and in the community; and improvements in the social skills of the patient. The purpose of these improvements is to change the patient's manner of relating to his or her socioeconomic environment. This change will then, it is hoped, produce a change in that environment and/or the patient's perception of it so that he or she will then experience less stress and anxiety; this in turn will improve his or her chances of recovery.

The cultural-historical dimension of psychotherapy is of central importance when the patient and therapist are of different cultures or subcultures. The culturally oriented aspects and impacts of psychotherapy include providing therapy completely or partially in the patient's native language; interpreting the patient's proxemic behavior according to the standards of his or her culture; understanding the patient's developmental history in light of the patterns of psychological development in the patient's culture; understanding key cultural symbols and role relationships so that interpretations of dreams, thoughts, and behaviors are not made entirely according to formulations developed in another culture; and understanding enough of the social system of the patient's culture so that directives can be given in a manner that is compatable with the demands of the patient's daily life.

The cultural-historical dimension of psychotherapy affects the treatment process in many ways. However, two of the most important areas that are affected are the transference relationship and behavioral directives. A positive transference relationship can be difficult to develop when the patient and therapist come from different cultures, and especially difficult when the patient is from a minority group. In such a situation the relationship between the therapist and the patient can be tainted with conceptions based on unpleasant experiences. Similar complications arise when behavioral directives are given. Depending upon the style of therapy employed, behavioral directives may be an important part of treatment. If such directives ignore the demands and requirements of the social and economic systems of the patient's culture, then it is difficult to imagine that they would be followed. In fact poorly formulated directives could be a major stimulus for the development of a negative transference and could lead to the premature identification of the therapist with non-native significant others in the patient's life, such as an overly strict school principal. Such issues early in the therapy tend to cloud the transference and can sidetrack it, leading it away from the identification of the therapist with the parents and other significant members of the patient's family, the object-relationships that usually contribute significantly to the patient's psychological conflicts.

The third element of treatment is environmental intervention. This involves the therapist in doing something to help promote a change in the patient's environment that will in turn reduce stress in the patient's life. This aspect of treatment can be controversial, since it involves such issues as involuntary commitment and the honoring of requests by the patient for the therapist to become involved in helping the patient achieve goals that lie outside the treatment setting. In the strictest of psychoanalytic approaches, the therapist would rarely, if ever, honor such requests. Instead, he or she should interpret them in the light of transference. But if, after such an examination, it can be determined that the request has a major impact upon the social, economic, or medical condition of the patient, it would then be referred to a social worker or a physician.

The situation is quite different on Indian reservations and in most other Third World settings, in which there are generally few environmental resources and economic and social opportunities. For example, the Navajo Reservation has an unemployment rate that frequently exceeds 50%,[4] an equally high percentage of substandard housing, and a significant population of individuals who are not sufficiently educated in the ways of Anglo-American culture to initiate contacts with social and welfare agencies. When these facts are combined with the remoteness of the Reservation, the frequent insufficiency of social and welfare agency staffing, and the lack of social workers in many Reservation communities, it becomes clear that what may be legitimately considered to be a manipulation of the therapist in another setting may be a genuine and realistic request for help. Therefore, although caution is always necessary, environmental manipulations are frequently an important element in treating Third-World patients.

[4] The level of unemployment on the Navajo Reservation has proven difficult to measure and has often been disputed. In 1982 the Navajo Tribe estimated that unemployment was as high as 70%. At the same time the Bureau of Indian Affairs said that it was only 35%. Given the fact that this was an election year, both estimates were issues of dispute. However more recently an estimate of 56% was released and considered to be fairly accurate by both parties. The problem in obtaining an accurate measurement is that there is much seasonal unemployment due to the fact that many Navajos work on off-Reservation farms in the summer and are not employed in the winter. Therefore, it would be most reasonable to state that reservation full-time unemployment is in the neighborhood of 35% to 40% and that another 20% to 30% of the Navajo population can, at best, find only seasonal jobs.

Environmental interventions can also be multidimensional in nature, even though they are primarily aimed at changing the patient's socioeconomic condition. For example, when a therapist suggests hospitalization to a patient, the psychological stress experienced by the patient is greatly reduced by the change in environment. In addition, hospitalization allows the introduction of medical therapies that cannot be performed on an outpatient basis. Finally, when the therapist becomes involved in helping reorganize the social institutions in the patient's culture, then he or she is participating in the process of cultural change and is becoming part of the cultural-historical process. A good example of such an intervention was the involvement of Bergman and Goldstein (1971) in the model dormitory project on the Navajo Reservation. Although environmental manipulation should never be approached lightly, when a social institution can be shown to contribute to the general level of psychopathology in a population, it may very well be appropriate.

TREATING NAVAJO "ALCOHOLICS"

The concept of care presented above is quite complex and raises questions of how it can be operationalized to provide evaluation and treatment services for a cultural entity as diverse as the Navajo. However, if a global approach is employed, and the problems of the drinker are viewed from the perspective of cultural systems, then some basic principles upon which specific treatment protocols can be based may be defined. Once the basic principles are defined, then it may be somewhat easier to make suggestions.

The first principle is that treatment modalities that are strongly biased toward either the medical or cultural-historical poles of the quadra-dimensional continuum are not likely to be sucessful. "Alcoholism" is an illness that creates problems in all four dimensions. For example, in the medical dimension, abusive drinking can be seen to create morbidity and mortality through disease and alcohol-related accidents regardless of how the drinking is patterned. The high alcohol-related morbidity and mortality rates among the Navajo clearly indicate that medical treatments for alcohol-related conditions cannot be overlooked just because they are not products of traditional Navajo culture. Failure to apply proven medical treatments for such conditions as delirium tremens and cirrhosis of the liver would only increase the medical impact of these conditions. Instead, it is necessary to adopt these treatments so that they are compatible with the evolving culture of the contemporary Navajo.

To use Spicer's (1972) terms, one might say that an "incorporative integration" of medical treatments for alcohol abuse is a major aspect of a balanced cross-cultural treatment system.

A similar point can be made about the evaluation and treatment techniques that arise in the psychological dimension. They also are often associated with the "medical model" and are therefore seen by some as being alien to the treatment of Third-World drinkers. Although there are many arguments about the psychological causes of alcohol abuse and arguments about the applicability of Western diagnostic categories to the mental illnesses and emotional disorders of native peoples, it has been shown cross-culturally that abusive drinking can either occur by itself or can occur in the company of more severe mental disturbances. I have treated Navajo drinkers whose drinking problems were accompanied by a variety of other mental disorders, ranging from "organic brain syndrome" (*bitsiighaan doo ahaliyáada*) to "schizophrenia" (*tsi' naaghá*). Of the 39 patients with drinking problems I have treated in Winslow in the past two years, only 15 have suffered from simple, uncomplicated alcoholism. The other 24 have had their drinking problems complicated by other mental disorders. These have included depression; anorexia nervosa; posttraumatic stress disorder; organic brain syndrome due to alcohol, drugs, epilepsy, or stroke or other cerebral vascular disorders; schizophrenia; and suicidal behavior.

These observations present clear implications for clinical care. Many treatment modalities for drinkers are designed for individuals who have mild to moderate "character disorders" (*tsi'na'ada*). These treatments involve confrontation, acceptance of the label *alcoholic* as part of the patient's identity, and the use of highly structured group treatment plans that place a great deal of responsibility upon the patient to demonstrate his or her "commitment" to sobriety. This type of care is well-suited to a number of alcoholic patients. However, it is highly unsuited to "schizophrenic" (*tsi' naaghá*), severely "depressed" (*yíní bi niiłhi*), "cyclothymic" (*tsi'nizhdiidááh*), "manic-depressive" (*yéego tsi'nizhdiidááh*), "organically impaired" (*bitsiighaan doo ahaliyáada*), or "seizure-disorder" (*na'anijitl'ish*) patients. These patients need different psychological and medical treatments than those who are suffering from simple, uncomplicated alcoholism. In fact confrontation can be contraindicated in patients who are suicidal and may produce violent behavior in decompensated schizophrenics and manic-depressives.

Therefore, it is vital that a proper diagnosis be the first step to treatment. Although there are few one-to-one equivalencies between traditional Navajo language categories and the standardized categories of *DSM-III*[5] (American Psychiatric Association, 1980), there is a rough equivalency that, in many cases, can provide a significant understanding of what is going on with the patient in both systems of thought. Such a dual viewpoint provides sufficient understanding of the patient's condition and its cultural meanings so that a culturally sensitive and appropriate treatment plan can be developed for the specific disorders, defined in both Navajo and psychiatric terms, indicated by the patient's symptoms. In this way, treatments that are contraindicated in either system of care can be avoided.

The argument presented here is that all four dimensions are necessary for proper diagnosis and treatment. The so-called "medical model" fell into disrepute because the standard treatment techniques of Western medicine and psychology, if applied without considerable sensitivity to the culture of the people being served, can produce inadequate care systems that local communities underutilize and do not understand. On the other hand, many of the so-called alternative programs fail to deal adequately with life-threatening medical and psychological conditions. Clearly the first step in the development of any cross-cultural treatment program is the placing of all four dimensions of evaluation and treatment in their proper relationship.

The concept of multidimensional treatment leads to a second principle. Just as the program as a whole must be broad enough to deal with variability on a cultural level, it must also be broad enough to deal with it on an individual level. It is very clear that the broad variety of psychopathology, medical conditions, levels of acculturation, and socioeconomic statuses exhibited by Navajos who drink abusively calls for the individualization of treatment planning. Navajo culture is different from the dominant society, but it is also a complex entity itself. Given this, it is in the best interests of the patient to develop and implement individual treatment plans that utilize and choose between the available culturally sensitive treatment resources in a manner that focuses on the individual patient's specific needs and limitations.

A third principle that arises from the foregoing discussion is that treatment cannot be limited to the patient. I have argued here that the recent increases in alcohol-related morbidity and mortality among the Navajo have resulted from a decades-long process of cultural change that has been uneven and inadequate to meet the needs of the growing

---

[5] The Navajo terminology employed here refers to Navajo language categories for the mental and emotional disorders that were elicited by myself. These Navajo categories are only roughly equivalent to the DSM-III terminology, presented in quotation marks next to the Navajo terms. For a fuller discussion see Topper (1978).

Navajo population. This process has left the Reservation with a depressed economy, an educational system that, through understaffing in dormitories, contributes to the development of psychopathology in the young, and new drinking behaviors in which the social cohesion function of drinking is largely inoperative.

These changes in Navajo culture, both long-term and recent, have created a situation in which alcohol abuse treatment cannot be seen apart from cultural changes that influence the rate and etiology of mental disorders and abusive drinking. Even if fully individualized multidimensional care were completely in place and if treatment resources were sufficient to meet the needs of all Navajos who drink abusively, the basic problems generated by changes in Navajo culture would mitigate against the eradication of abusive drinking by treatment alone.

The Navajo Tribe has recognized in its Comprehensive Alcohol Plan that a portion of the struggle to contain abusive drinking on the Reservation must be waged in the arena of cultural change. The plan focuses on a wide area of prevention services. These services are vital if young Navajos are to learn what alcohol is, what it can do to the human body, and what the general parameters of non-abusive drinking are. However, many tribal administrators are beginning to feel that non-abusive drinking cannot be taught by educational means alone. Children and adolescents learn by example as well. Prohibition on the Reservation does not allow for the development of role models for non-abusive drinking and older patterns of kin-based drinking are now fading away. At present, drinking is frequently a clandestine activity and there are few public environments where drinking behavior can be openly controlled according to social guidelines. Drinking has permeated a number of social environments in which the primary drinking rule has become "Don't get caught." This has led to rapid intoxication, since alcoholic beverages are consumed quickly to "get rid of the evidence." However, even more problematical is the fact that this type of drinking has taught young people that others will control their drinking behavior and creates a condition in which increasing risks are taken by drinkers when they are not caught by the Tribal Police. The net effect is a loss of the internal locus of control and steady increases in alcohol-related risk-taking, morbidity, and mortality. Finally, the phenomenon of bootlegging on the Navajo Reservation has created a number of law enforcement problems and has led to a loss of tax revenue that could be used to fight alcoholism.

The legalization of alcohol use would not improve this situation overnight. Phillip May (personal communication) has stated that the legalization of drinking on Indian reservations generally creates some initial adjustment problems, but then leads to the evolution of more socially controlled drinking. Legalization of alcohol among the Navajo would be an important step in the development of social cohesion functions that would lead to control of drinking in non-traditional environments. In such places, people would be able to interact in a more relaxed manner over longer periods of time. The removal of the threat of arrest along with an increase in interpersonal interactions would gradually create social pressures for the development of internal loci of control for many drinkers and would thereby generate pressure for the evolution of patterns of social control in these newer drinking environments.

In addition, two other cultural changes could considerably improve the prospects for reducing the amount of abusive drinking on the Navajo Reservation. The first involves the development of a mixed economy and the reduction of unemployment. The outcome of any psychotherapeutic treatment is to have the patient achieve a more independent lifestyle. The current level of unemployment on the Navajo Reservation makes treatment difficult because the unemployed patient faces the same means–goals disjunction after treatment that very likely contributed to the degree and timing of his or her emotional illness. The Navajo Tribe (1982) is currently seeking to develop the reservation economically through a number of different programs. In the long run, the success of one or several of these avenues of development will probably be one of the most important factors influencing both the ability to achieve long-term beneficial treatment outcomes and the reduction of alcohol-related morbidity and mortality on the Navajo Reservation.

A second, long overdue cultural change involves the improvement of the boarding school dormitories. Bergman and Goldstein (1971) effectively demonstrated a method for reducing the psychological impact of the boarding school environment on Navajo children. Unfortunately, the "model dormitory" system was never adopted. As long as the population of the Navajo Reservation remains sparsely settled and the percentage of paved roads remains low, boarding schools will be the means of education for a significant number of Navajo children. The question therefore becomes one of how to generate more adult involvement with boarding school children in order to reduce some of the adverse impacts that the understaffing of these schools have been shown to have on psychological development. There are several possibilities for effecting change. They range all the way from the implementation of model dormitory programs to the use of the "big brother/big sister" concept to increase the amount of parenting each child receives. Regardless of the approach taken, however, it is clear that a concerted effort to improve the quality and quantity of adult–child interactions in the boarding schools is needed if children raised in these institutions are to develop more stable and less substance-dependant personalities.

CONCLUSION

The treatment of "alcoholism" among the Navajo Reservation is not a simple matter. Just thinking about abusive drinking among the Navajo requires one to go beyond most of the popular theoretical conceptions of the causes of abusive drinking among minority peoples. Given this complexity, this chapter has taken the position that the treatment of abusive drinking among the Navajo calls for an individualized multidimensional approach toward the patient and his or her problems. Such an approach follows neither the "medical model" nor any "alternative model" of patient care. Instead it attempts to take what is most effective from many concepts of evaluation and treatment and combines them. Finally, the approach presented here recognizes that both the patient and his or her culture are loci of treament.

The net effect is an approach that views a system of treatment as an ongoing set of patient-oriented and culture-oriented interventions. It does not look for quick solutions to conditions that have developed over many years, but focuses instead on alleviating the conditions that contribute to abusive drinking while it treats the illnesses of specific patients. Over time, such an approach looks forward to reducing the incidence of abusive drinking both by treating patients and by assisting in the ongoing process of cultural change. In this way, both immediate and long-term improvements can be realized in the medical, emotional, and cultural health of the people being served.

REFERENCES

Aberle, D. B. *The Peyote religion among the Navajo*. Chicago: Aldine, 1963.

American Psychiatric Association. *DSM III Diagnostic and Statistical Manual of Mental Disorders*. Washington, DC: APA, 1980.

Bergman, R., & Goldstein, G. *The model dorm: Changing Indian boarding schools*. Window Rock, AZ: Navajo Area Indian Health Service, 1971.

Broudy, D. B., & May, P. A. *Demographic and epidemiologic transition among the Navajo*. Window Rock, AZ: Navajo Area Indian Health Service, 1982.

Kluckholn, C., & Leighton, D. *The Navajo*. New York: Natural History Press, 1958.

Kunitz, S. J., & Levy, J. E. Navajos. In A. Hammond (Ed.), *Ethnicity and medical care*. Cambridge, MA: Harvard University Press, 1981.

Levy, J. E. *South Tuba.* Paper presented at the meeting of the American Anthropological Association, Chicago, 1960.

Levy, J. E., & Kunitz, S. J. *Indian drinking: Navajo practices and Anglo-American theories.* New York: Wiley (Interscience), 1974.

Levy, R. I. The psychodynamic functions of alcohol. *Quarterly Journal of Studies on Alcohol,* 1958, *19,* 649–659.

Navajo Tribe. *Economic recovery on the Navajo Reservation: The view from Window Rock.* Window Rock, AZ: Navajo Tribal Chairman's Office, 1982.

Spicer, E. H. Types of contact and processes of change. In D. Walker (Ed.) *The emergent Native Americans.* Boston: Little, Brown, 1972.

Steward, J. H. *Theory of culture change: The methodology of multilinear evolution.* Urbana: University of Illinois Press, 1955.

Topper, M. D. *The determination of Navajo drinking patterns by police policy and behavior in Tuba City, Az., Cortez, Colo., Monticello, Utah, and Kayenta, Az.* Evanston, IL: Northwestern University Press, 1970.

Topper, M. D. *The daily life of a traditional Navajo household: An ethnographic study in human daily activities.* Evanston, IL: Northwestern University Press, 1972.

Topper, M. D. Drinking patterns, culture change, sociability and Navajo adolescents. *Addictive Diseases: An International Journal,* 1974, *1,* 585–618.

Topper, M. D. The cultural approach, verbal plans,and alcohol research. In M. Everett, J. O. Waddell, & D. B. Heath (Eds.), *Cross-cultural perspectives on the study of alcohol.* The Hague, Holland: Mouton, 1976.

Topper, M. D. Not Navajo life: Mental illness, psychotherapy, and clinical anthropology among the Navajo. In J. L. Steinberg (Ed.), *Cultural factors in the rehabilitation process.* Los Angeles: California State University, 1978.

Topper, M. D. Drinking as an expression of status: Navajo male adolescents. In J. O. Waddell & M. Everett (Eds.), *Drinking among Southwestern Indians: An anthropological perspective.* Tucson: University of Arizona Press, 1980.

Topper, M. D. *The effects of economic change on Navajo family psychology.* Window Rock, AZ: Navajo Area Indian Health Service, 1981.

Topper, M. D. The etiology and treatment of psychological problems caused by delayed adolescent separation–individuation among the Navajo. In *The listening post.* Albuquerque: Indian Health Service, 1983.

White, L. A. *The evolution of culture.* New York: McGraw-Hill, 1959.

Witherspoon, G. *Navajo kinship and marriage.* Chicago: University of Chicago Press, 1975.

V

Case Studies
Spanish-Speaking Populations

14

Mexican-Americans in California
Intracultural Variation in Attitudes and Behavior Related to Alcohol

M. JEAN GILBERT

There is no *one* set of beliefs, norms, and behaviors associated with the use of alcohol among Mexican-Americans in California. To understand why this is so, it is necessary to consider some of the characteristics of the Mexican-American population that makes up nearly one fifth of the state's peoples. This group includes an estimated 1 to 3 million resident aliens, as well as many times that number who are second-, third-, and fourth-generation Americans. Successive waves of heavy immigration from Mexico, beginning in about 1910 and continuing to the present, have brought hundreds of thousands of persons from all parts of rural and urban Mexico to California. And in the last two decades, as second- and third-generation Mexican-Americans have reached maturity, considerable out-marriage has taken place, resulting in families of mixed heritage.

For the most part, therefore, the state's Mexican-Americans are at least bicultural; that is, they most share to a greater or lesser degree,

M. JEAN GILBERT • Spanish Speaking Mental Health Research Center, University of California, Los Angeles, California, 90024. Support for the research on which this chapter is based was given through grants from the National Institute of Education, the National Institute of Mental Health, the California Office of Drug and Alcohol Programs, and the National Institute of Alcohol and Alcohol Abuse.

depending on class, generation, occupation, and geographic region, both mainstream American and Mexican culture. The complexity of the situation is captured in the phrase "to a greater or lesser degree": The retailer in Santa Barbara whose forefathers predated the Anglo immigrants in California and whose mother and wife are Anglo is embedded in an entirely different cultural complex than that of a second-generation Mexican-American leadman in a Saticoy lemon-packing shed whose wife, born in Mexico, but brought to the United States at age 10, is a school aide. Both have, however, to a greater or lesser degree some cultural commonality with the newly immigrated, undocumented factory worker living with his cousins in the *barrio* of East Los Angeles. However, there is much that these men do not share. And it is probable that they do not share a set of beliefs and attitudes toward the use of alcohol.

As has been stated, no doubt in other places in this volume, a cultural group's rules prescribing who may drink what, where, and how often are linked to other norms regulating status and social and sex roles, as well as their corresponding rights and obligations. In a situation in which many persons participate in varying degrees in two cultures and in which the norms and behaviors relevant to each culture also vary with respect to class and other structural factors, the complexities are very great indeed.

The goal here, therefore, will be to describe alcohol-related beliefs and behaviors that appear to be widely shared by different groups within California's Mexican-American population, but also and most importantly, to demonstrate how factors associated with the intermingling of culture and class modify these configurations. I have chosen to focus on three separate but interrelated aspects of the Mexican-American alcohol landscape around which a majority of the alcohol-related norms coalesce. First, I will be concerned with the social context of Mexican-American drinking: What is considered normal or nondeviant drinking and drinking behavior for whom, where, how much, and when. Next, I will discuss the variation across Mexican-Americans with respect to the operation of alcohol norms related to sex roles; what is presently occurring among Mexican-Americans in the arena of alcohol beliefs and behaviors linked to gender offers a fascinating microcosm in which to observe the subtle effects of acculturation on sex roles. Following this, I will examine Mexican-American family structure and interactional patterns as they related to the abuse of alcohol.

The interview and observational data upon which the discussion is based came from several sources: a cross-cultural, ethnographic study of alcohol-involved and non-alcohol-involved couples currently in progress (Gilbert, 1981b); a three-region (East Los Angeles, rural Fresno County,

San José) study of drinking practices and beliefs of Spanish-speaking Californians, which combined ethnographic and survey methods (Alcocer & Gilbert, 1979); and an ethnographic study of Mexican-American familial interaction and exchange, which, although not designed for that purpose, yielded rich data on relationships among families and alcohol-involved family members (Gilbert, 1980).

THE SOCIAL CONTEXT OF ALCOHOL USE

Interview and observational data both conclusively show that a majority of Mexican-Americans view drinking as appropriate behavior when celebrating a happy or important life event, socializing with friends and family, or relaxing and taking part in recreational activities. Although the specific kinds of activities that fall into these categories may vary across class and generational groups, it is clear that, for Mexican-Americans, drinking is closely associated with "time out" and social interaction for enjoyment and is positively valued in these contexts. However, it is equally clear, as will be demonstrated, that different sets of sanctions operate to modify alcohol intake and behavior in differing contexts of socializing and relaxing.

Three-quarters of the 608 Mexican-American men and women interviewed in three widely separated California locales—East Los Angeles, San José, and rural Fresno County—reported celebrating as the most important reason for drinking (Alcocer & Gilbert, 1979). Such family-oriented ritual events as weddings, birthdays, and baptisms are occasions that are typical of such ritual celebrations. Most Mexican-Americans in California are Catholic, and many, particularly those who are Mexican-born, celebrate religious events with parties to which family and friends are invited and at which liquor is served. It is perhaps this aspect of Mexican-American drinking that led one Los Angeleño to comment, while describing a wedding party, "drinking is part of our way of life," or the remark by a Santa Barbara informant (born in Mexico) that "a *bautismo* [baptism] is just an excuse for a *borrachera*" (drinking spree). Among many immigrants from Mexico and some more traditional Mexican-Americans, it is incumbent upon the *padrino* (godfather) to supply liquor for the post-ceremonial party or dinner honoring a new baby's entrance into the religious life of the Church, thus cementing the *compadrazgo* (fictive kin) relationship between the *padrino* and the parents of the newborn. Another ritual event calling for celebration is a couple's re-enactment of their marital vows upon the occasion of the 25th or 50th wedding anniversary. It is customary for male members of the wedding party to supply the alcoholic beverage. In the two ritual celebrations described above, the provision of alcohol is a normatively prescribed aspect of a specific participant's role.

Among second- and later-generation Mexican-Americans, not unlike Anglo-Americans, celebrating often takes on a more informal, secular cast, augmenting or supplanting occasions that are associated with a religious ritual: to properly launch a new boat, "warm" a recently purchased house, mark a job promotion, or honor a college graduate. At large gatherings, usually in a hired hall, beer and jug wines are most often offered and frequently a smaller supply of tequila is available for those (usually men) who prefer stronger spirits. At smaller home gatherings, especially those of middle-class Mexican-Americans, California wines, champagne, mixed drinks, and blended whiskeys are served. At celebrations of almost every sort, ritual or secular, there is a wide variety and great abundance of food. Thus, the occasion is viewed as one of conviviality, rather than as a drinking occasion. Moreover, celebrations typically include persons of all ages, although children and youth in their early to mid-teens are not usually offered alcoholic beverages. Light drinking (one or two drinks) on the part of women is considered appropriate; men may drink more heavily and may gather in smaller drinking groups toward the end of the evening.

The presence of women and children, as well as a wide range of relatives and friends, often operates to limit the repertoire of alcohol-related behaviors allowable in such a context. When an individual is seen to be exceeding or about to exceed such allowable limits as indicated by his or her behavior, family members and (less often) friends will attempt to curb alcohol intake. A brother may gently caution his sister to "ease up on the beer for awhile, Angela," as was the case at a San Jose baptism where a woman's laughter became raucous and frequent. A young, middle-class Mexican-American hostess's intoxicated and wildly dancing father was the target of good-humored gibes at a house-warming—until the man's wife and son urged him off the dance floor and homeward amidst the disparaging remarks of his embarrassed daughter. Thus to a large extent, "drinking to celebrate" is most often done in a context in which setting and attitudes force rather narrow limits on consumption and drinking comportment. It is likely that the wide approval or acceptance of celebratory drinking by Mexican-Americans of varying generation and class is indicative of an underlying assumption that such drinking will be monitored by the use of the kinds of social controls described here.

In California where the climate is conducive to out-of-doors recreation, a substantial amount of drinking takes place in public. Light drinking (one or two glasses) of beer, and much less often, wine, is practiced by Mexican-Americans in a variety of public settings—*mercados* (marketplaces), public *fiestas, tardeadas* (afternoon dances), swap meets, beach picnics, backyard barbeques, and soccer games—all easily observable instances in which incidental drinking accompanies other activities.

Heavier drinking is often sanctioned in settings more private in nature and with only adults present. Typical primarily, though not exclusively, of U. S.-born Mexican-Americans, are informal couple parties that are age-homogeneous. These may be comprised of neighbors, work associates, friends, or fellow club members. The occasion may be a Superbowl Sunday, a minor holiday, the successful completion of a church fundraising event, or just a desire to socialize with friends. Food is plentiful and accompanies such mixed drinks as margaritas as well as the (nearly) ubiquitous beer. The cocktail party familiar to middle-class Anglo-Amerians and the cocktail hour preceding a dinner party is typical only of U. S.-born, middle-class Mexican-Americans and in fact may include as many Anglos as Mexican-Americans.

Despite the differences in occasion and setting across Mexican-American subgroups, social occasions and celebrations are the most widely accepted settings for nondeviant drinking among all Mexican-Americans. Interview data bear out these observations. A majority of the respondents in the California Spanish-speaking study (Alcocer & Gilbert, 1979), both men and women, Mexican-born and U.S.-born, made clear that light drinking, to be sociable at a party or when relaxing and seeking recreation with friends, is acceptable and appropriate behavior. Although few people interviewed approved of drinking to the point of drunkenness in such settings, about one third of all respondents thought it all right to get high. The basic acceptance of alcohol, "just for sociability," as one Chicano termed it, is reflected in the kinds of behavior proscribed or tolerated in social gatherings such as those described above.

If alcohol ceases to produce amicable social interaction and is perceived to be the cause of behavior that results in conflict among people, moves are made to reduce the alcohol intake of offending individuals, usually by encouraging them to leave. Boisterousness, for example, is frequently tolerated (not without humorous comment), but hostility and aggressiveness are quickly squelched. A particular type of behavior occurring in mixed-sex settings, inappropriate sexual overtures, often is perceived as the result of an individual having "had a couple too many," as one person charcterized this situation. At a work-related Halloween party in a private home in San Jose, for example, a group of wives and dates felt they finally had to "talk to" another female guest after watching her "throw herself around at the men" for the better part of the evening:

The ensuing dialogue resulted in the woman leaving with her date. And again, at a private party in the Montebello suburb of Los Angeles, when a young wife was the recipient of the aggressive attentions of a male guest, both the husband of the woman and the wife of the overly ardent man came in from the patio and put a stop to the behavior.

More widely divergent sets of norms related to alcohol consumption and drinking behaviors are applied by different groups of Mexican-Americans, however, in commercial settings strictly oriented to the consumption of alcohol: bars, *cantinas*, cocktail lounges, and nightclubs. Specific drinking establishments often draw their clientele solely from one segment of the Mexican-American population, and it is in such settings that clear intra-ethnic differences in the organization of alcohol-related activities are observable. A description of several of these drinking environments will allow the reader to understand these differences.

In rural Fresno, for example, where the seasonal needs of the vast agri-business of the San Joaquin Valley draw thousands of migrant workers annually, the small towns are filled with *cantinas*, which cater to the needs of male migrant workers, many of whom are single or who travel without their families. A large number of the *campesinos* are undocumented workers from Mexico, although many are also native-born Californians or migrants from Texas. The *cantinas* sell beer as their major beverage and feature *norteño* (northern Mexican) music on their jukeboxes. Single women, themselves often migrants, serve as barmaids in these *cantinas;* unescorted women from Fresno or the local area join the men in the evening for dancing and drinking. Unlike the bars in the region that do not cater mainly to migrant farmworkers, few couples frequent these establishments together. Men arrive singly or in groups. Very heavy drinking is not discouraged or negatively sanctioned, and overt sexual overtures toward women employees are tolerated and sometimes encouraged. It is not unusual for patrons to drink to the point where their speech becomes slurred and their walk unsteady or for individuals to occasionally pass out—companions who are not far from these conditions themselves will see that incapacitated persons are looked after and taken home. Patrons are encouraged to buy drinks for themselves and for women in these settings. Interaction between barmaids, bartenders, and patrons is frequent, informal, and expected.

Clearly, such establishments are almost exclusively male-oriented and serve the drinking needs of a very limited segment of California's Mexican-American population. Let us look for a moment at another kind of male-oriented commerical drinking setting observed in urban San Jose and Los Angeles, which serves a different, more acculturated segment of Mexican-American men—for want of a more appropriate name, I will call it the "wet T-shirt" bar. This type of bar serves an ethnically mixed clientele of younger Chicano, U.S.-born Mexican-American, and Anglo men. Beer is the primary beverage consumed and informal entertainment, often with a Master of Ceremonies, is offered. The major focus of the evening is on heavy drinking and watching female contestants in a "wet T-shirt" contest. Drinking contests are held among the patrons with pitchers of beer awarded the winners. Onlookers shout obscene comments in English and Spanish focused on the anatomical attributes of the contestants. The MC may admonish the clientele to "Drink it up to keep it up" and "Stay in a party mood!" References to the relationship between drinking and sexual activities are frequent, and it is obvious that such associations are shared by Anglo and Mexican-American men alike, although it is noteworthy that the activities are always officially conducted in English.

In the two kinds of male-oriented drinking settings described above, both those patronized by Spanish-speaking farmworkers and those with a clientele of acculturated Hispanic and Anglo populations, permissive norms governing drinking comportment are operative. About the only restricted behavior is physical aggression—if fighting breaks out among the patrons, the disputants are asked to leave or, if necessary, are forcefully removed from the premises. And it should be clear that the strong associations between sexual permissiveness and drinking operate cross-culturally among Anglo and Mexican-American men, as well as across subgroups of the Mexican male population. Such activities and norms contrast sharply with those observed in the three kinds of drinking settings described below.

Scattered along commercial thoroughfares and in shopping centers throughout urban neighborhoods populated primarily by working-class *Mexicanos* and Mexican-Americans are multitudes of small beer bars. Other bars are located in industrial areas, drawing their patronage almost exclusively from the work force of a single industry: the "cannery bars" in San Jose are a typical instance. Here male groups gather in the late afternoon after a day's work for what might best be termed "respite" drinking—drinking as an earned "time-out" after a day's work. A single observation in Los Angeles typifies the activities seen at many of these bars. Groups of two or three men began arriving at the neighborhood bar near a furniture manufacturing plant at about 5:00 P.M. Pitchers of beer termed "fringe benefits" were served at the two tables available, each round bought by a different person. Men on barstools called out comments in Spanish to drinkers seated across the room, and the Spanish-speaking barmaids joked with the drinkers. The woman bartender laughingly issued warnings when drinkers ordered additional pitchers, telling the men not to "get wasted" as they had the day before. Several of the groups took turns playing pool, betting pitchers of beer. As the late afternoon progressed, most of the patrons left and a core group of drinkers remained. One man attempted to dance with a barmaid, but she retired behind the bar telling him he was a *borracho* and couldn't dance, the rest of the men jokingly encouraged him to try again. His stumbling walk and inadvertent upset of a chair were met with laughter and jokes by other drinkers and the women employees. No negative sanctions were attached to this behavior by anyone present. It was typical of this kind of bar, however, that *most* men did not appear intoxicated (slurred or loud speech, unsteady gait or stupor), although drinking was steady for one and one-half to two hours for several groups. The easy interaction indicated that the groups were acquainted with each other and with the employees, which implied a regular patronage. At such a bar, anything but good-natured or joking "passes" at female employees, such as the one described above, are not usually condoned, and a patron is asked to leave if his persistence becomes offensive or takes a serious cast. This happens infrequently because a Mexican-American man does not wish to lose his dignity before his co-workers and other regular patrons he may know well and who may also know his family and friends.

The three settings described above are all primarily settings for male drinking, with women incorporated into the drinking environments in very different ways: as employees or as the object of overt sexual focus. Cocktail lounges and nightclubs, usually patronized by U.S.-born Mexican-Americans and most typical of urban areas, offer a contrast to these primarily male-oriented settings. Cocktail lounges, often attached to large motels or dinner houses, attract both men and women and are frequented by white-collar or professional Mexican-Americans who come to these settings in same-sex pairs or groups, as couples, or in mixed-sex groups after a day's work or possibly later in the evening to listen to music, drink, and converse. Mixed drinks or wine are the usual beverages served, and the drinks are more expensive than in most other settings. The crowds here are very mixed as to ethnicity, and Mexican-Americans may be in ethnically homogeneous groups or in groups incorporating Anglos. Interaction between patrons at different tables or between groups seated at tables and persons at the bar, save for an occasional greeting, is virtually nonexistent, but groups may talk and laugh animatedly among themselves. The interaction between customers and employees is restricted to the issuing and taking of orders—that is, strictly instrumental. Rarely occurring boisterous, loud behavior, arguing, or verbal overtures to women patrons all bring negative response—icy stares, turned backs, and the like. For example, in a cocktail lounge in Monte Vista, a suburb of San Jose, Mexican-American and Anglo patrons alike briefly interrupted their conversations to gaze at two men, business associates from the tenor of the conversation, arguing heatedly at the bar. A third businessman took the most vociferous disputant to a table, attempted in a quiet voice to calm him, and at the suggestion of the bartender, ordered coffee.

A variety of different nightclubs are frequented by Mexican-Americans, from the very sophisticated Hollywood dance clubs catering to elite crowds of movie and TV industry people to the clubs in East Los Angeles that feature contests among *mariachi* singers that draw only *Mexicano* couples. However the most typical urban or suburban nightspot is the club that offers recorded or live disco, *salsa*, or rock music. Here the crowd is comprised of young *Mexicano* and U.S.-born Mexican-American, Anglo, and black men and U.S.-born Mexican-American, Anglo, and black women. Few *Mexicanas*, that is, Mexican-born women, frequent these clubs. Most patrons are young working-class men and women, white-collar clerical workers, and students. Dance partners are chosen across ethnic and racial lines.

The primary focus in these clubs is on dancing, meeting people of the opposite sex, and drinking. Young women and men arrive in couples or groups at around ten in the evening to "check out the action," that is, to see what kind of crowd the club has drawn that night. If, after a drink or two, the atmosphere is pronounced sleazy or boring, a group may move on to another club, hitting two or three by "last call" at 1:45 A.M. Mixed drinks are most frequently ordered at these spots by Mexican-Americans of both sexes, with more women than men tending to ask for fruit drinks. Drinking is usually heavier among Mexican-American men, three to seven drinks per evening, than among their female counterparts, two to three drinks. Many clubs serve weak, heavily iced drinks, both because they are appreciated by the dancers returning from hot, crowded dance floors and because the club's margin of profit is thereby increased. In these nightclubs, poly-drug use is not at all uncommon, with men and women of all ethnic groups retiring to the ladies' or men's room or to a parked car for a toke of marijuana or a snort of cocaine.

Since the major objective of men and women coming to such a club is to meet eligibles, overtures in the form of requests to dance or the purchase of a drink are not censured. Individual women fend for themselves in terms of selectively encouraging or discouraging would-be partners, and men are frequently ignored or rudely put down by women if they cannot dance well because they are intoxicated. Mexican-American women rarely get so intoxicated they cannot dance, but if this does happen a female companion will sit with the woman at a table, accompany her to the restroom, or take her home. Men will either ignore women who are intoxicated or press an advantage; most of the time when the latter occurs, friends move in protectively in the manner described above. Fights among men occur most often outside these establishments, since when fighting begins the disputants are quickly "bounced."

In the cocktail lounges and nightclubs just described, the drinking and behavioral norms followed by Mexican-American patrons cannot be said to be specifically Mexican-American. The attitudes governing what is permissible in cocktail lounges might best be characterized as class-related, as these comparatively heavy constraints on drinking and drinking behavior are typical of what might be expected in commercial drinking establishments catering to a middle-class clientele in most regions of the United States and Mexico. In the dance clubs, the mixing of young people from a wide range of class and ethnic backgrounds also results in another sort of normative homogenization, more permissive but brought to bear through peer pressure and considerations of personal safety.

In general, drinking by Mexican-Americans at commercial settings such as the several described above is heavier than in noncommercial drinking settings, such as homes, halls, and parks. I observed that when a group left a home-based celebration to continue the party at a bar or club, heavier drinking began. Some self-selection is no doubt taking place here, that is, the heavier drinkers want to continue the party, the others want to go home. Indeed, in the study mentioned earlier, in which over 600 Mexican-Americans were interviewed (Alcocer & Gilbert, 1979), more people considered it appropriate to drink heavily ("drunk" or "high but not drunk") at a bar than at any other place: 30% of the women interviewed thought it appropriate to get high, but only 7% thought it was all right to get drunk in such a setting; 35% of the men thought high was all right and 12% did not balk at drunk.

What is clear from this examination of the social contexts in which non-deviant drinking by Mexican-Americans takes place is that the norms that guide Mexican-American drinking behavior are in fact very different from situation to situation, depending on who is present in the drinking setting (age range, sex, class, and ethnicity of participants) and the type of activities in which the drinkers are participating. Thus, what might be considered inappropriate or deviant quantities of alcohol consumed or inappropriate behaviors in one setting among a particular group defined by age, sex, ethnicity, and nativity may not be so considered in another. If a Mexican-American confines all of his or her drinking activities to middle-class cocktail lounges or family get-togethers, for example, that person would most likely regard the drink-related activities occurring in beer bars or *cantinas* as inappropriate drinking behavior.

ALCOHOL-RELATED NORMS AND ATTITUDES RELATED TO SEX ROLES

In virtually every study focused on Mexican-American drinking patterns inside and outside of California, men's drinking habits have been found to be very different from women's (Cahalen, Roizan, & Room, 1976; Johnson & Matre, 1978; Paine, 1977; Maril & Zavaleta, 1978; Trotter, 1982). Overall Mexican-American women are much more likely to be abstainers, to engage only in very light social drinking in restricted environments, and to report fewer problems related to alcohol than Mexican-American men. Although this sex difference also exists in the larger population, research suggests that sex role differences are clearer among Mexican-Americans. However, there are also indications that the sex gap for Mexican-Americans is narrowing. Many alcoholism service providers I interviewed noted an increasing consumption of alcohol by Mexican-American women; in the words of one Mexican-American counselor, "women of Mexican heritage aren't supposed to drink, but they are catching up to men fast these days." A housewife in a small Southern California town put it cogently in a culturally specific context: "It used to be that a woman would ask her *comadre* in for a cup of chocolate—now she asks her in for a beer!"

In the Alcocer and Gilbert study of drinking practices (1979), 19% of the women, but just 13% of the men claimed to be abstainers, 34% of the women and 52% of the men judged themselves to be light drinkers (one to two drinks on occasion), and 42% of the men and 29% of the women, social drinkers (three to four drinks). In another California survey (Cahalen *et al.*, 1976), 20% of the Mexican-American women and 12% of the men reported being abstainers. These figures therefore indicate that, although most Mexican-American women in California are not teetotalers, their drinking habits differ significantly from men's. What then are the beliefs and attitudes that underlie these differences and, if the differences are lessening, what factors are associated with the changes?

Answers to these questions become evident when one seeks to understand the different ways in which drinking is integrated into the lives of Mexican-American men and women. Earlier in this chapter, in the description of neighborhood beer bars, I referred to a type of drinking among Mexican-American men that I characterized as "respite drinking," that is to say, drinking as a respite from labor or after a hard day's work. The connection between a man's right to drink in this manner and his obligation to support himself or others was made so consistently in interviews and during setting observations that this relationship cannot be ignored. *"Yo trabajo, tengo derecho de tomar"* (I work, I have the right to drink), asserted one respondent, and another, *"Yo soy el hombre de la casa, si quiero tomar, tomo cuando me de la gana"* (I am the man of the house, and if I want to drink, I drink when I feel like it). Interesting in this connection is an answer given frequently by Mexican-Americans, particularly immigrants and working-class U. S.-born Mexican-Americans, when asked to give the age at which drinking may be appropriately begun by a man: "When he is old enough to earn a living, he's old enough to drink." Thus drinking is a right seen to be earned by masculine self-sufficiency and assumption of the provider role. In line with this association is a corollary perception of deviant or problem drinking occurring when drinking is seen to interfere with the performance of the provider role or the breakdown of self-sufficiency. When a man can continue to provide, despite heavy drinking and the disruption of family relations, an alcohol problem may not be recognized. The remarks of a young Mexican-American professional cast light on this circumstance, "I guess my father would have been considered an alcoholic, but no one thought of it that way," he mused at the end of an interview, "he drank at bars almost every night and passed out when he came home, but he never missed a day's work and met his family responsibilities in that way, so no one did anything about it."

Respite drinking takes place at home or at a bar following the workday, or it may be part of a lunch break. Fifty-nine percent of the men interviewed in one survey (Alcocer & Gilbert, 1979) but just 25% of the women reported that drinking during lunch break was appropriate behavior, although few thought that more than one or two drinks was acceptable at this time. It is a frequent practice for fieldworkers to be provided with beer from trucks brought to the field or for cases of beer to be available when it is desirable for workers to put in overtime. The remarks of one fieldworker shed light on this practice: "Beer is something you do for relaxation, so if we have beer in the fields, that makes us feel that maybe we aren't working so hard or that we aren't just working. I guess that doesn't make too much sense, does it?" It certainly does make sense, if alcohol is regarded as a reward for labor or as a "time-out" a man earns through the sweat of his brow, as clearly appears to be the case among much of the working-class Mexican-American population.

In the bars in which men gather after work, there are rarely women patrons. Additionally, as revealed in the Alcocer and Gilbert study (1979), fewer women (42%) than men (54%) consider it appropriate to relax at a bar following a workday, and whereas 81% of the men interviewed felt that it was acceptable for a man to drink at home after work, 71% of the women felt that it was appropriate to do so. However, as Mexican-American women enter the work force in greater numbers and in white-collar positions, this difference in attitude may be expected to decrease. Certainly, the growing number of middle-class Mexican-American women observed in cocktail lounges at the Happy Hour suggests that at least some women may be appropriating this once-male prerogative, although their numbers are still comparatively few. More typical than barroom respite drinking perhaps is the practice of several working female house-hold heads I observed. These women would typically stop off after work at a neighborhood convenience store for a six-pack, then consume one or two beers while preparing supper.

The descriptions of drinking settings in the earlier part of the chapter illustrate another important sex difference for Mexican-American men. Drinking figures importantly in same-sex social contexts as well as in contexts involving male/female interaction; for women it does not. Women tend to restrict their drinking to such situations as family- or home-based parties where strict norms operate to control drinking behavior. They also drink in settings where dancing and interaction with men are the major activities—although they may attend nightclubs in same-sex groups, interaction among women is not central. This same sex-related difference, consistent across all but the most acculturated groups of women, has been reported for Mexican-Americans outside of California; for example, Trotter (1982) notes it in a Texas study.

Drinking together has a different *meaning* for Mexican-American men than for Mexican-American women: it is a significant bonding mechanism, intensifying relationships and underscoring a shared masculinity. This sex-related difference was brought home to me in a most vivid way in two quite unrelated incidents. While riding on a bus in a California city, I eavesdropped on a conversation between two Mexican-American male passengers, strangers to each other until fate placed them in adjacent seats on a crowded bus. After a mumbled greeting and a few desultory remarks, one volunteered to the other the he had a bad headache and that the bumpy bus ride was not doing it any good. *"Crudo?"* smiled the other, using the slang word for hangover. The first ruefully returned his smile, nodding mock-mournfully. Both laughed and embarked on a discussion of appropriate cures for a hangover, finally mutually pronouncing a popular restaurant's *menudo* (tripe soup) as being the best medicine for this malady. As the passenger without a hangover prepared to disembark, he warmly clapped the other's shoulder and wished him better luck—next time. Both laughed again, and this brief episode of male communion ended. What had been exchanged in this small time was, I believe, a sense of shared masculine weakness and pleasure.

The second incident involved an upwardly mobile Chicano professional who, having acquired a graduate education and an Anglo wife, found himself in a social and occupational world isolated and very different from the *barrio* in which he had spent his young manhood. His complaint was that he desperately missed the intense relationships with male drinking companions. "When I try to get together for a drink with my Anglo friends," he related with amazement, "they want to bring their wives!" Drinking together, then, is an important aspect of Mexican-American male interaction, whether it be with a group of co-workers, neighbors, or friends or with a single long-term companion.

For acculturating and middle-class Mexican-Americans, the increasing participation of women in the work force and the more frequent inclusion of women in public drinking settings is having a distinct effect on norms related to drinking behavior for both men and women. Changes in the content of women's roles, with wives taking over some of the economic responsibilities of the family, often result in increased assertiveness in their attempts to control their spouses' drinking behaviors. "It used to be that a man put the bread and butter on the *mesa*," said one Chicano alcoholism-service provider. "Now the man brings home the bread and the woman brings home the butter, and she won't put up with his doing as he pleases." It is worth noting, however, that this apparently increasing "right to intervene" is, like the man's "right to drink," linked in the minds of many to the distribution of economic power within the family.

Additionally, there is a positive correlation between women's participation in public drinking activities and acculturation, and this greater participation of women, most especially escorted women, is accompanied by systematic changes in the drinking comportment of men. Where women are not escorted by husbands or dates, fighting among men occurs much more frequently, sexual overtures to women present are not uncommon, and boisterous or stuporous behavior is also tolerated or ignored. None of these behaviors is permitted for long in places where escorted women are drinking. And in both public and private drinking settings where Mexican-American wives or dates participate in the drinking, women, with the clear approval of others, often actively attempt to curb the quantity of liquor consumed by their escorts. Such inhibiting interaction is rarely observed in settings dominated by unaccompanied men—either because no women are around to act as inhibitors or because those women present are not seen to have the right to curb men's behavior. The irony may therefore be that acculturation operates, indirectly, to limit consumption among Mexican-American *men*, but is accompanied by the erosion of cultural norms restricting Mexican-American *women's* participating in drinking activities.

THE DISRUPTION OF KINSHIP TIES

As Ablon has noted (1971), Hill (1958) aptly described alcoholism as an intra-familial stressful event because it destroys family role patterns and relationships. However, family roles and relationships are interpreted differently across ethnic groups; thus, the meaning and effect of alcohol-related behavior in the family context may differ across groups as well. The questions here, then, are How do problems with alcohol cause demoralization within *Mexican-American* families? What are the situations or events that signal family disorganization, and how are they interpreted and acted upon by family members? To begin to answer these questions, let us look first at the ways alcohol abuse affects the functioning of the close kinship networks in which the nuclear family is embedded.

My first insight into the disorganizing effects of alcohol abuse on Mexican-American kinship relations came during research focused on second-generation extended family interaction and exchange networks (Gilbert, 1978). In the course of talking with 119 adults in both rural and urban settings about their exchange and interaction with extended family members, I was struck by the frequency with which informants cited the alcohol involvement of a relative as a causal factor in the attenuation of breaking off of a kinship bond. To clarify the nature of this disruptive process, it is important to understand some of the norms that govern Mexican-American kin ties. First, a norm of genealogical closeness is clearly operative in the groups of California Mexican-Americans whose extended kinship relations have been carefully scrutinized (Gilbert, 1978; Keefe, 1979; Keefe, Padilla, & Carlos, 1979). This appears to be true among both immigrant and native-born Mexican-Americans of both the middle- and working-classes and has been noted for Mexican-Americans in locales other than California, such as the Detroit Mexican-Americans studied by Ramirez (1980). That is to say, the strongest and most frequent interaction and exchange takes place within the immediate family: parents, siblings, and offspring. Although certain individuals may maintain strong interactive and exchange relations with one or two *local* kinsmen outside this close group, the vast majority of mutual aid and interaction takes place with its confines. Second, the strongest and most durable kin bonds, even within the close kin group, are those characterized by a high degree of reciprocity, that is, a relatively equal exchange of goods, services, and interpersonal interaction over time. When a given kinsman fails to hold up his end of the relationship, reciprocity breaks down, and social distancing frequently occurs.

Discussions of the breakdown of relationships with alcohol-involved kinsmen made it clear that affective and instrumental bonds between more distantly related kin were rather easily eroded as a result of alcohol-related behavior. When, for example, a cousin's problems with alcohol resulted in repeated and nonreciprocated requests for loans of money, transportation, and intervention in marital difficulties, one small-town woman flatly described her feelings: "We no longer allow him to set foot in this house! Let his brothers take care of him, if they want to."

Brothers will, and sisters, too, expend a tremendous amount of time, energy, and compassion in dealing with an alcoholic sibling. I learned of several instances in which a sibling with an apparent alcohol dependency, alienated from his or her spouse and children, was taken into a brother's or sister's home, nurtured and "preached at," as one informant put it, with the reported result that the abusive drinking ceased. The length of time and amount of resources a sibling will expend on such a rescue operation varies with the closeness of the sibling bond, the amount of strain placed on other emotional commitments, and the economic resources. One man, a retail sales manager in an urban center with a close and long-term mutually supportive relationship with an older brother, recounted his conflicting loyalties upon turning the now-alcoholic brother out of his house: "My brother always gave me a hand when I needed it, and I always helped him in return. It is hard now for me to turn my back on him, but I have to do it or my children will suffer—maybe my wife will leave me." Individuals or families with scant resources of their own often find it especially difficult to sustain nonreciprocal kin ties, even with close kin. In speaking of the breach between herself and her heavy drinking sisters who lived in a nearby town, one welfare-supported single mother made these remarks: "My sisters embarrass me, and they don't help me even when they could; they don't help my mother either. They are no good to anyone, so why should I help them?" Thus, the long-term and progressive nature of alcohol involvement, with its erosion of an individual's emotional, social, and economic resources, cripples his or her ability to sustain a mutually supportive kinship bond.

These processes operate in a similar manner to erode the kin relations of alcohol-involved members of the larger society, but the attenuation of kin links may have a more devastating impact on Mexican-American alcohol abusers and their families than on their Anglo counterparts. The greater importance of the extended family among Mexican-Americans has frequently been alluded to in the literature (Murillo, 1971; Moore with Cuellar, 1976; Mirandé & Enríquez, 1979), although this notion has been less frequently sustained by empirical, cross-cultural research. Nevertheless, the empirical studies that do exist suggest that although Mexican-Americans may not rely more on their kinsmen for instrumental and affective aid, they rely *more exclusively* on kinsmen than do Anglo-Americans. That is, although both Anglo-Americans and Mexican-Americans rely on their kinsmen for social support, services, and material aid, Anglo-Americans are much more likely to augment kinship support with assistance from friends, neighbors, and work associates (Keefe *et al.*, 1979; Gilbert, 1981a). In the light of these data, it is not difficult to understand that the breakdown of kin ties under the contraints and burdens imposed by problem drinking may represent for the alcohol-involved Mexican-American the disintegration of most or all of his social world. The most critical instances of this type of family demoralization can be observed in the extreme isolation of migrant and immigrant alcohol abusers and their families. Having left many of their closest kinsmen behind in Mexico or other regions of the United States and lacking integration into other social support, they are totally without interpersonal resources when the few family ties they may have locally available disintegrate.

It is important to point out however that the interpersonal networks of Mexican-American families also vary in important ways by generation and class. Native-born Mexican-Americans, especially those who have attained middle-class status, more frequently augment their kinship ties with mutually supportive friend, neighbor, and work-associate links than do immigrants (Gilbert, 1980, 1981a). These ties give them additional interpersonal support and oftentimes serve as sources of information on community resources, including alcoholism services.

In sum, although the moderate use of alcohol is favorably seen by Mexican-Americans as a social or even ritual device for lubricating and strengthening bonds between kinsmen in happy or celebratory contexts, the long-term results of abusive use of alcohol clearly contribute to the disintegration of familial ties.

DIFFERING MARITAL NORMS AND STRATEGIES FOR COPING WITH ALCOHOL-ABUSING SPOUSES

Perhaps no aspect of alcoholism and family life has received greater attention than the effects of abusive drinking on the marital dyad. The extraordinary attention devoted to alcoholism and spousal relations is no doubt in part due to the assumption made early-on by such theorists as Bacon (1945) who held that marriage is particularly vulnerable to the corrosive effects of alcohol-related behavior because such behavior directly impacts on the affective interpersonal relations that cement the union. Although this no doubt holds true to some extent for marriages in most cultures, this notion clearly reflects the marital norms of complex, Westernized societies in which romantic love and companionate and egalitarian role relations, as well as joint spousal activities, figure importantly in the criteria by which marital relations are evaluated. Scant attention has been focused on the way alcohol-related behavior patterns affect marriages in cultures in which the components of marital roles are differently emphasized. Mexican-American marriages in Southern California exhibit highly variable spousal role norms, ranging from those found in Mexican peasant culture to those similar to the configuration described above. The way in which drinking is managed in these diverse marriages offers insights into how differences in the interpretation of rights and obligations accruing to marital roles affect drinking patterns and shape spousal responses and strategies for dealing with an alcohol-abusing mate.

The discussion that follows centers upon the manner in which variance in marital norms affects *wives'* responses to their husbands' abusive drinking. Despite the growing consumption of alcohol by Mexican-American women, there are virtually no data on the effects of women's abusive drinking on Mexican-American marriages, and my own research has failed to produce substantive information on this issue—although I have numerous descriptions of divorced or separated women whose drinking patterns are considered problematic by relatives. Thus, it may be that, as in the larger population, married women's drinking is underreported due to a greater shame associated with their drinking (Gomberg, 1982), husbands are more likely to terminate marriages with alcoholic spouses than are wives (Paolino & McCrady, 1977), or divorcees are at higher risk for alcohol problems (U.S. National Institute of Alcohol and Alcohol Abuse, 1979).

Traditional Mexican marriages, particularly those in rural areas, emphasize sex-segregated activities and sexual division of labor. Males work and socialize in same-sex groups outside the home and have almost exclusive authority over the articulation of the household with the larger social world. Women, whose sphere of influence is confined primarily to the household, also socialize in same-sex groups, usually inside the home. Wives are frequently abstinent and, save for family ceremonial occasions, are rarely present when their husbands are consuming alcohol. The emphasis in these marriages is on role complementarity:

Dominating the husband role configuration is the provider function; dominating the wife role is the nurturing function.

Numerous interviews with Southern California Mexican-American wives whose marriages closely conform to the pattern just described yielded striking similarities in the ways women in such marriages interpret and cope with husbands who drink heavily. First, since their husbands' drinking is conducted primarily outside the home and within the male social context, heavy drinking is not necessarily seen as deviant, since drinking is viewed as simply part of what men do when they get together. Further, since the wives are not present at the majority of these drinking scenarios, they have little basis for comparing their spouses' consumption with that of other men. Most interesting, however, are the strategies used by these women to cope with their husbands' heavy drinking. The strategies center on keeping the husband's consumption of alcohol beyond the boundaries of the home rather than attempting to control the quantity of alcohol consumed or the occasions for drinking. Some of these women reported that although their husbands may come home intoxicated, they do not allow alcohol in the house nor do they allow their husbands' drinking companions to enter their homes. The daughter of one of these women commented that "my mom is very strong-minded—you might say that she 'wears the pants in the family where alcohol is concerned.' If my father's drinking buddies come around, she'll chase them out past the fence!" Another woman reported that, should her husband come home drunk and in possession of alcohol, she would "make him put it outside in the yard or on the windowsill," or would do so herself. Further, although these women will frequently absorb much verbal and sometimes physical abuse from their spouses, they hustle their children away from the view of their intoxicated mates and will not tolerate the children's talking against the fathers because, as one woman put it, "I can say anything I like, but he is their father, and he provides the roof over their heads so they must respect him." Because the home is the arena in which women have responsibility and a degree of authority confirmed by cultural norms, they may rightfully and acceptably act to control behavior within its confines.

Frequently, such marriages continue for years in this "containment" pattern and dissolve only when the husbands cease to carry out the provider function or opt to abandon so restrictive a domestic environment. Women in these marriages may enlist the aid of relatives in attempting to ameliorate the effects of their husband's drinking, but rarely seek help outside the kin circle.

Quite a different set of strategies is employed by Mexican-American wives in marriages wherein greater normative emphasis is placed on joint spousal activities and in which the character of the emotional/interactive bond between husband and wife figures more importantly than the provider/homemaker role configuration. These marriages demonstrate less strictly defined division of interactive spheres, with the wives assuming greater responsibility for relations between the household and the wider social system than in traditional marriages. These marriages are most typical of U.S.-born, middle-class Mexican-Americans. In such marriages, women do participate with their husbands in social activities, including drinking occasions other than those confined to kin-related activities. They are thus more easily able to assess the quantity of alcohol consumed by their spouses and to compare their behavior with others. Additionally, they may themselves have had sufficient personal experience with social drinking so as to better be able to assess the relationship between quantity of alcohol consumed and behavioral manifestations. These women may attempt to restrict their mates' consumption at social occasions by urging them homeward or commenting on their consumption. If these tactics are unsuccessful over time, wives may refuse to accompany their husbands to social events because, as one woman put it, "I was tired of the embarrassment of getting him into the car and dragging him home." Such a strategy does not necessarily curtail the husband's drinking, but it does tend to restrict such drinking to same-sex settings and it certainly narrows the wife's opportunities to participate in the couple-oriented, extra-familial activities that she values highly as part of her expectations of mutuality in the marital context, producing a different perception of the drinker's infringement upon her marital rights than occurs in a traditional marriage.

Further, because she herself may be employed, she may not be as dependent on her husband as provider. And because she relates to a wider social world, she is likely to discuss her feelings and circumstances with friends or work associates whose comments and information may serve to heighten her perception of the problematic nature of her husband's drinking. Because of this awareness she may seek aid from an outside helping source, although most probably she will first ask close kin to intervene. Should such tactics not produce results, she is more likely than her traditional counterpart to terminate the marriage herself.

The two polar marital configurations sketched above should not be construed as the only or even typical models of the perceptions, behaviors, and responses of Mexican-American wives to the drinking behaviors of their husbands. Rather, they illustrate the range of marital norms operating in Mexican-American marriages in Southern California and the behavioral possibilities resulting from the interaction of diverse marital norms and drinking behavior. The reader can imagine the complexity engendered in the immense variation that occurs between these polar models.

PERCEPTIONS REGARDING THE ETIOLOGY OF DEVIANT DRINKING

In general, within this diverse population, drinking is identified as deviant when it interferes with good social relations or the performance of role-related obligations, however those relations or obligations are defined. Among all but the most formally educated Mexican-Americans, alcohol problems are not viewed as part of a progressive disease syndrome with clearly defined symptoms and stages. The term *alcoholic* or *alcoholico,* if used, describes an individual whose drinking is causing negative social consequences for himself and for others. Although some individuals may see heavy drinking as contributing to poor health, alcohol abuse is not widely recognized as a medical problem. However, it is interesting to consider several reports of cases in which a heavy drinker's consumption was diagnosed by a doctor to be life-threatening: many of these individuals claimed or were reported by kin to have quit drinking altogether, "cold turkey." A knowledge of such cases may be one of the reasons why many Mexican-Americans view abusive drinking as a failure of will. Aside from this perception of problem drinking, the other commonly expressed idea is that deviant drinking results from emotional or situational stress related to life problems. Although many Hispanic service providers emphasize the stresses associated with acculturation as significant in the etiology of Mexican-American alcohol abuse (Alcocer & Gilbert, 1979), acculturation stress was rarely noted by informants in the general population who chose rather to emphasize family, financial, and job-related problems as casual factors. Difficulties in each of these specific areas may certainly be related to acculturation as it is manifested in changing norms and expectations, but such an encompassing and theoretical explanation is not easily grasped by persons who are wrestling with the complex expression of culture change in everyday relations and decisionmaking.

In recent years, alcoholism treatment modalities have shifted from an individual-only treatment concept to models emphasizing the individual in his entire social surrounding; family systems approaches and formal interventions reflect this newer perspective. The extreme variation in interactional and drinking environments of Mexican-Americans documented in this chapter should warn against either a stereotypic

traditional concept of Mexican-American drinking patterns or the imposition of therapeutic models based solely on Anglo patterns when
working with Mexican-American abusive drinkers and their families.

REFERENCES

Ablon, J. Family behavior and alcoholism. In M. W. Everett, J. O. Waddell, & D. B. Heath
    (Eds.), *Cross-cultural approaches to the study of alcohol*. The Hague, Holland: Mouton,
    1971.
Alcocer, A., & Gilbert, M. J. *Drinking practices and alcohol related problems of Spanish-speaking
    persons in three California locales*. Sacramento: California Office of Drug and Alcohol
    Programs, 1979.
Bacon, S. D. Excessive drinking and the institution of the family. In *Quarterly Journal of
    Studies on Alcohol* (Ed.), *Alcohol, Science and Society*. New Haven, CT: Summer School
    of Alcoholism Studies, 1945.
Cahalen, D., Roizen, R., & Room, R. Alcohol problems and their prevention: Public
    attitudes in California. In R. Room & S. Sheffield (Eds.), *The prevention of alcohol
    problems*. Sacramento: California Office of Alcoholism, 1976.
Gilbert, M. J. Extended family integration. In J. M. Casas & S. E. Keefe (Eds.), *Family and
    mental health in the Mexican-American community* (Monograph Number Seven). Los
    Angeles: University of California, Los Angeles, Spanish-speaking Meantal Health
    Research Center, 1978.
Gilbert, M. J. *Los parientes: Social structural factors and kinship relations among second generation
    Mexican Americans*. Doctoral dissertation, University of California, Santa Barbara, 1980.
    [*Dissertation Abstracts International*, 1981, *42*, 4959A-END. (University Microfilms No.
    DA8207651)]
Gilbert, M. J. *The transition to parenthood: Mexican American and Anglo American first time
    parents and their networks of support*. National Institute of Mental Health Grant No. MH
    31882, 1981a.
Gilbert, M. J. *Cultural determinants in alcoholism help seeking*. National Institute of Alcohol
    and Alcohol Abuse. Grant No. AA05172, 1981b.
Gomberg, E. L. Special populations. In E. L. Gomberg, H. R. White, & J. A. Carpenter
    (Eds.), *Alcohol, science and society revisited*. Ann Arbor: The University of Michigan
    Press, 1982.
Hill, R. Generic features of families under stress. *Social Casework*, 1958, *39*, 139–152.
Johnson, L., & Matre, M. Anomia and alcohol use: Drinking patterns in Mexican American
    and Anglo neighborhoods. *Quarterly Journal of Studies on Alcohol*, 1978, *39*, 894–902.
Keefe, E. Urbanization, acculturation, and extended family ties: Mexican Americans in
    cities. *American Ethnologist*, 1979, *6*, 349–362.
Keefe, S. E., Padilla, A., & Carlos, M. The Mexican American family as an emotional
    support system. *Human Organization*, 1979, *38*, 144–152.
Maril, R. L., & Zavaleta, A. N. *Applications of anthropological research on Mexican American
    drinking patterns in the lower Rio Grande Valley of South Texas*. Paper presented at the
    Annual Meetings of the Society for Applied Anthropology, Merida, Mexico, April
    1978.
Mirandé, A., & Enríquez, E. *La Chicana*. Chicago: University of Chicago Press, 1979.
Moore, J. with A. Cuellar. *Mexican Americans*. Englewood Cliffs, NJ: Prentice-Hall, 1970.

Murillo, N. The Mexican American family. In N. N. Wagner & M. J. Haug (Eds.), *Chicanos: A social and psychological perspective.* Saint Louis, MO: Mosby, 1971.

Paine, H. J. Attitudes and patterns of alcohol use among Mexican Americans. *Journal of Studies on Alcohol,* 1977, *38,* 544–562.

Paolino, T. J., & McCrady, B. S. *The alcoholic marriage: Alternative perspectives.* New York: Grune & Stratton, 1977.

Ramirez, O. Extended family phenomena and mental health among urban Mexican Americans (Monograph Number One). Reston, VA: Latino Institute, 1980.

Trotter, R. T. *The Mexican American experience with alcohol: South Texas examples.* Paper presented at the Annual Meetings of the American Anthropological Association, Washington, DC, December 1982.

U. S. National Institute of Alcohol Abuse and Alcoholism. *Alcohol and health; Third special report to the U. S. Congress from the Secretary of Health, Education and Welfare, June 1978.* E. P. Noble (Ed.) (DHEW Publ. No. HDM 79-832). Washington, DC: U. S. Government Printing Office, 1979.

15

Mexican-American Experience with Alcohol
South Texas Examples

ROBERT T. TROTTER, II

The Lower Rio Grande Valley of South Texas provides a fascinatingly complex research site for the anthropological investigation of alcohol-related behavior. The area is multi-ethnic, multi-cultural, and multilingual. It supports a complex mixture of urban and rural life-styles.

Three major cultural systems predominate in the Valley. These include the Mexican National cultural system, which is the dominant cultural system on the Mexico side of the border, south of the Rio Grande River. More precisely, it is the border expression of Mexican National culture and differs in some ways from the expression of that culture in the interior. The other two cultural systems, the Mexican-American and Anglo-American cultures, are on the United States side of the border. They too are influenced by the border dynamics of the region and differ from the non-border expressions of each system.

Each major cultural system has multiple divisions, which separate the ethnic and cultural groupings into more homogeneous belief and behavioral patterns. I call these groupings life-style subdivisions. These divisions are created by economic, occupational, linguistic, and educa-

ROBERT T. TROTTER II • Department of Psychology and Anthropology, Pan American University, Edinburg, Texas 78539.

tional patternings. They combine class with cultural ecological and linguistic variables to create a series of divisions that reflect more accurately the cultural reality than more commonly used social science divisions of class, caste, and socioeconomic status.

The existence of these divisions has both practical and theoretical implications for the alcohol field. Cultural differences are directly related to differences in the accepted levels of consumption of alcohol, type of behavior associated with drinking, differences in drunken comportment, and the whole area of alcohol consumption. Since all successful models of alcoholism treatment depend on communication and on the values and attitudes toward alcohol that are built into cultural and subcultural systems, these divisions are also important in setting up appropriate therapeutic systems for the individuals who follow these life-styles.

REGIONAL CHARACTERISTICS

The research for this chapter focused on the Mexican-American cultural system. Mexican-Americans make up approximately 80% of the more than one-half million people living in the Lower Rio Grande Valley. The remaining 20% are predominantly Anglo-American, with less than 0.5% of the population black, Native American, or Asian. The area contains the two poorest standard Metropolitan Statistical Areas (SMSAs) in the United States, based on *per capita* income. Over 50% of the resident households fall below the federal poverty level. The area is the home base for an estimated 180,000 Mexican-American migrant and seasonal farm workers. It is also a seasonal in-migration area for approximately 100,000 winter residents, most of whom are elderly Anglo-Americans from the midwestern United States, and Canada.

Mexican-Americans in the Valley can be heuristically divided into six life-style groups, based on a combination of income, cultural–ecological, and linguistic variables. Two groups—the migrants and the non-migrant poor—make up the poorest elements of the Valley life-style subdivisions. Both groups engage in seasonal and periodic employment, primarily in agriculturally related occupations, or as unskilled labor. Linguistically, both groups tend to be Spanish dominant and often have very limited or no English proficiency.

Nearly all the most recent Mexican National immigrants fall within these groups, especially among the non-migrant poor. The bulk of both groups are second-, third-, fourth-generation or more native U. S. citizens. Educational levels tend to be below fifth grade for older persons and below eighth grade for younger persons.

People and household units move back and forth between migrant and non-migrant groups on either a permanent or temporary basis. Both groups tend to live in *colonias* or in *barrios*. Colonias are small rural communities that are unincorporated, lack most utilities (e.g., potable water, drainage, trash disposal), but the land there is cheap enough so that a resident can own his own lot and build a house on it. The houses are often built in stages that reflect the economic success of the owner. Barrios are neighborhoods within the incorporated urban centers in the Valley, which are predominantly low-income areas, although some have achieved middle-class status due to the widespread success of barrio residents. In both one often sees homes that differ significantly from others in size and in quality of construction. This reflects a greater level of success for the household, compared with the others. It also reflects a cultural value of remaining in the colonia or barrio, rather than following the middle-class pattern of suburban flight for the upwardly mobile. Adherence to this value has important implications for the development of leadership and influence networks in the region.

The major differences between the two groups are due to cultural–ecological factors. The migrants engage in an annual migratory cycle that takes them away from the Valley for 3- to 9-month periods, whereas the non-migrants remain in the Valley doing seasonal labor. The agricultural specialization and the location of work tend to determine the length of time spent outside the Valley. However, after a short (1- to 3-year) adjustment period, most migrants follow a firmly established migratory cycle. They develop personal relationships with growers and return year after year to the same crops, farms, or areas. This establishes a life-style that is analogous to the migratory cycles of nomadic groups in other areas of the world. Included in this life-style is the clear belief that the lower Rio Grande Valley is home, a permanent residence from which they temporarily move periodically. In addition, groups working in specific "upstream" locations tend to be members of the same extended family and/or residence networks. Thus, many of the social patterns and networks that bind the migrants together when they are at home remain in force when they are upstream. These consistencies in life-style, networks, and life experiences create an identifiable social group. The non-migrant poor overlap somewhat in networks with the migrant, but not in out-of-Valley work experience or life-style.

The third group of Mexican-Americans in the Valley can be labeled the working class or the stable poor.[1] These are individuals whose earn-

---

[1] I am indebted to Dr. David Alvirez, Professor of Sociology, Pan American University, for pointing out the separate life-style subdivisions of this group in the Valley when I

ings fall within or just above poverty level incomes, but who have stable, permanent, year-round jobs. Their occupations include such semi-skilled jobs as janitors, hospital aids, and orderlies, some agricultural jobs (such as grove care personnel), and some jobs in the light manufacturing industries that have recently moved into the Valley. Linguistically, this group, like the former two, tends to be Spanish dominant, but generally has a somewhat greater proficiency in English. Educational levels are higher, with most having between an eighth grade and a high school graduate level of attainment. The key to these individuals' improved social and economic position within their communities is the steady nature of their employment, rather than their higher income levels. Many migrant families earn more during the season than do working-class families. The stable poor, however, have incomes that are relatively immune to seasonal fluctuation or weather conditions that can cause a disastrous year for both the migrant and non-migrant groups. This stability of income allows for better planning, establishment of credit (however minimal), and a permanent residence. Thus, their children can complete their schooling uninterrupted. This has implications for upward mobility. A Chicano scholar pointed out that of the three groups this is the only one in which individuals have realistic aspirations for upward social mobility in succeeding generations.[2] The Garcia family is one such family. The father works in a local *bodega,* or packing shed. All eleven of his children are completing or have completed high school; two are registered nurses, one is a licensed vocational nurse working on her R.N. degree, another is a dental assistant, and two are housewives; one child is still in high school, another in college, and so on. All are achieving, or have achieved, a middle-class life-style. This is primarily due to the stable employment of the father. As can be seen this group maintains a strong positive orientation to education and generally has more prestige, as a family unit, within the local social system than the other two groups, even though many of the households are found within the urban barrios. Few reside in the colonias.

The next subdivision is the middle class. This is perhaps the most heterogeneous of all the subdivisions within the community. Unlike other areas in the United States, the middle class in the Valley is difficult to designate upper or lower middle class because of an overall compression of the upper income groups, due to its recent, rapid growth. The recent genesis of this middle class does not permit subtle distinctions based on occupation, education, or residence. This group is linguistically

---

asked him to review this particular typology of Mexican American subgroupings.
[2] Dr. David Alvirez, personal communication, November 26, 1982.

heterogeneous, including households and individuals who are Spanish dominant, others who are completely bilingual, and still others who are English dominant, with extremely limited proficiency in Spanish. Education levels are generally at the high school graduate level and above, including a significantly growing number of individuals with advanced and professional degrees. Although Valley towns were segregated during the pre- and the immediately post-World War II era, there are virtually no census tracts that are not at least 50% Mexican-American. This holds from the poorest to the wealthiest. The middle class does not live in the barrios, but instead resides in more loosely organized neighborhoods, or housing tracts, which are virtually indistinguishable from suburban neighborhoods elsewhere in the United States. Many Valley communities are becoming increasingly suburban, without even having been distinctly urban.

Like the first two groups, the middle class can be viewed as consisting of two life-styles, one urban, the other "kickers" or *los rancheros*. Kickers are individuals who are involved in the South Texas ranching complex. The word, derived from the pejorative label *shit kickers*, are easily visible due to their Western-style dress, pickup trucks, Country Western and Tex-Mex music, and involvement with ranching. This particular group is especially interesting in respect to alcohol, since it is strongly oriented to and, at times, organized around alcohol consumption, especially beer.

The final group is variously described as "the old families" or "the rich." The Valley was first settled by the Escandon expedition of 1750, and some families in the area have maintained their land and/or their wealth for many generations. These families are active in local and regional politics and form a small but significant elite of ranchers and owners of citrus and commercial farms. Their children are frequently lawyers or physicians, or enter similar professions. They are bilingual, and a significant number of the families have retained ties, including land holdings, in Mexico. Some also include extended family networks whose members are prominant in regional politics in Mexico. As with many elite groups, the families are relatively inconspicuous in the community, except within their own social network. They influence all the life-style groups.

RESEARCH METHODOLOGY

The evaluations presented in this chapter result from data collected during longitudinal research contact in the field. I have been living, working, and doing research at the same field site continuously for the past 10 years. This allows for the development of a much more complex methodological structure than short-term research.

My approach, called "spiral methodology," begins with extensive ethnography that provides the descriptive base for a valid research spiral. This stage is followed by "focused ethnographies." These are studies of specific problem or content areas within the overall cultural system. Focused ethnographies are intensive investigations of narrow subject areas that allow a part of the total system to be described and analyzed, but always within the context of the overall frame created by earlier research. They provide the structure and the variables for the next stage, which is generally some form of survey. Surveys test validity, reliability, and distribution of variables across the broader cultural groups being studied. This is particularly important in the complex societies being investigated here. The survey results can then be taken back to the community for commentary, discussion, and even debate or rejection. The community reaction to the survey focuses on further ethnographic exploration, and further refines typologies and concepts. This initiates a second loop of the spiral—toward more specific, yet broader coverage.

The outcome of this process is an accurate, detailed ethnography coupled with methodologically sound generalizations across a complex, heterogeneous, social system. This approach is being followed with the research on alcohol use in the lower Rio Grande Valley, although because of the nature of the approach, it is by no means complete.

To date the general ethnography of the region has been pursued for over 10 years. One ethnography has focused on alcohol-related behaviors in the conventional health care system, on the folk medical system (and on the *curandero's* treatment of alcohol-related problems in particular), and on bar behavior, primarily in predominantly "kicker bars." Surveys of alcohol use and abuse have been conducted on the college population (Trotter, 1982) and on a general population in one city (Maril & Zavelta, 1979). An ethnography of migrant drug use, just completed, is complemented by a survey of migrant college students.

Material has been drawn from all such sources and summarized as trends and tendencies within or across the life-style groups presented above for this chapter. This summary should be taken as a broad-brush approach, presenting normative patterns while omitting some important elements of variation within and between some of the subgroups. The spiral methodology is excellent for balancing normative and variational data, but only when it has been carried out to a degree that does not yet exist for the Valley for alcohol-related data.

OVERVIEW OF DRINKING PATTERNS

The general thrust of the research completed thus far has been to determine culturally normative drinking patterns, to discover emic views of and values toward alcohol use and abuse, and to make recommendations about the development of culturally appropriate treatment of alcohol-related problems. However, before those subjects can be covered, one of the persistant Anglo myths about Mexican-American drinking needs to be reevaluated. This is the myth of the poor Mexican male who instead of buying food and clothing for his children goes out, and because of uncontrolled machismo, buys drinks for his equally profligate friends. Some of the roots of this stereotype can be seen in the ethical perspective of the temperance movement described by Ames (Ch. 3). This is compounded by the fact that many of the Mexican-American males' preferred drinking locations in the Valley are publicly visible, as opposed to the more culturally "hidden" drinking of Anglos, which takes place within the home. The resulting stereotype of Mexican-Americans is that they are frequent, heavy, often boisterous and abusive drinkers. Included in the stereotype is an assumption that alcohol abuse and alcoholism must be rampant, especially among the poor. The stereotype is not supported by the data collected in the Valley. In fact, as will be shown below, Mexican-Americans as a group tend to be more conservative in their drinking patterns than are Anglos in the same environment.

The most significant finding about Mexican-American drinking patterns is the persistent differences in sex roles relative to alcohol consumption. Two surveys (Maril & Zavaleta, 1979; Trotter, 1982), as well as the ethnographic data, confirm that there are significant differences in the amount and type of alcoholic beverages consumed by Mexican-American males and females, as well as differences in preferred drinking locations, the social context of drinking, drinking patterns, and attitudes toward drunkenness.

The data point to a general conservatism among Mexican-Americans, vis-à-vis drinking. The percentage of the male population of drinkers is the same, or slightly lower than males in the United States as a whole. One study (Trotter, 1982) indicated a tendency for Mexican-American males and females to have their first drink, and to begin drinking on a regular basis, approximately a year or more later than their Anglo-American counterparts.

The percentage of drinkers among the female population is considerably less than among women in the United States as a whole. Not surprisingly, there are stronger sanctions against women drinking than there are against males. The data also indicate that although Mexican-American males, as a group, engage in more drinking episodes per unit of time than do their Anglo contemporaries, they drink less per occasion, so the total amount of alcohol consumed by both groups is about the same. Mexican-American females, however, participate in significantly fewer drinking occasions than do Anglo females. They also consume fewer drinks on those occasions. In the Valley, both Mexican-American and Anglo females drink less often and consume less alcohol per occasion than do their male counterparts. In fact, in terms of individuals who drink, more than one third of the Mexican-American females questioned did not drink at all (Trotter, 1982).

The ethnographic research indicates that the use of alcohol by women is constrained by group pressures revolving around the twin concepts of virtue and respect. An individual, particularly a female, can "demonstrate" a lack of virtue through the inappropriate use of alcohol in the community. Inappropriate use of alcohol by either sex also demonstrates a lack of respect for family and family members, and especially for parents. Both of these conditions act to impose social controls on drinking in the small, tightly knit communities and in the extended family systems that predominate in the Valley.

The situation of women, relative to drinking behavior, is especially crucial to the use of culturally shaped social roles as an explanation of behavior. A significant number of Mexican-American women in the Valley are part of a "protective environment." The "protection," in this case, refers not only to strong negative sanctions against drinking, but also to restrictions on other behavior that would facilitate the individual's access to places where drinking might occur. Thus, it is not surprising that Mexican-American females tend to be the most conservative group in the Valley in drinking behavior.

In addition to lower levels of alcohol consumption, the role differentiation between males and females in the Valley also leads to different preferences in drinking settings and the choice of drinking companions. Both males and females tend to drink more frequently outside the home than in it, and this is especially true for young people. Unmarried children who smoke or drink in front of parents are often thought to be extremely disrespectful, and to shame their family. There are few social pressures on drinking at friends' houses or in public drinking establishments. For males, there is a strong positive sanction for drinking with members of the same sex, which may act as a "pull" factor for drinking outside the home, complementing the "push" factor of disrespect. For females, the choice of drinking scene is more limited. Both can drink at friends' homes or in public establishments, but for females there is a restriction in terms of which public establishments are acceptable and which are not. Generally, as one informant put it, it is acceptable for a female to drink in a public establishment as long as the major reason for people being there is to dance. But a girl can easily "lose her reputation" by being seen drinking in a bar. A bar is any establishment where the primary purpose is drinking, even if it has a dance floor. Male and female role differences not only influence drinking, but also the choice of drinking companions. There is a strong tendency for Mexican-American males to drink most frequently with other males, unless they are engaged in courtship or family celebrations. For example, in one survey (Trotter, 1982, p. 319), nearly two thirds of the Mexican-American males stated that they most frequently drank with small or large groups of the same sex. This compares to less than one fourth of the Mexican-American females stating that they most frequently drank with individuals or groups of the same sex. These data suggest that female drinking patterns are primarily associated with or even partially depend upon relationships with the opposite sex, whereas those of males more closely depend upon relationships with friends of the same sex.

Overwhelmingly, Mexican-American males tend to drink beer, with hard liquor second, and wine a distant third. Women, on the other hand, strongly prefer mixed drinks or wine rather than beer. Beer is considered a non-feminine drink, as well as a less reputable drink (one informant said only bar women should drink beer). Males tend to drink in social settings (bars, friends' homes) and with individuals (male friends) that would encourage the consumption of beer because of its lower cost and its association with masculine images. Televised sport events, regular bar associations, along with a special social occasion called the *pachanga*, provide important male focal points for drinking in the Valley.

Pachangas are a special drinking scene in the Valley. They are generally all male secular rituals that occur periodically within Mexican-American male networks in the Valley. When asked to define a *pachanga*, one informant said "That's easy. The definition is meat, beer, and politics." *Pachangas* are drinking scenes in which a group of males with close familial, friendship, and political ties get together for a barbecue, generally with quantities of meat and beer. The men stand or sit outside around the grill, occasionally grabbing bits of meat with the tortillas provided for that purpose, and drink beer. Most *pachangas* function to reinforce existing social ties and simultaneously serve as a focal point for political, business, or social strategy sessions, depending on the composition of the group. Among middle-class people, there has been some effort to include females in the *pachangas*, but most attempts have failed. The existing social setting in which barbecue and beer coexist with women's participation are familial parties or celebrations; the function of the gathering is different, and in many cases, there is a mild to strong sanction against women drinking. Therefore, the conflicting role models for *pachangas* (unlimited male beer drinking) and familial parties (limited female drinking) have not made mixed-sex *pachangas* comfortable.

Females, however, tend to drink in settings and with drinking companions that favor the consumption of mixed drinks and include the presence of male companions, either family members or dates. This helps maintain sanctions for low levels of consumption. It creates the interesting condition within the broad cultural system of making female drinking and the control of drunken comportment dependent upon cross-sexual role patterns, while making male drinking and drunken comportment dependent upon same-sex models.

In the Valley, males are generally given significantly more license in drunken comportment than are females. There are strong sanctions against female drunkenness; drunk females tend to attempt to mimic sobriety in their drunken comportment. The exceptions to this are females who have accepted a deviant role, occasional all-female situations where the absence of males allows a different, frequently humorous expression of drunkenness, and occasional family parties where stressful situations (accident, death, etc.) are used to excuse a rare bout of inebriation for an adult female.

Male drunken comportment, on the other hand, does not change according to the sex of the participants in the drinking scene and is used in some situations to express aggression (mostly verbal, but occasionally physical) that could not otherwise be expressed in a social context. Males can normally be confident that behavior that would be socially unacceptable when they are sober will be excused when they are drunk. Many times you hear someone describing the socially negative things a male said or did, excusing and absolving the person by saying "but, *pobrecito*, he was drunk and didn't know what he was doing." A woman's behavior in a similar circumstances is much more likely to be castigated than absolved.

This overview of drinking behavior and drunken comportment indicates that significant differences exist between the sexes. These can be viewed as role responses to the values of respect and modesty within the culture. Another convention should be added to the above. Very few young Mexican-Americans drink in public in front of their parents. To do so is to show extreme disrespect for both parents and family; it constitutes a public flouting of parental authority. To be drunk in front of them would be improper in the extreme.

Females virtually never drink in front of their parents, regardless of marital status, having children of their own, or any other indicator of adult status. Males, if invited to do so by their fathers, are sometimes given freedom to drink with parents after they are married and have children of their own. However, individuals of both sexes refrain from drinking in front of their parents, at least in mixed-sex groups, all their lives.

MIGRANT AND NON-MIGRANT POOR

As summarized earlier, the migrant or non-migrant poor person is normally only minimally bilingual/bicultural. Hispanic values for drinking predominate. Several authors have pointed out that this group has the most sexually dimorphic drinking patterns within the cultural system (Maril & Zavaleta, 1979; Paine, 1977). Both men and women view male drinking as an expression of manliness and maintain strong negative sanctions toward female drinking. However, contrary to some Mexican-American stereotypes, there is little association in this group between deviant behavior and heavy drinking (Johnson & Matre, 1978). This group has the highest level of abstinent females: up to more than 85% not having consumed any alcohol for the past 12 months (Meril & Zavaleta, 1979). Most males drink (over 90%).

Drinking scenes for this group are *cantinas* (inexpensive bars located close to the barrios and virtually exclusively male), *pachangas*, and dances. Drinking also occurs at family gatherings, although the primary focus of these parties is the social celebration, not the drinking. This contrasts sharply with the middle-class "cocktail party." Celebrations in the barrios are primarily family and fictive kin gatherings having the purpose of social reinforcement of kinship ties. Cocktail parties are more "alcohol focused" and serve primarily a non-kin social function, especially to aid the social advancement of the participants.

The primary alcoholic beverage consumed at all migrant and non-migrant poor functions is beer. The only exception is dances. Drinking is less negatively sanctioned for females at dances. The normal dance is one in which the participants bring their own bottles and buy setups (cups, ices, mixer) from the proprietors of the dance hall. Clear-cut distinctions are made between dance halls and bars. The dance hall personnel sell beer, wine, and setups, but no hard liquor. Since everyone brings a bottle, it is considered a modified family gathering. The socially relaxed atmosphere gives married women greater license to drink than in their own homes or in their parents' home. This type of setting is also observable at some Christmas parties given by some companies for their employees, as well as at major celebrations (e.g., weddings and *quincianeras*, fifteenth birthday, coming-of-age celebrations).

A major variation in drinking between migrants and the non-migrant poor is seen during the migrant season. Most of the migrants from the Lower Rio Grande Valley participate in the midwestern migrant stream. Some participate in a western migrant stream, and a few in an eastern stream. The nature of the migration patterns causes differences in the migrants' drinking patterns while they are "upstream."

Migrant camps on much of the Eastern Seaboard have historically been singles camps. These camps have barracks-type housing for single males who have been recruited and transported to the harvest site by a crew chief. The crew chief contracts directly with the grower and has considerable authority and control over the laborers. The crew chief often provides all meals for the workers, as well as all beverages. The cost of these essentials are deducted from the migrants' wages and constitute an additional source of income for the crew chief. Since the crew is normally transported in a bus owned by the crew chief, the workers have limited mobility and limited knowledge of the area in which they are working. This factor, combined with the recruitment of single males, the isolation of the camps (local townspeople seldom welcome the migrants), and the control of the crew chiefs, leads to high levels of drinking as recreation.

Migrants in the western stream tend to be the most mobile of the three streams and many travel in family groups. Nearly every family owns a car. The abuses common in the eastern stream are limited by this mobility. The camps tend to be mixed family and singles. Drinking is certainly one recreational activity, but with mobility, other activities are available. The amount of recreational drinking appears to be less than in the eastern stream, but greater than in the midwestern stream.

The midwestern stream is the most conservative, in terms of the levels of drinking as a recreation. Until recent years, the camps have been predominantly family camps. They also tend to be somewhat smaller than either the eastern or the western stream camps. The leaders in the midwestern camps are called *troqueros* (truckers). They own a produce truck, and like the crew chiefs, they develop long-term relationships with one or more growers and contract to bring workers to the fields. Unlike the crew chiefs, the majority of the workers they bring are members of their own family, their extended family, or neighbors from their home base. Exploitation of these individuals would be inappropriate and would make life difficult for them when they returned each year to the Valley. It would also make it difficult for them to recruit workers for the next season.

Drinking as recreation in the midwestern stream is also mitigated by the attitudes of the midwestern growers. Many, but not all prohibit drinking in their camps and reinforce this prohibition by expelling individuals who break the rule. Some drinking in the camps is ignored, but there are also instances when the rules are enforced.

In talking with a grower and *troquero* in Michigan, it became obvious that the two of them had cooperated in making their camp a settled, stable environment. The *troquero* had brought along his extended family, plus the unmarried brother of his son's wife. This was the first time the brother had been away from his family. He celebrated his freedom by drinking excessively and becoming increasingly belligerent. On occasion, he refused to work, which increased the workload of the other members of the family. He was joined in his rebellion by one of the nephews of the *troquero,* and the two became a focal point for dissension in the camp. The grower ignored the increasingly visible misbehavior on the grounds that it was the *troquero's* privilege to maintain order in the camp, especially since productivity had not declined. However, the *troquero* was in a situation in which his position was being undermined, and the work was being threatened by dissension. He was exhorted by members of the family to both punish and to ignore and forgive the young men's behavior. His solution was to quietly let the grower know that he would appreciate the enforcement of the no drinking rule. The two young men were sent home at the request of the grower. The *troquero* saved face, since he did not have to throw his nephew and daughter-in-law's brother out of the camp. The grower accepted the role of the "heavy," and at the same time allowed the *troquero* to reinforce his position by "saving the whole family from being dismissed." Since the relationship between the grower and the *troquero* had existed for more than 10 years, there was never any intention on either side of letting things get so out of hand that it would be severely damaged.

Ironically, the existence of these family camps and their conservative drinking environment is being threatened by recent legislation designed to improve living conditions for migrants. The laws address much needed improvements in the construction and sanitary facilities available in some migrant camps (not all have poor living conditions). The laws generally have different standards for family camps as opposed to barrack-style, single-sex camps. Each family dwelling must have separate toilet facilities, whereas the barracks can have a single facility if it is adequate for the number of persons in the barracks. The difference in expense has caused a number of growers to switch over to barracks and single-sex camps. This, in turn, has increased the prevalence of drinking as recreation, and its associated problems. For many Mexican-Americans, restrictions against bringing their families along has been a severe dislocation, causing many to shift to new migrant sites. Some have made the change over to all-male camps. However, when they return to the Valley, their drinking patterns appear to revert to those of the non-migrant residents of the area.

THE WORKING POOR

The primary condition for membership among the working poor is steady employment. The wages and total income may actually be below that of migrants, or even the non-migrant poor, but the income is predictable. Many of the occupations involve contact with Anglos, hence the group tends to be more bilingual/bicultural than the former two. The steadiness of employment also improves the probability of the children completing their education. Many children of the working poor move into middle-class occupations as a result of their higher educational levels.

The drinking patterns of this group are more age dependent than the former two groups. For individuals who are middle aged, or older, the basic drinking patterns found among the others prevail. For the younger adult and the teenager, drinking patterns have moved toward a more liberal configuration, like that of the middle class described below. More of the young women drink, and they drink in some of the establishments that were proscribed for their mothers. However, the twin values of virtue and respect are still strong and heavily enforced. At least 50% of the women in the group are abstinent. The preference for alcoholic beverages is the same as that of the former two groups.

THE MIDDLE CLASS

The middle class shows the most diverse drinking patterns. The presence of a Mexican-American middle class in the Valley is relatively recent. It began after World War II and has shown explosive growth since the return of the Viet Nam War veterans. Drinking patterns in this group range from the norms described for the groups above to drinking norms that are indistinguishable from national norms.

The least conservative middle-class drinking patterns consist of the use of alcoholic beverages by both sexes in relatively similar amounts, drinking scenes, and degrees of abstinence or use. As with the working poor, these patterns appear to be linked; the more conservative practices are found in the older age levels of the society. It is not usual to see young females in bars. Although many go in pairs or small groups, with the avowed purpose of dancing, not drinking, their level of alcohol consumption is above that of women at family gatherings or dances. They also participate in some all-female drinking occasions. Some will form all-female groups to go to a bar for a few drinks after work. This latter pattern would be extremely rare or non-existent among all three of the earlier described groups.

There are differences in other parameters of middle-class drinking. Beer is the preferred beverage for males, but liquor, especially some of the prestige liquors (e.g., scotch, bourbon), have greatly increased in popularity. Wine is more frequently consumed than in the groups, especially when people are dining out or having others over for dinner. For women, beer does not have the same level of proscription, although it is still considered primarily a male beverage (as in much of the United States). Women tend to drink "light" beers and prestige beers, such as Coors and Michelob.

In some middle-class homes, unmarried adult children can freely drink with parents without threatening shared values of respect or virtue. However, even as the middle class becomes less conservative, the changes are relative. As shown elsewhere (Trotter, 1982), there is still a persistent difference between the sexes in levels of abstinence (about 30% of the females and 10% of the males). These differences also hold for amount of alcohol consumed, settings where drinking is considered proper, and choice of drinking companions. Males still tend to drink with all-male groups more often then with mixed-sex drinking companions. They hold *pachangas* and drink in bars far more often than middle-class women. The women still tend to drink in mixed-sex environments and to be concerned with their reputations, in terms of how much and where they drink. Respect is still a key element in teaching about and controlling drinking and drunkenness in the community.

THE ELITE

The Valley elite are only now beginning to be studied. From the minimal data available, their drinking patterns appear to be most similar to the more bicultural elements of the middle class. This group belongs to the local Country Clubs, frequents other socially restricted environments, and is rarely visible in the public locations visited by the middle class. From ethnographic observation, it appears that the preference for type of beverage has significantly shifted from beer, for the males, to liquor and wine for both sexes, at least in publicly visible locations. The *pachanga* is still a key social institution for this group, and there is still a menu of "lots of beer, barbecue, and politics." In fact the *pachanga* appears to be used by this group as a temporary social-leveling mechanism so that politics can be pursued with members of all the other social groups without loss of face or dimunition of social status, since everyone at a *pachanga* is "family."

At least in publicly visible environments and at cocktail parties, the elite appear to have virtually no differences in the number of abstainers or the beverages consumed by the two sexes. Women do appear to drink less than their male counterparts. Obviously, further research on this group would be useful, since so little has been published.

TREATMENT CONSIDERATIONS

This chapter has presented a general description of some of the normative drinking patterns found in Mexican-American communities in the Lower Rio Grande Valley of Texas. The current estimates for the region indicate that the alcoholism rates in the Valley are neither higher nor lower than in other areas of the United States. There may be fewer female alcoholics, at least among some of the migrant and non-migrant poor, due to the high levels of abstinance in those groups. Among individuals who drink, alcoholism exists at the same levels as elsewhere.

There is a wealth of anecdotal data on the need for bilingual/bicultural treatment systems for alcoholics. These data are strongly supported by the information collected by the author, as part of an evaluation of a local halfway house program (Trotter, unpublished).

The Midway House program has a success rate of approximately 60%. Success is defined by the staff as permanent abstinance. The program is eclectic, and closer to a truly bilingual/bicultural program than any other the author has seen.

The program is directed at dealing with the realities of language and world view brought into the program by each client. Clients undergo an informal, but rigorous cultural assessment, and their treatment is individually tailored to fit their preferences and orientation. One of the keys to treatment in Midway House is the avoidence of grouping clients. The clients are certainly brought together into functional groups, but the orientation toward individualized treatment keeps these from becoming "groupings." This is a vastly different bicultural approach from the ones popular in the late 60s and the 70s, when bicultural counseling was synonymous with cultural sensitivity training and therapy sometimes focused on Aztec art, Chicano poetry, and lectures on the history of discrimination against *Mejicanos* in the United States. In the Midway House model, language, values, and ideals are not something to be taught wholesale, from a single perspective, and then called culturally sensitive therapy. Instead, language, values, and ideals are starting points on a broad cultural spectrum. The world is heterogeneous and knowledge of that spectrum allows the counselor to determine how best to provide an individual client with the unique path for his or her recovery and reentry into society. In the process, a bilingual client may attend both Spanish and English groups, whereas monolingual clients may attend only one. Therapy modalities are mixed, and different options are available, depending on the language abilities and educational backgrounds of clients. Some therapeutic modalities have been created by the staff.

One such modality is the deliberate use of periods of boredom as part of the therapy. The assumption is that life is sometimes boring and that too many programs have people so tightly scheduled that they are on a therapeutic "high" in a program. When these people are occasionally bored in the real world, they cannot cope. The Midway House program deliberately introduces boredom, then teaches mechanisms for coping with it. Midway House also has the only Spanish language Alcoholics Anonymous group that has lasted more than six months. It has been in existence for over six years and is successful because the material has been linguistically modified to match border language and values. The diversity of the program is both a necessary condition and a mirror of the complex cultural environment of the Lower Rio Grande Valley. Its success can be measured by the fact that the rate of success of its clients is identical for both Mexican-Americans and Anglos: The model for cultural intervention in alcoholism is truly bicultural.

The complexity and diversity of norms for drinking that have been presented here should probably be taken as a model for Mexican-Americans and other ethnic groups in the United States. Working with parts of groups may temporarily obscure the heterogeneity, but it quickly resurfaces in most urban environments. The result of this complexity, when alcoholism must be dealt with, leads to the necessity of developing diverse, complex treatment systems. This is counter to the current trend of cost-effective treatment systems that are highly standardized. But as the Midway House example demonstrates, treatment systems must reflect the total cultural complexity of the regions they serve if they are to be successful. And again from the Midway House example, when they truly encompass that diversity, they turn out to be more successful than monocultural treatment systems.

SUMMARY

There is a significant heterogeneity within the Mexican-American culture. This diversity is tied to cultural–ecological patterns of subsistence/employment, language orientation, and residence patterns and can be termed life-style subdivisions. This condition is typical of such urban/industrial societies as that of the United States; it plays a significant role in our understanding of patterns of alcohol use and abuse. As demonstrated, levels of alcohol use, beverage preference patterns, and drunken comportment are all tied to these life-styles subsets. This diversity makes it necessary to develop complex treatment systems that are reactive to individual cultural patterns and needs of clients, rather than the earlier model of cultural sensitivity training in the guise of alcoholism counseling. As the Midway House model demonstrates, treatment systems must reflect the total cultural complexity of the regions they serve. They must use sensitivity to cultural orientations, but treat alcoholism, not simply promote one vision of cultural awareness. When treatment systems truly encompass regional cultural diversity, they are apparently more successful than monocultural treatment systems in the same setting.

REFERENCES

Johnson, L. V., & Matre, M. Anomie and alcohol use; drinking patterns in Mexican American and Anglo neighborhoods. *Journal of Studies on Alcohol*, 1978, *39*, 894–902.

Maril, L. M., & Zavaleta, A. N. Drinking patterns of low-income Mexican-American women. *Journal of Studies on Alcohol*, 1979, *40*, 480–484.

Paine, H. J. Attitudes and patterns of alcohol use among Mexican Americans; implications for service delivery. *Journal of Studies on Alcohol*, 1977, *38*, 544–553.

Trotter, R. T. II Ethnic and sexual patterns of alcohol use: Anglo and Mexican American college students. *Adolescence*, 1982, *XVII*, 305–325.

Trotter, Robert T. II *Project evaluation: Midway House, Inc.* available from Midway House, 1605 N. 7th St., Harlingen, TX, unpublished.

16

Alcohol and Hispanics
in the Northeast
A Study of Cultural Variability
and Adaptation

ANDREW J. GORDON

INTRODUCTION

This chapter summarizes the results of an ethnographic study comparing alcohol use in three Hispanic groups in a city in the northeastern United States (see Gordon, 1978, 1979, 1981).

When treatment organizations and clinicians address the issue of Hispanic drinking in the Northeast, they frequently approach the subject as if Hispanics' drinking were guided by cultural rules that apply to all Latin Americans in the United States. This would seem to be logical, since U. S. Hispanic society is generally distinguished by minority group status, lower income, a separate language, and such cultural features as

ANDREW J. GORDON • Division of Sociomedical Sciences, School of Public Health, Columbia University, 600 West 168th Street, New York, NY 10032. The research on which this paper was based was conducted while I held a National Institute on Alcohol Abuse and Alcoholism Postdoctoral fellowship (Grant Number T32AA0713105) sponsored by the Department of Anthropology, Brown University. Data relevant to the Dominican Republic was collected while I was a Fulbright fellow there in 1980 and as Research Associate in 1983 at El Museo del Hombre Dominicano, Santo Domingo, Dominican Republic.

Catholicism and a recent colonial and agricultural past. Hispanics in the Northeast—meaning the five New England states and New York, New Jersey, and Pennsylvania—share even more. Almost all have migrated to the region since World War II in search of work and a better life in the industrial cities. Despite these similarities, however, I found considerable variability in behavior related to alcohol use.

The differences have been obscured in part because clinicians have broadly stated that Hispanics avoid treatment for alcohol problems (see Aguirre-Molina, 1980; Caste, 1979; Caste & Blodgett, 1979; Christmas, 1978; Gomez, 1981; Melus, 1980; USDHEW, n.d.). The purported reasons are the shame about being dependent on alcohol, cultural insensitivity on the part of treatment providers, and the limited number of services. I found that Hispanics do find assistance for alcohol problems, and in different ways for the separate national groups. Help is not overlooked; it is merely found outside the clinical and social service setting.

Also clinicians have overstressed the unvarying importance of the masculine ideal, or machismo, in Hispanic drinking in the Northeast. Avoidance of treatment is attributed to male pride and the inability to accept weakness. Frequent bouts of heavy drinking are described as a result of norms of male behavior (Abad & Suarez, 1975; Caste, 1979; Caste & Blodgett, 1979; Galbis, 1977; Lopez-Blanco, 1980; Sanchez-Dirks, 1978; Simpson & Simpson, 1978–79). My findings revealed an absence of a standard of male behavior deterring men from finding help, or promoting binge drinking. Furthermore, I did not find a standard of drinking conforming to typical conceptions of macho behavior, such as preoccupations with seducing or dominating females or in demonstrating physical prowess. In almost all aspects of behavior related to drinking—amount, frequency, choice of beverage, context of use, attitudes toward drinking and treatment—the variability is impressive.

I examined alcohol use in three Hispanic groups—Dominicans, Guatemalans, and Puerto Ricans—in the medium-sized northeastern city of Newtown (a pseudonym). The Dominican population is 7,700, the Puerto Rican 3,500, and the Guatemalan 1,000. I limited my study to alcohol use among men because their consumption far exceeds that of women. Newtown has a population of about 156,000, and these separate Hispanic groups each lived in their own enclaves on the south and west sides of the city.

Puerto Ricans are the largest Hispanic group in the Northeast. They number about 1.3 million. Since most had migrated to the U. S. mainland in the first two decades after World War II, much of the population is not itself immigrant, but rather children of immigrants who, themselves, have children. Dominicans are the second largest Hispanic group in the Northeast. They number 500,000 in the New York metropolitan area (Kayal, 1978) and about 50,000 through the northern states. Dominicans started to arrive in the United States shortly after the Dominican dictator Trujillo was assassinated in 1961. Each year has brought thousands more. The recent migration of Central Americans—from Guatemala, El Salvador, Honduras, Nicaragua, and Panama—has been brought about by political upheavals and these migrants will most likely have a significantly larger presence in the future.

In presenting the case of Dominicans, Guatemalans, and Puerto Ricans in Newtown, I will address (1) treatment for deviant drinking. Here I will pay particular attention to indigenous treatment as found in faith healing and storefront churches. Indigenous treatment is to be distinquished from formal services found, for example, in hospitals, clinics, social service agencies, and private medical practice; (2) modal drinking practices and beliefs about alcohol use characterizing each group; and (3) the role of alcohol in the distinctive adaptation that each migrant group has made in the Northeast.

Data are garnered from my active participation in the social life of Dominicans, Guatemalans, and Puerto Ricans. I made dozens of observations of drinking and treatment settings at different times of the day, the week, and the year. Attempts were made to gather data from people representing a diversity of backgrounds, occupations, and ages and from households of differing compositions. Those who asked me about my fieldwork were told quite openly that I was studying alcohol problems. Few were interested, however. They simply assumed I was a professor doing some obscure research. In some instances, when my declared intentions would have a chilling effect, as in bars, I phrased my interests more broadly by stating that I was studying health problems in order to better deliver services.

DOMINICANS

TREATMENT

The staff of hospitals, of mental health centers, and of social service agencies report very few, if any, cases of Dominicans bringing a drinking problem to their attention. But Dominicans do recognize when drinking clearly deviates from norms and when behavior suggests a dependence on alcohol. A group of six comparatively older men are quite well known for their compulsive drinking. Family and friends of these six and of others with less obvious, but still deviant drinking often seek help from indigenous sources of healing.

The local Catholic Church and its theological offshoot, *El Movimento de La Renovacion Carismatica* (The Movement of the Charismatic Renovation), provide a therapeutic resource for deviant drinkers. On Saturday evening a core group of about 40 to 60 of the movement's most faithful participate in a special prayer service. Two large extended families, some single males, and some married couples, all of whom have "found Jesus," attend the service. Many of the males have reportedly given up a life of sin that included heavy weekend drinking. Many of these men were urged to join the movement by their family. Their pattern was to take part, cautiously, at first. Later they began regularly to attend Saturday meetings, Sunday masses, and weekend retreats.

The Saturday services dramatically depart from conventional Catholic liturgy, since they are led by participants, not by the priest. The meetings are fervently spiritual. Guitars and tambourines provide accompaniment for a song asking for God's help, for his humility, and for his companionship: They sing *ayudame, humildame, acompaname,* while holding hands in fellowship. They are emotive and fully appreciate their anti-establishment stance, as when they declare, "We read from the Bible where we want, not where Rome wants."

El Movimento has influenced all aspects of the church's activities and the church, in turn, has influenced the Dominican community. The more traditional Sunday Mass (with an attendance of 500, mostly Dominican) stresses social justice and the fraternity of mankind. A sermon is as likely to question American political involvement in Central America as it is to address the divine love of Jesus. Traditional charismatic themes of the apocalypse and the coming of a new age, *El Siglo del Senor* (the Century of the Lord), are merged with ideals of self-improvement and the solidarity of the community. Interspersed with church liturgy are references to a new spirit on Broad Street, the principal thoroughfare of the Dominican enclave, as a sign of a new and prosperous age.

An ethic of community responsibility is also expressed in norms that proscribe drunkenness and the excesses of weekend drinking. When asked about their church's general stance on drinking and drunkenness, men report that "it's okay to have a few beers but not to get drunk—it is not good to be drunk, he who gets drunk is not an *amante* [lover, one faithful to] God."

Dominicans report that an atmosphere of *familiarización* has returned to their lives as a result of the church. *Familiarización,* in the sense used in the Dominican Republic, refers to warmth and companionship, as found on festive occasions involving drinking. However, the *familiarización* common in Newtown and associated with the church is also found in the church's services, its retreats, and its *cursillos* (courses of religious instruction). Participants describe these events as an exhilarating positive change from their workaday world and Saturday night drinking.

DRINKING PRACTICES AND BELIEFS ABOUT ALCOHOL USE

Dominicans usually report that they drank less after migrating to the United States; however, changes in the context and style of drinking strike them as more significant. In Newtown, drinking is done on weekends, principally on Saturday night, and not on weekdays as in the Dominican Republic. Less drunkenness and fewer fights characterize drinking in Newtown. Also, alcohol use is far less often associated with the traditional male image of the *parrandero*. A *parrandero* is one who follows a life of rum, women, and song. Part of the *parrandero* style involves drinking outside the home, with extramarital liaisons or with prostitutes; in Newtown, however, drinking mostly occurs at home, at the social club, and with the family.

Avoidance of excessive drinking reflects a perceived social and economic advancement. To get drunk, to pursue the *parrandero* image, or to drink during the work week violate the ethic and ideal of being an *hombre serio*. An *hombre serio* is literally one who is serious; more figuratively, he attends to the needs of wife and family and is conscientious about work.

The words one frequently hears to describe the experience and goals of Dominicans in Newtown are *economizar, sacrificar,* and *progresar*. This reflects a pattern of dedication to hard work. Often, when both husband and wife can find employment, they work full time. Buying property is a major goal. As a result of a municipal program to sell abandoned houses, many Dominicans have bought houses cheaply and then put them in good repair.

To affirm their newfound position, Dominicans follow drinking etiquettes expressing their upward mobility. They are determined to buy liquors that command the greatest prestige: Chivas Regal, Johnnie Walker Black Label, and Haig and Haig Pinch. With the exception of beer in the hot weather and Dominican rum, Dominicans generally drink the best of scotches. Even when they drink beer, it is the foreign and higher priced brands, usually Lowenbrau or Heineken. They are fully aware of their conspicuous consumption and boast, "We Dominicans drink only good liquor."

Not only is there close attention paid to what is consumed, Dominicans stress the need "to remain conscious and comprehensible" so as to avoid misunderstandings. The ideal is to drink *suave*, at a measured pace, not letting alcohol impair self-control. The norm of moderate drinking is protected and enforced in the Dominican social club. Club rules dictate that six members oversee the conduct of their parties, which often attract several hundred people. If someone has too much to drink and is likely to disturb others, then it is expected that this person's companions will take him home. Failing that, one of the six will arrange for this person to be accompanied off the club premises.

In drinking too much, one is at risk of crossing the line separating appropriate conduct from *indecente* behavior. Being *indecente* means insulting people, fighting, talking rudely to women. Dominicans as a group distinguish themselves from other Hispanics, such as Puerto Ricans with whom they are confused, by their respective drinking behaviors. Dominicans claim that unlike them, Puerto Ricans "drink every day from morning until night," "they drink and smoke marijuana," "they drink only beer," and "they don't think of the future."

ADAPTATION AND ALCOHOL

The upward mobility and industriousness of Dominicans has also been reported by Hendricks (1974) and Gonzalez (1971), both of whom have done fieldwork on Dominicans in New York City. Dominican progress has been characterized by the settlement of entire families, not just single males. This has facilitated the formation of a stable household economy to which the male will contribute.

Unlike other migrant groups, Dominicans do not and are not expected to contribute to kin beyond the household unit. The household concentrates on its own progress, on accumulating capital. Although most do send money to parents and siblings in the Dominican Republic, they keep most of their earnings for their own use.

Opportunity for work and the ability of men to live up to the traditional role of provider and family head has reinforced the reasons for personal sacrifice and moderation in the use of alcohol.

Changing economic opportunities for women have also produced an ethic of progress and influenced the level of drinking. Wives are far more financially independent than they were in the Dominican Republic. Often they find jobs as easily as or more easily than men. Since they may be economically independent from their husbands—if not through their own work then from welfare subsidies—they need no longer accept the profligate ways of a *parrandero* husband. They threaten to leave if men behave in the traditional manner, and they often make good their threats. Although it is sometimes with reluctance and nostalgia, men do accept their new roles and consign drinking sprees and extramarital affairs to their past in the Dominican Republic.

GUATEMALANS

TREATMENT

Guatemalans seeking help for a problem with drinking turn to the one Spanish-speaking Alcoholics Anonymous (AA) group in Newtown. It has 20 regulars, of whom only two are not from Guatemala. One is from neighboring El Salvador, the other from neighboring Mexico. The beliefs and routines of AA are intact, but the AA group takes on the characteristics of indigenous treatment because of the persuasive influence of Guatemalan culture.

A distinctively Guatemalan characteristic is apparent in the *tribunas* (personal testimonials about drinking) of the AA members. They use native slang, *caló*, or as they put it, "the language of thieves." Attitudes, names, and places peculiar to Guatemalan culture and to the hangouts of drinkers are continuous and explicit as well. Much of the *geist* of their lives is captured when they talk about their lives as *picaros;* that is to say, following an adventurist and debauching life punctuated by odyssean wanderings. Their travels frequently include several Central American countries and a number of U. S. cities, such as Los Angeles, Chicago, and New York. The *picaro* is also the central character type of the picaresque novel. He travels from place to place, flouting social conventions by going from feminine embrace in a brothel to the same in a convent, from maid to married mistress, all the while changing locales and vocations.

These *tribunas* are richly expressive in Guatemalan culture, so much so that Dominicans and Puerto Ricans arriving at the AA meeting for a trial visit quickly find themselves confounded and excluded. They do not return.

The use of AA is a widespread phenomenon among Guatemalans. Of the 22 Spanish-speaking groups in New York City, 8 are predominantly Guatemalan in membership. This number far exceeds their comparatively small percentage among New York Hispanics. As a result of their close association with AA, Guatemalan members unhesitatingly accept the beliefs widely accepted by AA members: Alcoholism is a progressive and chronic disease beginning with blackouts (temporary lapses of recollection of what happened while drinking), continuing on through the pattern of associating with social inferiors, to severe symptoms of withdrawal, and finally to "touching bottom," the depths of dissolution.

All members of this group recall their drinking histories in like fashion. In their *tribunas* they say they started out with most promising futures in the military, business, or health care; they were teachers and topographers or followed other careers requiring skill and study. They report that the disease of alcoholism eroded their careers, led to extramarital affairs, to the deterioration of their marriages. It is not possible to discern if their respective biographies conform to this unitary pattern. It seems plausible that individual differences may be unwittingly submerged or altered so each may have his share in a past full of common experiences, disappointments, and expressions that are characteristically Guatemalan.

The AA group defines deviant drinking for the Newtown Guatemalan community. They all know of its existence; many come to visit, at least once, and when they do veteran AA's go to the front of the room and instead of discussing their lives, they explain alcoholism as they see it.

> It's progressive . . . we were like you, wondering whether we are alcoholic . . . if your drinking harms your health, your job, your wife, your family, then you're an alcoholic . . . you'll be back, but then after much sadness.

These declarations do spread knowledge of alcoholism from the point of view of AA. They suggest the future course of those who will continue drinking. Consequently, Guatemalan drinkers grow very articulate about alcohol use and its potential hazards. With an apparent clinical sophistication, they offer opinions of what an alcoholic is, what a social alcoholic is, the difference between one who has to drink and one who drinks occasionally, the fact that alcoholism is a disease.

DRINKING PRACTICES AND BELIEFS ABOUT ALCOHOL USE

Many Newtown Guatemalans have followed the life of a *picaro*, unencumbered by family, home, or possessions. They move frequently, allegedly in search of better pay. Excess cash is spent on alcohol and no money is saved.

Drinking almost always leads to drunkenness, and in drinking contexts there are no interdictions against heavy drinking or drunkenness. On the contrary, the drunk is glamorized as a victim of circumstances. They explain their heavy drinking by saying they are all too sentimental and nostalgic for the wives and families they left behind in Guatemala. It is not uncommon to hear people respond to a *como esta* (how are you) by answering in *caló "un poco jodido de chupar"* (a little wrecked from drinking). Using alcohol to cope with sadness is a fundamental expression of these men, who also turn to poetry and melancholy songs when drinking.

Drinking in bars is the norm. One bar in Newtown, Taglio's, serves as a gathering place for all the Guatemalans in that city, and it is known as a meeting place for Guatemalans throughout the Northeast. There, one may seek out others, develop friendships, and if needed, find temporary assistance in a loan or lodging or obtain information about jobs. The emphasis, however, is on easygoing camaraderie. The sole social expectation and standard is that one ought to reciprocate a purchase of drinks.

Beer is consumed almost exclusively. The pace of drinking is especially quick. Men sit at crowded bar tables and buy rounds for each other. The haste in reciprocity far outstrips the capacity to consume. In the initial 25 minutes in Taglio's, people will have had four beers purchased for them, while maybe they will have consumed only the second beer. By the end of the evening, at the 1 A.M. closing time, people stagger out, many after having fallen asleep at the table; those still capable buy six packs to drink at home.

Their style of drinking contrasts sharply with that of the Dominicans. Guatemalans drink beer heavily throughout extended weekends lasting from Thursday night through Sunday night. Their drinking is done in bars and mostly among men. Dominican drinking is different in almost all respects. They drink "the better" scotches in relatively moderate fashion. Saturday night is the time for drinking; and they are with their families at home or at their social club. The moderating influences on drinking in the Dominican community come from the ethic of progress and sacrifice, the theological movement, and the altered family structure. The sole influence on Guatemalans' drinking is exerted by AA.

ADAPTATION AND ALCOHOL

The reasons given for migrating are similar for Dominicans and Guatemalans. Both groups come to the United States in search of a better job, better pay, and a superior life for their families. Both say Newtown is a good place to live, there being plenty of opportunities for work in this relatively tranquil city. In the vast majority of cases, the similarities stop there. Guatemalan men, unlike Dominican men, arrive without their families. Due to the vicissitudes of illegal migration, Guatemalan men travel first to Mexico and then illegally cross the border into the United States. They come with a plan to acquire a substantial bankroll in order to return to Guatemala and start a business. But their goals do not remain fixed in their minds. Not being a part of a family unit, they make short-term investments in friendship and in alcohol while they satisfy immediate needs and maintain friendships in case of future need. Their long-range plans of returning to Guatemala become vague. Other women and alcohol undermine their best intentions; the men forget their wives and their future in Guatemala.

The adaptation of Guatemalans is also influenced by national traditions. Traditional cliques of men continue to be tied by the valued ideal of *amigoismo* that is typical of Guatemalan agrarian society.

Alcoholics Anonymous is also a cultural tradition in Guatemala and all through Central America. In Guatemala there are 250 groups with 10,000 members. Its popularity is, in part, the result of political support. In the early 1960s, the president and vice-president of Guatemala made public statements on the value of AA, encouraging drinkers to sober up and making it clear that this was patriotic.

PUERTO RICANS

TREATMENT

Even though clinicians remark that Puerto Ricans are reluctant to use treatment services, the case of Newtown reveals they make greater use of services than either Dominicans or Guatemalans. The identification of pathological drinking in a Puerto Rican family is more likely to result in the individual seeking out treatment for alcoholism specifically.

Apart from formal treatment services, Puerto Ricans find help with their alcohol problems in the Protestant Pentecostal Church. There are four in Newtown, and their memberships are almost entirely Puerto Rican. I was able to interview six self-identified alcoholics from these churches. Each experienced a miraculous deliverance from heavy drinking. Simply put, Jesus took away the desire to drink. The pattern of cure is very similar in all cases. Typically, arrival to the church, acceptance of Christ, and dedication to a life without vice is preceded by a long period of alcohol-induced illness or stress. The conversion is catalyzed by seeing a vision of Jesus "in a long white robe," or "hearing his voice calling me to a church I didn't know existed," or as in the case of one, "the exorcism of the Devil who took the form of alcoholism." The spiritual cure is apparently effective. In these six cases, abstinence has continued for periods of six months to fourteen years. Their commitment to daily church attendance suggests it may well continue.

When pastors and congregants describe alcoholism, it is defined as a sin on a par with smoking, homosexuality, and drug use—all problems the acceptance of Jesus will solve. From the Pentecostal view, the body is the Holy Temple for the spirit, and it should not be defiled by alcohol or even coffee.

Other Hispanics reject the Pentecostals. Dominicans contend that being Pentecostal means being indifferent to material prosperity and consigning oneself to poverty. Guatemalans report negative experiences with the church in their own country. This has led them to eschew both the Pentecostals and the Catholic Church.

The failure of Puerto Ricans to be involved in AA is not applicable outside of Newtown. In New York City, the 14 Spanish-speaking AA groups not populated by Central Americans are mostly Puerto Rican in composition.

The failure of AA for Newtown Puerto Ricans cannot solely be attributed to the overwhelming presence of Guatemalans. The fate of a Spanish-speaking AA group in Boston suggests reasons for AA's variable success with Puerto Ricans. The AA group there was actively promoted by an American minister and a Hispanic AA member. The Puerto Ricans they recruited were for the most part indigent and without social and psychological supports. When meetings were scheduled, continuing small obstacles prevented the formation of an AA group. On one occasion, the key to a meeting room was forgotten; on another occasion, too few arrived because of a lack of money for transportation. Another time, someone had visitors from out of town and could not attend. It is plausible that a degree of social and economic stability, or at least a core group of reliable individuals, is critical to launch and sustain an AA group. Puerto Ricans who had drinking problems in Newtown did not, for the most part, meet these criteria.

One other alcoholism treatment resource has been identified. In the Puerto Rican community of Hartford, Connecticut, Singer (1984) and Singer and Borrero (1984) have found that *centros* (centers) for *espiritismo* (spiritism) are being used by Puerto Ricans seeking help with problems in alcohol use. It is likely that the spiritist *centros*, quite common in other major urban centers, are also addressing alcohol problems in other Puerto Rican communities.

Drinking Practices and Beliefs about Alcohol

The period of residence on the U.S. mainland is longer for Puerto Ricans than for other Hispanics, and acculturation has been more thorough. Consequently, Puerto Ricans have more fully adopted U.S. drinking customs and added them to their traditional drinking customs. Thus in Newtown, Puerto Rican drinking behavior is an amalgam of drinking practices. They follow the pattern of weekday drinking typical of the American workingman. Many of them speak rather good English and they freely fraternize in the non-Hispanic workingman's tavern. Weekday drinking among Puerto Ricans does not affect the importance of their traditional weekend fiesta drinking more commonly seen in a rural society. Additionally, many males well into their 30s use cocaine and marijuana; and to a lesser extent, they sniff glue and use barbiturates and methadone in conjunction with alcohol.

The regularity with which Puerto Ricans use drugs clearly sets them apart from other Hispanics. The Puerto Rican custom of making a *bomba* (becoming utterly stoned and narcotized) has no parallel among other Hispanics in Newtown; and there is no analog to the *atomico* (one who puts whatever is available in his mouth to get high). The influence of the drug culture is explicit. Puerto Ricans wear tee shirts emblazoned with pictures of marijuana leaves or with slogans advocating drug use.

The number of individuals known to drink excessively in Newtown is large and recognized by social service providers to be between 75 and 100. This constitutes about one-tenth of the adult male population. The rate of Puerto Rican alcohol and drug use has been high wherever it has been studied, as in Pennsylvania (Kessler, Rosario, & Lindgren, 1977), in New York City (Haberman & Sheinberg, 1967; Haberman, 1970; Fitzpatrick, 1973; Rodriguez, 1980; Strug, Walsh, Johnson, Anderson, Miller, & Sears, 1984; Valez-Santori, 1980), and in New Haven (Abad & Suarez, 1975; Caste & Blodgett, 1979).

Adaptation and Alcohol

The perceived possibility and will to improve one's social and economic position have been undermined by barriers to work. Unemployment rates range from 30% in New York City (Rodriguez, 1980) and New Haven (CSAC, 1973) to 25% in Massachusetts (MSAC, 1972). Migrants arrive from Puerto Rico with little preparation for the mainland's job market. When here, they suffer racial prejudice, they live in Puerto Rican ghettos, and their youth receive an inferior education.

The problems for Puerto Ricans are considerably more profound than for other Hispanics because of their dependency on welfare, which has roots deep in Puerto Rican society. The pattern of relying on Federal subsidies to households in Puerto Rico continues on the mainland. Unfortunately, the men maintain the traditional expectations that they ought to fulfill the traditional *padre de familia* role; however, the possibilities are severly limited. As a result, the dignity or *dignidad* (a central idiom in Puerto Rican culture) of men is assaulted. Men increasingly use alcohol and other drugs to cope with their problems. But alcohol, rather than tempering discontent, leads to a high incidence of family violence by Puerto Rican males when they are drinking, as shown by police reports.

DIFFERENCES AND THEIR IMPLICATIONS

DRINKING ETIQUETTE

There are clear differences in the expectations and norms of drinking in each of the national groups. Dominicans drink in relative moderation on Saturday nights among their families, either at home or their social club. The choice of alcoholic beverage is high-priced scotch, reflecting upward mobility. Standards of drinking dictate drinking slowly and carefully, so as not to lose control. Guatemalans, on the other hand, drink in excess over extended weekends. They congregate in bars, generally with male-only company. The standards of drinking are to continue consuming until it is no longer possible to consume. They rarely feel the need to stay in control. Their style of drinking, they feel, is the inevitable expression of a people whose lives have turned out tragically. In comparison, Puerto Ricans use drugs along with alcohol far more frequently than do either Dominicans or Guatemalans. Alcohol and drugs are not associated with a specific social position or context. The associations are diverse, such as a continuing rural pattern of binge and fiesta drinking on weekends and holidays; the daily drinking of the U. S. workingman; and the style of the drug culture that glamorizes pills, marijuana, inhalants, and other drugs.

TREATMENT

Claims that Hispanics do not seek help for problems with alcohol use overlook treatment that does not take place in formal clinical settings. Those who seek help have a very different point of view. For instance, all Guatemalans know about AA, even if they do not drink. If a Guatamalan AA member suspects a compatriot to be alcoholic, then he will be encouraged to go to AA rather than to a clinic or a social service agency.

Additionally, indigenous treatment will rarely be recognized as a form of alcoholism treatment because indigenous treatment fails to separate alcohol as a cause of problems from alcohol as an aspect of a problem. Those in formal treatment tend to see alcohol as the real problem underlying difficulties in one's family life, in work, or in acceptable behavior.

For example, when Dominicans attend church and urge their family members to do the same, it is not specifically for a problem with alcohol use. Rather, it is for a way of life that is considered irresponsible, out of phase with progress, and, most importantly, without the strength of purpose provided by believing in Jesus.

Protestant Pentecostals promise deliverance from a variety of spiritual transgressions such as the use of alcohol or drugs, smoking, dancing, and welfare. Treatment in this case is a total change in the person's life; the absense of alcohol is just one part. Alcohol is not regarded as the problem. The issue is one of life-style, and the failure of a person to lead a life guided by Jesus is the problem.

MACHISMO

The deterrent to treatment for alcoholism that is assumed to come from machismo does not appear to impede the use of indigenous services.

Neither do the traditionally held associations of machismo—authority over females, sexual prowess, or competition among men—appear to influence drinking behavior. The norms of drinking comportment for Dominican men stress self-control and moderation. For Guatemalan men, the norm is to be convivial, expressive, and often melancholy. For Puerto Rican males, drinking and taking drugs is simply a matter of becoming and staying high.

Although macho behaviors do certainly exist among Dominicans, Guatemalans, and Puerto Ricans, they do not occur with greater frequency during drinking, nor do they appear to underlie the impulse to drink. Insofar as fighting may be a sign of machismo when drinking, it is well to point out that fights occur with greater regularity among most ethnic groups simply because of alcohol's disinhibiting effects.

The Differences in Light of Adaptation

An examination of drinking behavior from the perspective of adaptation has strongly suggested that differences in these groups accrue from far more than continuities in national culture. Immigration policy, along with the structure of economic opportunity, plays a crucial role in the migrant groups. Dominican families have been able to immigrate together, to set up a household, and to work toward a common objective of economic improvement. The Dominican male with an intact family has a social group to which he is responsible, as well as wives and children and extended kin who expect him to act responsibly. With a good measure of economic opportunity, Dominican men adopt an ethic stressing progress, personal sacrifice, and austerity—all of which implies moderation in alcohol use.

The pattern and policy of migration may undermine reaching the very goals that inspired migration. The Guatemalan migration is an example. Because of the dangers of illegal entry, wives and children often remain in Guatemala, and their husbands fend for themselves in the United States. In the absence of a family and household, there is little incentive for saving and planning for the future. Guatemalan men eagerly take advantage of the opportunities for work. Extra cash goes into socializing with other men, and drinking. Plans for improving one's lot are sacrificed to short-term satisfactions, and heavy drinking becomes the norm.

Immigration policy has not affected the adaptation of Puerto Ricans, since they are U. S. citizens, but economic policy has been influential. The program of U. S. development in Puerto Rico rapidly transformed the island from an agrarian to an urban society, wherein many households became dependent on federal subsidies. This chartered a course for some Puerto Ricans who later moved to mainland cities to live off welfare payments and unemployment benefits. Social norms in this sector fail to inspire social progress, moderation in alcohol use, or avoidance of drugs.

Findings in this study indicate that patterns of Hispanic drinking are the result of a complex of experiences of Hispanic men—as family members, workers, and individuals with perceived life chances. These experiences are shaped by state-level institutions, such as those responsible for immigration and the creation of economic opportunities. It would be fruitful, then, for those concerned about alcohol problems to modify institutional policies so as to ease adaptation and enhance social well-being.

It would be equally fruitful to address problems with alcohol by identifying and reinforcing indigenous modes of treatment. However, professionals and policymakers in health and social services are strongly inclined to pursue clinical and administrative approaches to problems. Attempts to alter institutional policies and to reinforce indigenous cultural therapies are generally not included in their range of activity.

ACKNOWLEDGMENTS

I would like to express my appreciation to A. M. Cooper of the Alcohol Studies Center at Brown University for bibliographic assistance. Also, I would like to acknowledge the support of the staff of the Smithers Alcoholism Treatment and Training Center in the preparation of this manuscript. Interested readers may find a fuller account of the results of this research in Gordon (1978, 1979, 1981).

REFERENCES

Abad, V., & Suarez, J. Cross-cultural aspects of alcoholism among Puerto Ricans. *Proceedings of the Fourth Annual Alcoholism Conference, National Institute on Alcoholism and Alcohol Abuse*, 1975, 282–294.

Aguirre-Molina, M. N. *Critical training requirements for alcoholism counselors working with first generation Puerto Rican males in New York City*, Doctorial dissertation. Columbia University, New York, 1980.

Caste, C. A. Cultural barriers in the utilization of rehabilitation programs by Hispanics in the U.S. *First International Action Conference on Substance Abuse Vol. 1 Alcohol: Use and Abuse*. Phoenix: Do It Now Foundation, 1979.

Caste, C., & Blodgett, J. Cultural barriers in the utilization of alcohol programs by Hispanics in the U.S. In J. Szapocznik, (Ed), *Mental health, drug, and alcohol abuse: Hispanic assessment of present and future challenges*. Washington, DC: National Coalition of Hispanic Mental Health and Human Services Organization, 1979.

Christmas, J. Alcoholism services for minorities: Training issues and concerns. *Alcohol Health and Research World*, 2, 1978, 20–27.

(CSAC) Connecticut State Advisory Committee. *El Boricua: The Puerto Rican community in Bridgeport and New Haven*. United States Commission on Civil Rights, 1973.

Fitzpatrick, J. P. *Puerto Rican Americans: The meaning of migration to the mainland*. Englewood Cliffs, NJ: Prentice Hall, 1973.

Galbis, R. Mental health service in a Hispano community. *Urban Health*, 1977, 6, 33–35.

Gomez, A. G. Ethnicity and transculturality: Its relevancy in training personnel to work with clients with dependency disorders. In A. J. Schecter (Ed.), *Drug dependence and alcoholism 2 Social and behavioral issues*. New York: Plenum Press, 1981.

Gonzalez, N. Peasants progress: Dominicans in New York. *Caribbean Studies*, 1971, 10, 154–171.

Gordon, A. J. Hispanic drinking after migration: The case of Dominicans. *Medical Anthropology*, 1978, 2, 61–84.

Gordon, A. J. *Cultural and organizational factors in the delivery of alcohol treatment services to*

Hispanos (Working papers on Alcohol and Human Behavior 7). Providence: Department of Anthropology, Brown University, 1979.

Gordon, A. The cultural context of drinking and indigenous therapy for alcohol problems in three migrant Hispanic cultures; an ethnographic report. D. B. Heath, J. O. Waddell, and J. D. Topper (Eds.), *Cultural Factors in Alcohol Research and Treatment of Drinking Problems. Journal of Studies on Alcohol* (Special Suppl. 9), 1981, 217–240.

Haberman, P. Denial of drinking in a household survey. *Quarterly Journal of Studies on Alcohol* 1970, *31*, 710–714.

Haberman, P. W. & Sheinberg, J. Implicative drinking reported in a household survey: A corroborative note on subgroup difference. *Quarterly Journal of Studies on Alcohol* 1967, *28*, 538–543.

Hendricks, G. *The Dominican diaspora.* New York: Teachers College Press, 1974.

Kayal, P. The Dominicans in N.Y. Part 1. *Migration Today* 1978, *616–23.*

Kessler, J. L., Rosario, B., & Lindgren, W. S. The needs of the Spanish American Community related to alcohol abuse treatment: An updated report and background statement. In R. T. Trotter II & J. A. Chazira (Eds.), *El Uso de Alcohol: A resource book for Spanish speaking communities.* Atlanta: Southern Area Alcohol Education and Training Program, Inc., 1977, 63–70.

Lopez-Blanco, M. Some considerations in counseling DUI offenders. In *Proceedings of the Second National DUI Conference, Rochester, Minnesota, May 30–June 1, 1978.* Falls Church, VA: Foundation for Traffic Safety, 1980.

(MSAC) Massachusetts State Advisory Committee *Issues of concern to Puerto Ricans in Boston and Springfield.* U. S. Commission on Civil Rights, 1972.

Melus, A. Culture and language in the treatment of alcoholism: The Hispanic perspective. *Alcohol Health and Research World*, 1980, *4*, 19–20.

Rodriguez, C., Economic survival in New York City. In C. Rodriguez, V. Sanchez Korrol, & J. O. Alers (Eds.), *The Puerto Rican struggle: Essays on survival in the U.S.* New York: Puerto Rican Migrant Research Consortium, Inc., 1980.

Sanchez-Dirks, R. Drinking practices among Hispanic youth. *Alcohol Health and Research World*, 1978, *2*, 21–27.

Simpson, L. V., & Simpson, M. L. The Spanish-speaking social service worker: Attitudes toward the alcoholic and alcoholism. *Drug Forum*, 1978–79, *1*, 399–347.

Singer, M. Spiritual healing and family therapy: Common approaches to the treatment of alcoholism. *Family Therapy*, 1984, *11*, 155–162.

Singer, M., & Borrero, M. Indigenous treatment for alcoholism: Espritismo in the Puerto Rican community. *Medical Anthropology*, 1984, *8.*

Strug, D., Walsh, E., Johnson, B., Anderson, K., Miller, T., & Sears, A. The role of alcohol in the crimes of active heroin users, Mimeo. New York City: Interdisciplinary Research Center, Division of Substance Abuse Services, State of New York, 1984.

(USDHEW) U. S. Department of Labor, Bureau of Labor Statistics. *Socio-economic profile of Puerto Rican New Yorkers.* New York: Middle Atlantic Regional Office, U. S. Dept. of Labor, n.d.

Velez-Santori, D. *Drug use among Puerto Rican youth: An exploration of generational status differences.* Ann Arbor University Microfilms International (#8125409), 1980.

17

Alcohol and Adaptation to Exile in Miami's Cuban Population

J. BRYAN PAGE, LUCY RIO, JACQUELINE SWEENEY, and CAROLYN McKAY

INTRODUCTION

Miami has a multicultural population in which Hispanics are a plurality (about 35%), and primarily Cuban (Metropolitan Dade County, 1983). Because Cubans who originally immigrated to other parts of the United States are constantly "trickling back" to settle in Dade County (Prohias & Casal, 1973), and Hispanic households tend to underreport in the Census, total figures are approximate, but 700,000 to 800,000 is a reasonable estimate of the Cuban population here.

Alcohol consumption among Miami's Cuban population is not well documented in public statistical sources. Cubans have tended not to present themselves in public facilities for treatment of either acute or chronic alcohol or other drug use problems (Page, Gonzalez, McBride, McCoy, & Forment, 1981). They have a public image of industry and sobriety, which seems logical, in light of their absence from treatment centers' records.

Our information on Cuban drinking patterns in Miami comes from

J. BRYAN PAGE, LUCY RIO, JACQUELINE SWEENEY, AND CAROLYN McKAY • Department of Psychiatry, University of Miami School of Medicine, Miami, Florida 33101.

the results of two anthropological studies conducted in Dade County between 1978 and 1983. The first study involved observing and interviewing 72 poly-drug-using male Cubans, age 11 to 54, in a street setting (viz., Page *et al.*, 1981). The second was a multi-ethnic study of prescription drug use among women that included 460 interviewees, 152 of whom were Hispanic. Of the Hispanic interviewees, 132 were Cuban. Because of the investigators' holistic orientation, both studies resulted in the gathering of sufficient data on alcohol use to constitute the basis of this chapter.

Our anthropological field work in Miami's Cuban population has produced data on patterns of alcohol use and related psychotropic drug use among Cuban men and women. These data, which include survey materials, observational notes, and intensive interviews, show that the stated public image holds for Cuban women (with modifications related to use of prescription drugs), but not necessarily for Cuban men.

ALCOHOL USE AND ABUSE AMONG CUBAN MALES

Self-reports of drinking frequencies among the male study subjects showed that over 50% claimed that they drank more than once a week, but nearly 40% drank less than once a week. The number of other drugs used by the study subjects had no statistically significant relationship with these drinking frequency figures. This may be due in part to the fact that minors have less ready access to alcohol than they do to illegal drugs. Nevertheless, the range of drinking frequencies in this group was reasonably well distributed, including abstinence (all boys under age 15) as well as light and heavy frequencies. Participant observation verified the self-reported drinking patterns of the male informants (Page & Gonzalez, 1980; Page, 1980).

Some of the older Cuban men who admitted to regular and heavy alcohol consumption were observably intoxicated (slurred speech, stumbling) on several occasions in the street setting. Despite their obvious loss of control, these individuals consistently denied that they were drunk. Denial is fairly common among alcoholics, and it therefore appeared at first that the Cuban heavy alcohol consumers were no exception. A later examination of Cuban traditional values revealed the cultural precedents for this behavior, however.

Younger Cuban males tended not to use alcohol as regularly as the older adult male study subjects. The youths' orientation toward intoxicated states also appeared at first to be less proscriptive than that of their older counterparts. They sometimes set out to "get all fucked up" (i.e., become so intoxicated that they are not fully conscious of their actions). If getting "fucked up" involved alcohol, the youthful abusers' parents or guardians looked upon the act with less prejudice than if it involved other drugs. As one 19-year-old youth explained it:

> He tells me every day that I get into my house . . . "You get caught stealing, I'm not going to get you out of jail; you get caught smoking grass, I'm not going to get you out of jail. If you get caught drunk, drinking beer, I'll get you out; that's normal, drinking beer, you know. You are having a party, you know, with friends and you drink some beer, okay, That's cool. If you get caught in a fight, I'll get you out. But for grass or stealing I'm not going to get you out."

For all its problematic potential, alcohol is the only drug available to Cuban young people that is tacitly sanctioned by Cuban parents. If children have trouble linked with alcohol, parents are much more likely to "look the other way" or "slap wrists" than for offenses involving marijuana, methaqualone, or cocaine.

The majority of the Cuban male interviewees were regular drinkers, but the range of alcohol consumption patterns demonstrated by the whole group included abstention and addiction. The relationship between these patterns and those found in the general Cuban population is tenuous, because the study subjects were all identified users of illegal drugs as well as alcohol. Nevertheless, observations in the broader Cuban community led to the conclusion that daily drinking to the point of intoxication was unusual among Cuban men. A consistently observed pattern among Cuban men outside the study population involved occasional excessive drinking at social gatherings.

CUBAN WOMEN AND ALCOHOL USE

The overall drinking pattern of the 132 Cuban women was relatively light. Compared to other race/ethnic groups in Miami, Hispanic women were strikingly disinclined to use alcohol on a regular basis, and 33 percent of the Cuban interviewees did not drink at all. Older Cuban women were less likely to use alcohol on a regular basis than younger Cuban women. Although 64% of the former group abstained from all alcohol use, of those who drank occasionally, only 3% (one individual) drank more than once a month. Furthermore, on their responses to the highly diagnostic items taken from the Michigan Alcohol Screening Test (Pokorny, Miller, & Kaplan, 1972), the 132 Cuban women showed no drinking pathology.[1]

---

[1] The Michigan Alcohol Screening Test (MAST) items used here included a "morning

Younger Cuban women allowed a wider range of social drinking contexts than did older Cuban women. The following young Cuban woman expressed her attitudes toward alcohol in terms of a transitional adjustment between her parents' drinking patterns and patterns learned in the United States:

INTERVIEWER: Did either of your parents drink?
SUBJECT: No!
INTERVIEWER: Do you ever drink?
SUBJECT: No. Only during family reunions, some wine, but I'm not the kind of person who drinks regularly. One a week, twice a week, if we get together I have a glass or two of wine, not much more.

Examination of the cultural background of Cuban drinking values helped to clarify the relationship between non-consumption of alcohol among older Cuban women and light or moderate alcohol consumption among younger Cuban women. Combined with other elements of Cuban cultural background, this information helped to explain the minor tranquilizer use of 132 female Cuban interviewees.

ALCOHOL USE AND CONTACT BETWEEN HISPANIC AND NON-HISPANIC CULTURES

To understand present-day patterns of alcohol consumption among Cubans in Dade County, it is essential to examine the process of adjustment to contact between the Cuban and North American cultures. Theoretical perspectives in the psychological literature (cf. Scopetta, Szapocznik, King, Ladner, Alegre, & Tiliman, 1977; Szapocznik, 1977, 1979; Szapocznik, Daruna, Scopetta, & Aranalde, 1979) have concentrated on various sources of stress related to acculturation (e.g., adjustment to speaking a new language, differences in interaction patterns, pace of daily activities). These perspectives, however, did not succeed in identifying specific features of acculturation that were statistically related with alcohol- or drug-using behavior (Page *et al.*, 1981). Obviously, situations of culture contact are stressful to the contactor, but the specific pressures of learning a new language or new interaction patterns do not appear to be sufficient to constitute correlates of measurable stress. For our purposes, then, we shall assume that Cuban immigrants who ex-

---

drinking" item, a "blackout" item, and several social environment items known to be highly diagnostic. No Cuban interviewee consistently responded in problematic fashion to these items. The three isolated positive responses were statistically insignificant and did not merit interpretation.

perienced exile perceive that process as stressful. We have ample evidence that this continues to be true among older Cuban Miamians (Page *et al.*, 1981).

Graves (1967) and Gordon (1981) examined situations of contact between Hispanic and non-Hispanic cultures in the United States, and defined key variables related to patterns of alcohol use in Hispanic populations:

1. Pre-contact patterns of alcohol and/or drug use among Dominicans, Guatemalans, and Puerto Ricans (Gordon, 1981).
2. Age and sex characteristics of the immigrant group (Gordon's comparison of Dominicans and Guatemalans).
3. Presence or absence of agencies of social control (e.g., families of procreation, other norm-setting groups) for the immigrant (e.g., Dominican or Guatemalan) or contactor (e.g., Mexican-American) group (Graves, 1967; Gordon, 1981).
4. Stratification of the host society to the extent that newcomers or outsiders (Mexican-Americans) do not have access to economic benefits (Graves, 1967).

These variables suggest direct and obvious logical linkages between the culture contact experience and observed patterns of alcohol use. The historical background of the Hispanic population's drinking tradition helps in the interpretation of observed patterns of drinking in the new environment. Graves (1967) demonstrates the relationship between alcohol use and access to economic benefits in a situation of culture contact. The composition of the immigrant group (age, sex, household makeup) will influence the immigrant group's drinking patterns. The presence or absence of socially controlling factors, such as accompanying spouses and children, influence male immigrants' drinking (Gordon, 1981). All these specific aspects of culture contact or immigration apply to the Cuban data presented here.

Historical Background of Cuban Drinking Patterns

The origins of the Miami Cubans' drinking patterns lie partly in Cuba's colonial history. Although it is doubtful that the pre-Colombian inhabitants of Cuba consumed alcoholic beverages, it is almost certain that the first Europeans to settle there brought drinkable alcohol with them. By the early sixteenth century, gambling and strong drink were established as the principal vices of the island's new settlers (Mouriño, 1947). Given Cuba's status as a check point for Spanish colonial shipping (Page *et al.*, 1981) and its growth as a sugar producer (Ortiz, 1963), alcoholic beverages were available in Cuba during its entire colonial period, which lasted until the beginning of this century. Despite this availability and the presence of a full range of drinking behaviors in the Cuban population, including both complete abstinence and extreme drunkenness (Gutierrez, 1948), the stimulating (or, what some medical researchers call potentiating) effects of alcohol were emphasized among Cuban consumers of alcohol. Cuban drinkers were not encouraged to drink to the point of drunkenness, but rather enough to encourage free social interaction and an atmosphere of *alegría* (happiness) or *vacilón* (merry-making).

Cuban old-timers recall the disparaging stereotypes assigned to immigrants in Cuba (especially Haitians) who had the reputation of being given to drunkenness (Page, 1982, p. 60). Especially abhorred were women who drank publicly and excessively (Castellanos, 1929) because this violated the ideal of the Cuban woman as the demure and viceless mainstay of the household.

According to traditional standards of drinking behavior, Cuban men had a somewhat greater latitude than women, being allowed to patronize taverns (the only women who would do such a thing were prostitutes) and to come home drunk on isolated occasions (a drunk woman was almost certain to have compromised herself sexually). Nevertheless, Cuban men also had to meet basic criteria of proper behavior while drinking: Slurred speech or speech more slurred than one's drinking mates' and loss of muscle motor control endangered a man's ability to assert himself in the heated debates and fast-flowing conversations and interactions characteristic of Cuban settings for public drinking. This need for control of one's physical and mental capacities did not prevent all Cuban men from drinking past the point of lost control, but it set behavioral limits within which most Cuban men remained when drinking.

Traditional Cuban recreational gatherings (*fiestas*) in which alcohol was present generally had an animated quality that reflected the stimulating function of that drug in these settings. Small doses of alcohol to loosen inhibitions and to encourage dancing and facile interaction were the rule. Only at these times were women permitted to drink recreationally in pre-Castro Cuba.

The values and behaviors described above characterize members of the middle and upper classes of Cuba before Castro, because members of these classes dominated the historical documentation on drug and alcohol use before the exodus to the United States. Those Cuban writers who deigned to chronicle the lives of rural or urban poor or working-class people did so in tones of disdain (cf. Ortiz, 1973; Castellanos, 1929; Gutierrez, 1948; Mouriño, 1947), and therefore patterns of recreational alcohol or drug use among these people were usually portrayed in terms of depravity or debauchery. The negative tone of Ortiz and other early Cuban writers appears to reflect class prejudice rather than objective observation. Even so, our observation of working-class Cubans in drinking establishments and our talks with them about their drinking practices give us reason to believe that the pre-revolutionary Cuban poor recognized and attempted to practice the same constraints on drinking behavior as did the Cuban middle and upper classes.

The post-immigration environment of Miami did not appreciably change the availability of alcohol for the newly arrived Cuban. Abundant liquor stores and bars catering to the growing population and the tourist trade offered a full range of alcoholic beverages, including the traditional rum.

The main cultural inconvenience for Cubans in Miami's beverage supply system was that the American "lounge-"style bar, with thick carpet, dark lighting, and a quiet, enclosed atmosphere, did not offer the Cubans the well-lit, animated, noisy, open drinking atmosphere to which they were accustomed. The high price of liquor licenses in Florida discouraged the establishment of Latin-style, hard-liquor-serving bars in Miami for several years after the first wave of immigration.

A more accessible way of starting a Cuban-style drinking place was the *cafetería* (coffee shop) in which customers could drink Cuban coffee, beer, or wine or eat a modestly priced meal. These are among the oldest Cuban-founded small businesses in Dade County, and they are still places where Cuban males gather to have a couple of beers and to discuss the issues of the day. Presently, however, Cubans and other Hispanics succeeded in establishing a wide variety of truly Latin-style drinking places, including everything from corner bars, through liquor-by-the-bottle dance halls, to feathers-and-sequins-show nightclubs in the style of Havana in the 1940s. Some of the more-acculturated Cuban young adults, however, have developed a preference for the Naugahyde lounge.

Use of psychotropic medications in Cuba before Castro was not widespread, but the cultural background of practices involving the use and dispensation of prescription pills is related to Cuban women's alcohol use patterns. In pre-Castro Cuba, physicians were often inaccessible to the general population both in rural areas and in working-class neighborhoods in the cities (Danielson, 1979). Cubans who had little or no access to physicians learned to resort to the local pharmacist for the prescription and the dispensation of medicines, based on a brief conference with the pharmacist. Some Cubans in Miami still practice this system of therapeutic drug dispensation (Page *et al.*, 1981; Gonzalez & Page, 1981; Scott, 1974).

The themes of control of intoxication for men, very restricted drinking for women, and informal distribution of medicines are important to our understanding of current alcohol and other drug use patterns among Cubans in Miami. Each theme relates to the analysis of specific observations and interviews conducted in this Cuban population.

DEMOGRAPHICS AND ECONOMICS OF THE CUBAN IMMIGRATION EXPERIENCE

The aggregate characteristics of the Cuban immigrants who came to the United States bear directly on the specific variables that Graves (1967) and Gordon (1981) have demonstrated to be related to alcohol use: (1) age and sex of immigrants, (2) presence or absence of agencies of social control, and (3) blocked access to the host society's economic benefits.

Between Castro's takeover in 1959 and 1982, approximately 1 million people left Cuba, choosing a life of exile rather than a life under a communist regime. Three quarters of these Cuban exiles currently reside in South Florida, mostly in Dade County.

Each stage of immigration had a distinctive socio-demographic character. The Cuban population in Miami began with the arrival of relatively well-educated and well-to-do refugees. The successful adaptation to exile of this early group of refugees contributed to a public perception of Cuban refugees as prosperous entrepreneurs (Alexander, 1966; U.S. News and World Report, 1967). Subsequent stages of Cuban immigration brought lower-middle and working-class Cuban exiles into South Florida, and researchers (Fleitas, 1976; Portes, Clark, & Bach, 1977; Prohías & Casal, 1973; Diaz, 1981) have pointed out that economic adjustments to life in the United States have been difficult for most Cuban refugees. Diaz (1981) estimates the median yearly household income of Cubans in Dade county to be at least $2,000 below that of white non-Hispanics in the same area. Prohías and Casal (1973) identify two of the predominantly Cuban zones of Miami as having low average education and income levels for heads of household.

Recalling also that many of the Cuban exiles were over age 40 *at the time of arrival,* we would expect large numbers of these older immigrants to be at a disadvantage in trying to learn English and in becoming self-supporting in this country. Page *et al.* (1981) and Gonzalez and Page (1981) have described some of these disadvantages, which include difficulty in learning English, difficulty in obtaining re-certification in this country, age discrimination, and the need of both spouses to work at lower status jobs. Older immigrants perceived these experiences as generally stressful, regardless of previous socioeconomic status, but the less educated and less employable of their number were at the greatest disadvantage in trying to support themselves in this country. Even after more than 20 years of exile, generalized immigration stress impinges on the lives of older Cuban immigrants.

Women accounted for nearly 60% of some waves of Cuban immigrants (especially the third and fourth in the 1960s). This relates directly to the question of employability, because service and light manufacturing were the only jobs available to Cuban women upon arrival. Household composition and disruption of extended family were also related to the proportion of women in the immigrant population. The ratio of women to men affected the continuity of alcohol use values because adult males were the principal agents for transferring Cuban values on alcohol consumption.

The demographic contours of the stages of Cuban immigration also affected agencies of social control for youth and heads of households. The vicissitudes of Castro's immigration policies, combined with the United States' reactions to these changes, reshaped Cuban exiles' nuclear families in many different ways. Some families sent children unaccompanied to the United States (during the freedom flights in the 1960s) and the parents arrived anywhere from six weeks to ten years later. Throughout the exodus, wives of political prisoners brought their children over early to be joined by their husbands as much as 20 years later. Opportunistic parents used the upset of immigration to break up marriages they perceived to be unsatisfactory. Some children lost one or both parents to revolutionary violence.

Cuban immigrants who came to the United States (mainly Miami) between 1959 and 1980 were not uniformly well educated or well connected. Their numbers included many poor, more women than men, some unaccompanied children, and many broken families. They did not all prosper after they arrived here, and many still live below the poverty level. The familial disruption and general stress they experienced during the immigration process comprise, along with economic upsets, the set of variables we have examined in terms of alcohol abuse.

HISTORICAL ANTECEDENTS AND PRESENT-DAY ALCOHOL USE

The cultural constraints that discouraged drunkenness among men in Cuba before the revolution were verified among the 72 male study subjects, through both observation and interview. Older male Cuban informants described a style of alcohol consumption in pre-revolutionary

Cuba that was moderate and closely linked with sociability. According to them, drinking was a part of a larger set of social activities among friends and other members of the community. Similar patterns obtained in Miami's Cuban neighborhood *cafeterías*. Participants in these settings seldom appear drunk, although they may spend hours there.

Exceptions test the rule, and among the male study subjects, there were some who occasionally appeared to be drunk. None except the youngest would admit to that level of intoxication. For example, one informant, observed several times in a state of advanced inebreity, was emphatic in his denial:

> Alcohol . . . I don't see that it has done me any damage. On the contrary, alcohol calms my nerves. I take a pair of drinks of alcohol. . . . *I'm not an alcoholic, nor has anybody seen me drunk.* . . . I have a drink of whiskey, I have two or three beers. . . . Why, being here I believe that it is even necessary because right now in my state, in the situation I'm in right now . . . here there's no place to go.

This informant's translated testimony sounds like the denial often articulated by alcoholics, but it is consistent with culturally defined attitudes regarding loss of control. There is also a note of personal desperation in this passage bespeaking an inability to find work because he lacks education or training. The same study subject later asserted implicitly that he was in control of his alcohol intake, at the same time indicating that he had established a level of tolerance common to alcoholics.

Other Cuban drinkers' testimony confirmed the pattern of denying intoxication among Cuban males. In fact, only 2 of the 72 users of illegal drugs interviewed and observed by Page testified that they might at times have a problem controlling their alcohol or other drug intake. The denial was consistent regardless of the variety of intoxicant; nearly all claimed that they controlled the intake of their drug of choice.

The youngest and the oldest male study participants had the highest observed frequency of episodes of intoxication. Older drunks denied their loss of control, whereas the youngest ones openly admitted it. This reflects differences in adherence to traditional values of control under the influence of drugs. The youngest inebriates had not internalized the proscription against losing control, and their system of values regarding intoxication represented the values of contemporary North American youth more than the Cuban drug-use values of the 1950s. Their older counterparts, on the other hand, had established abusive drinking patterns that resulted in intoxication in spite of their values forbidding such behavior. These old-timers had apparently achieved a disease or an addictive state that overpowered any value-based proscription against losing control. Hence they got drunk, but consistently denied that they had done so.

The age group in between the youngest and the oldest male Cuban study participants, older youths and young adults, tended to be wary of alcohol and its effects, expressing especial distaste for the hangover and accompanying side effects. Light drinkers expressed the following attitudes (translated) concerning alcohol:

SUBJECT: What I don't like is to drink much, I don't like to drink whiskey or vodka or anything like that. Beer, nothing more.
INTERVIEWER: Always beer?
SUBJECT: Yes, I don't like to mix drinks or any of that.

SUBJECT: I'm not a lover of drinking.
INTERVIEWER: Have you drunk little in your life?
SUBJECT: Yes, I never have had experience with drinking, I don't like it.
INTERVIEWER: Not even beer or wine?
SUBJECT: No.

INTERVIEWER: Did you drink MD 20–20?[2]
SUBJECT: Yeah, I've done that a couple of times. Makes me feel real bad.
INTERVIEWER: That's good reason for not liking alcohol. It's awful stuff.
SUBJECT: Yeah, It makes you vomit . . . disgusting.

Most of these study subjects desired control of their drug experience, and they were familiar enough with alcohol to realize that it could cause loss of control. They represent a syncretism of drug-use values: Control remained a positive value for them, but they accepted the wider repertoire of psychotropic drugs and altered states of consciousness valued by North American drug users.

In the drinking patterns reported by 132 Cuban females, there is clear evidence of the influence of traditional values. The Cuban traditions that forbid women to drink regularly are responsible for the lack of regular drinkers among the older female interviewees. The most frequent drinker among 34 Cuban women over 60 years old drank two to three times a month.

The following comments (translated) on alcohol use by older Cuban women illustrate their restrictive attitudes toward drinking:

I don't dance. I don't drink. I don't smoke.

---

[2] MD 20–20 is a cheap fortified wine, containing 20% alcohol (hence the numbers in the brand name). "Mad Dog," as it is sometimes called, is available in pint bottles, which are easily carried and quickly consumed.

I don't need to drink to have a good time.

In my entire life I've never sat in a bar, not even to have a Coca-Cola. No sir.

The younger, more acculturated Cuban women, like the young adult Cuban men, demonstrated syncretic attitudes toward alcohol. They retained traditional restrictive values, but modified them to permit slightly higher drinking frequencies.

THE RELATIONSHIP BETWEEN DEMOGRAPHIC FEATURES OF IMMIGRATION AND PRESENT-DAY ALCOHOL USE

The previous discussion has led us to accept the relationship between immigration and stress only in a general sense. Nevertheless, some features of the process of contact between Hispanic and non-Hispanic cultures are related to drinking behavior. Three of them, exclusion of the contactor population from economic benefits of the host culture, age and sex characteristics of the contactor group, and presence or absence of agencies of social control, are primarily demographic in nature.

Taken singly or in combination, the demographic factors would seem to predispose large segments of the Cuban immigrant population to heavy alcohol use. Many have not attained the prosperous life-style of the American Dream (Page *et al.*, 1981). Their inability to achieve the recognized life-style in the United States is stressful. The 72 male Cuban study subjects came from neighborhoods that Prohías and Casal (1973) identified as below Dade County medians in education and household income. Poverty was a common theme in their conversations and interviews. If drinking were an accepted strategy among Cubans for mediating stress due to perceived economic deprivation, as it was among Graves' interviewees, we might expect a large proportion of Dade County's Cuban population to drink heavily. Even though they are excluded from the economic benefits of the host society, the Cubans in Dade County do not have a reputation for drinking heavily. Exclusion from economic benefits, as a single variable, does not have the power to predict the observed conditions.

Age and sex characteristics of the Cuban immigrants contribute to their exclusion from economic benefits in the United States. A large segment of Dade County's Cuban population was female and over 60 at the time of our study (Metropolitan Dade County, 1983, p. 79). Similarly, the 132 female Cuban interviewees included 34 elderly women, 24 of whom lived in households that had incomes under $9,000 a year. This low level of income among the elderly Cuban women who responded to our interview reflected the fact that 42% of them lived alone, and an additional 37% lived with one other relative. This not only lowered the probability of multiple contributions to the household by extra wage earners, it also reflected a lack of extended family resources for these people. Most of those who lived alone did not have access to other relatives in South Florida. The employability of these women, as indicated by education levels, was low. Over one-half (55%) had not had more than an eighth-grade education. Even those who had lived in the United States for 20 years had not obtained well-paying jobs during their working years here. In addition, their English facility was generally low, which further hindered their employability. These characteristics have "trapped" elderly Cuban women in an unaccustomed life-style, which we may define as stressful.

If drinking were an acceptable strategy for mediating stress among Cuban women, we might expect to see some older Cuban women consuming alcohol regularly. In fact, we do not, because of their cultural background of Cuban drinking values. Elderly Cuban women are often under the general stress of forced immigration combined with the specifically alcohol-related condition of living in poverty. Furthermore, they lack recourse to relatives or friends to cope with their predicament. Increased drinking is sometimes a response to such a convergence of difficulties (Schuckit & Miller, 1976). Nevertheless, these women hardly drink at all. The cultural background variable demonstrates its explanatory power here. Elderly Cuban women were brought up to believe that drinking was not for them. This belief does not forbid the use of pills to alleviate nervous conditions, however, and the Cuban women in our study, especially the elderly, have apparently adapted to their chronically nervous state by a chronic and frequent use of minor tranquilizers.

The rate of recent minor tranquilizer use among Cuban women is significantly higher ($p = .01$, Chi-square) than for either black or white non-Hispanic women. As the age in the sample of Cuban women decreased, the rate of recent tranquilizer use decreased. Recalling that older Cuban women tended not to drink, and that younger Cuban women drink somewhat more frequently than older Cuban women, we hypothesized that the minor tranquilizers were substitutes for alcohol among the older women. These women perceived life in Cuba as non-stressful, and there they had needed neither alcohol nor pills to mediate stress on a regular basis. They perceived life in Miami as constantly stressful, but they could not use alcohol or rely on interpersonal resources (family and friends) to mediate stress. They traditionally identified regular alcohol use with disreputable women, and their families and other social support resources were no longer available. The availability of minor tranquilizers, which were identified with culturally acceptable female behavior and traditional pharmaceutical practice in Cuba, made "nerve" pills the major alternative coping strategy for many Cuban women.

Younger, more acculturated Cuban women illustrated the variability among the 132 female interviewees. They tended to drink somewhat more frequently than their older counterparts, but they used tranquilizers less frequently. The younger Cuban women were suspicious of pills as stress mediators, in spite of their mothers' encouragement to use Librium or Valium or Tranxene when they felt anxious (Gonzalez & Page, 1981). They also did not perceive the immigration experience as especially stressful on a day-to-day basis, so they did not feel the need to use pills to cope with daily pressures. Because they had the advantage of education over the older study subjects, Cuban women under age 60 were less likely to experience the poverty and lack of alternatives that afflicted the elderly women.

The demographic feature of the immigration process that was most strongly related to male alcohol and other drug use involved the absence of agencies of social control, especially the disruption of the nuclear family of orientation. The migration process had a variable impact on the families that came to the United States. Many households lost one or both parents as a result of immigration. This kind of loss is statistically related to drug use (Page *et al.*, 1981, p. 319) regardless of the sex of the lost parent. Our interpretation of this result, based on an analysis of life-history materials, is that the loss of the parent made the individual more likely to come under the influence of older peers outside of the home. According to the interview materials, Cuban children who, through the attenuation of the nuclear family, lacked positive supervision or a nurturing environment in the home were drawn to street life. In the street setting, they came under the influence of older peers who acted as their instructors in drug use.

Drinking alcohol was one of the drug-use patterns learned in the street setting, and it varied according to the cultural orientation of the older peer instructors. If they were men from the streets of Havana, the alcohol- and drug-use values they taught emphasized self-control; if the older peers were slightly older than the initiates, the values taught might emphasize oblivion-seeking or some mix of self-control and mind-expansion.

The syncretism of alcohol- and drug-use values that took place in

Miami's streets was highly complex, involving a mixture of traditional Cuban values (not dissimilar to the middle-class values described earlier), an Age of Aquarius ethos (a product of the 1960s revolution in drug-use values), and modern, nihilistic forms of oblivion-seeking. According to our experience, the values that encouraged oblivion-seeking among the very young drug users were the most destructive. Cuban boys who sought to get as drunk or as crazy as possible on alcohol or drugs or both commonly fought or drove cars while under the influence of alcohol and methaqualone. This behavior contrasted sharply with that of the older, more traditional Cuban interviewees, who drank beer or smoked marijuana while engaging in rapid-fire discussions of sports or politics.

OTHER PATTERNS OF ALCOHOL USE

Cuban males who used only alcohol (observed in the context of the study of drug users, but not formally interviewed) exhibited a pattern that also constituted a health risk to themselves and the general population. In festive gatherings, they reminisced about their lost homeland and, reacting to the pain that these memories brought, they drank to the point of drunkenness. After they had become drunk, they would attempt to do things for which they no longer had a capacity, such as swimming, driving, or operating machinery. Their denial of intoxication was so strong that their families' warnings could not dissuade them from these activities. The risk of accident from these activities, both to the nostalgia drinkers themselves and to the general population is obvious.

Interviews with clinicians specializing in Cuban families (Page, 1980; Page et al., 1981, p. 150) brought out a pattern that tied together several of the issues discussed here. Among Cuban families who presented for treatment, the presenting problem was often marijuana experimentation by the children. As the clinicians probed the family's behavior, they found that the father of the youthful marijuana smoker was drinking heavily and/or that the mother was using minor tranquilizers heavily. The clinicians' interpretation of this scenario was that the family only recognized the youthful marijuana use as problematic, even though the familial problem was more broadly based. The clinicians eventually brought to light the other, perhaps more serious, problems of heavy drinking by the father and/or pill taking by the mother, but they noticed consistent patterns of denial by the adult male drinker and the adult female pill user. The clinicians' experiences further verify the traditional values of controlling drugs' effects among males and seeking alternative stress mediators among females. The effects of nuclear family disruption, either by the absence of one parent or the reaction to the social and economic pressures of immigration, are reflected in the childrens' involvement in the use of illegal drugs. Unfortunately, these clinical perspectives are probably skewed by the tendency of Cuban families not to present for family therapy. Furthermore, those families who present do so in reaction only to use of illegal drugs by or extreme antisocial behavior of their children. They do not attach as much importance to youthful experimentation with the most dangerous of drugs, alcohol.

SUMMARY

The patterns of drinking presented here are generally moderate in nature. Cultural values regarding states of intoxication proscribe discernible drunkenness in both men and women. The stress that accompanies immigration has interacted with pre-migration proscriptive values in identifiable ways. Cuban women, traditionally denied drinking as a coping alternative, often respond to generalized immigration stress, as well as specific disruption or economic status, through the use of minor tranquilizers. This trend is modified in younger, second-generation Cuban women through partial adoption of North American values, which allow social drinking among women. Cuban women who are under 60, are U. S. citizens, and have some college education are less likely to use minor tranquilizers regularly and more likely to use alcohol occasionally than their older, less educated, non-citizen counterparts. Therefore, patterns of chronic excessive drinking are very unlikely among Cuban women.

Among Cuban men, the traditional values that prescribe self-control and proscribe outward evidence of intoxication have undergone little modification in the Miami setting. With few exceptions, older Cuban males consistently deny losing control of their motor functions regardless of their state of observable intoxication. Younger Cuban males appear to respond especially to the stress of nuclear family attenuation by trying a wide variety of street drugs, in addition to alcohol. Although less likely to deny states of extreme intoxication than older Cuban males, the young are eventually inculcated into a street subculture that proscribes the loss of motor control due to drug or alcohol consumption in this setting. The value that defines the limits of street comportment directly descends from traditional Cuban values prescribing self-control among males. It is not, however, simply a case of intoxicated males perceiving themselves to be in control of the effects of alcohol and other drugs. In most cases, those who believe in maintaining control do so by limiting dosage. The unreliability of street drugs leads to "accidents" of dosage among poly-drug users, causing loss of control in acute situations. Heightened, exile-related stress can push dosage past recognized limits among alcohol-only users in Miami's male Cuban population. Older Cuban males, because they lack alternatives for coping with exile-related stress, are perhaps more likely to overstep the bounds of the male self-control prescription in Miami than they might have been in Cuba. They repre-sent, in our opinion, the group at greatest risk of alcohol problems in Miami's Cuban population.

REFERENCES

Alexander, T. Those amazing Cuban emigrees. *Fortune,* 1966, Oct., 144–149.

Castellanos, I. *La delincuencia femenina en Cuba.* Havana: Imprenta "Ojeda," 1929.

Danielson, R. *Cuban medicine.* New Brunswick, NJ: Transaction Books, 1979.

Diaz, G. M. (Ed.) *The evaluation and identification of policy issues in the Cuban community.* Miami: Cuban National Planning Council, 1981.

Fleitas, R. F. *Adjustment without assimilation: The Cubans in the United States 1959–1976.* Master's thesis, University of Miami, Coral Gables, Florida, 1976.

Gonzalez, D. H., & Page, J. B. Cuban women, sex role conflicts and use of prescription drugs. *Journal of Psychiatric Drugs,* 1981, *13,* 47–51.

Gordon, A. J. The cultural context of drinking and indigenous therapy for alcohol problems in three migrant hispanic cultures. *Journal of Studies on Alcohol,* 1981 (Suppl. 9) 217–239.

Graves, T. D. Acculturation, access, and alcohol in a tri-ethnic commuinity. *American Anthropologist,* 1967, *June–August,* 306–321.

Gutierrez Delgado, L. *Estadística moral de Cuba.* Havana: Seoane, Fernández, Cia., Impre-sores, 1948.

Metropolitan Dade County, Department of Human Resources. *Dade County characteristics.* Miami: Human Resources, 1983.

Mouriño Hernandez, E. *El juego en Cuba.* Habana: Ucar, Garcia, Y Compania, 1947.

Ortiz Fernández, F. *Contrapunteo Cubano del tabaco y el azucar.* Havana: Universidad Central de las Villas, 1963.

Ortiz Fernández, F. *Hampa Afro-Cubana: Los negros brujos.* Miami: Ediciones Universal, 1973.

Page, J. B. The children of exile: Relationships between the acculturation process and drug use among Cuban youth. *Youth and Society,* 1980, *11,* 431–447.

Page, J. B. A brief history of mind-altering drug use in pre-revolutionary Cuba. *Cuban Studies,* 1982, *12,* 57–71.

Page, J. B., & Gonzalez, D. H. Drug use among Miami Cubans: A preliminary report. *Street Pharmacologist,* 1980, *III,* 1–4.

Page, J. B., Gonzalez, D. H., McBride, D. C., McCoy, C. B., & Forment, C. *The Ethnography of Cuban drug use.* Report submitted in fulfillment of NIDA Grant No. 1 RO1 DA 02320-02, 1981.

Pihl, R. O., Marinieri, R., Lapp, J., & Drake, H. Psychotropic drug use by women: Characteristics of high consumers. *The International Journal of the Addictions,* 1982, *17,* 259–269.

Pokorny, A. D., Miller, B. A., & Kaplan, H. B. The brief MAST: A shortened version of the Michigan Alcoholism Screening Test. *American Journal of Psychiatry.* 1972, *129,* 342–345.

Portes, A., Clark, J. M., & Bach, R. L. The new waves: A statistical profile of recent Cuban exiles to the United States. *Cuban Studies,* 1977, *7,* 1–32.

Prohías, R., & Casal, L. (Eds.) *The Cuban minority in the US: Preliminary report on need identification and program evaluation.* Boca Raton: Florida Atlantic University, 1973.

Schuckit, M. A., & Miller, P. L. Alcoholism in elderly men: A survey of a general medical ward. *Annals of the New York Academy of Sciences,* 1976, *273,* 558–571.

Scopetta, M. A., Szapocznik, J., King, O. E., Ladner, R., Alegre, C., & Tiliman, W. S. *Final report: Spanish family guidance clinic-Encuentro.* Report to NIDA on Grant No. 1(8) DA 01696–03, 1977.

Scott, C. Health and healing practices among five ethnic groups in Miami, Florida *Public Health Reports,* 1974, *89,* 524–532.

Szapocznik, J. *Role conflict resolution in Cuban mothers.* Doctoral dissertation, University of Miami. (Ann Arbor, MI: University Microfilms, 1977.)

Szapocznik, J. *Acculturation, biculturalism and adjustment among Cuban Americans.* Paper presented at the Annual Meetings of the American Association for the Advancement of Sciences, Houston, Texas, January, 1979.

Szapocznik, J., Daruna, P., Scoppetta, M. A., & Aranalde, M. A. The characteristics of Cuban immigrant inhalant abusers. *American Journal of Drug and Alcohol Abuse,* 1979, *4,* 377–389.

U.S. News and World Report. Cuban success story in the U.S., 1967, *62,* 104–106.

VI

Case Studies
Asian Groups

18

Japanese-American
Drinking Patterns

HARRY H. L. KITANO, HERB HATANAKA,
WAI-TSANG YEUNG, and STANLEY SUE

There is a common belief that the Japanese, as well as most other Asian groups in the United States, do not drink as much as their American and European counterparts and are therefore relatively immune to problems associated with alcohol. The existence of a strong family system reinforcing a moderate drinking style is cited as one factor behind the low rates of alcohol problems, and a physiological reaction to alcohol, known as the "Oriental flushing reflex," is thought to be another important variable. Research evidence to validate these and other impressions is scarce, but the observation that Japanese-Americans do not appear in any significant numbers at alcohol treatment programs or that there are so few visible Japanese drunks is often cited as supportive evidence.

We will attempt to ascertain the validity of these perceptions by presenting data on alcohol consumption among Japanese-Americans. We will also present information on alcohol consumption in Japan, with a warning that the Japanese in Japan and the Japanese-American are

HARRY H. L. KITANO • School of Social Welfare, University of California, Los Angeles, California 90024.   HERB HATANAKA • Special Service for Groups, 1313 West 8th Street, Los Angeles, California, 90017.   WAI-TSANG YEUNG • School of Social Work, University of Hong Kong, Shatin, Hong Kong.   STANLEY SUE • Department of Psychology, University of California, Los Angeles, California, 90024.

different populations in spite of their common ancestry. For example, the Japanese in America are a small, powerless minority group with a long history of facing racial prejudice and discrimination. They have had to adapt to majority group power, and questions of acculturation, integration, pluralism, and identity remain as constant issues. Conversely, the Japanese in Japan are in the majority and reflect dominant group status. Their defeat during World War II and the Allied occupation brought in strong American influences, but even then the Japanese were numerically superior and were able to retain many of their cultural traditions. Therefore, Japanese-Americans reflect an immigrant group exposed to a different culture, whereas the Japanese in Japan present a more orderly cultural progression in spite of the effects of World War II.

Our data on Japanese in Japan rely primarily on a review of research reports; the data on the Japanese-American rely heavily on a survey of alcohol drinking patterns of a randomly selected 295 heads of households residing in Los Angeles. We will also present information on the Japanese in Hawaii.

ALCOHOL BELIEFS AND DRINKING PATTERNS IN JAPAN

The Japanese have had alcoholic beverages, primarily saké, for most of their history, but prior to World War II Chafetz (1964) reports that alcoholism was rare. Shintoism and Buddhism were generally tolerant of drinking in moderation and provided clear prescriptions for drinking by age, sex, and role. Hayashi (personal communication October, 1982) remembers two common sayings: "Saké is beneficial to your health more than any other drug," and "Saké is the water which makes people go insane," as indicators of the Japanese philosophy of tolerance and moderation regarding alcohol consumption.

Since the end of World War II, changes in alcohol drinking patterns have been observed. For example, Nukada (1972) reports that per capita alcohol consumption rose from 2.3 liters in 1934 to 5 liters in 1975, with particularly steep increases for women and youth. Cirrhosis deaths also rose from 5,678 in 1950 to 13,564 in 1972. Wynder and Hirayama (1977), in comparing consumption of alcohol in Japan and the United States, wrote that in 1971, the U.S. population over age 15 consumed 9.31 liters per capita, whereas in 1974 the Japanese consumed 6.65 liters per capita. About 46% of all alcohol consumed in the United States was whiskey; in Japan saké consumption was high in the rural areas, and beer was the preferred drink in the urban areas. However, young drinkers in Japan are turning toward whiskey, usually with water (*mizuwari*) or ice.

A survey by the Health and Welfare Ministry (*Hokubei Mainichi*) shows widespread drinking among Japanese males. In 1981, over 37% reported that they drank every day, compared to the 32.1% who reported that they drank every day in 1966.

Although there are some minor discrepancies in the reported research, two generalizations appear appropriate concerning alcohol consumption in Japan: First, alcohol consumption has increased and second, although it remains primarily a male activity, more women and youth are drinking.

Drinking Styles

An important question in dealing with alcohol consumption relates to the various styles of drinking. Ohashi and Nishimura (1978) describe four popular styles of drinking in Japan.

1. *Banshaku,* or drinking at dinner. This type of drinking is done primarily by family men who wish to relax after a day at work. This type of drinking probably goes far back in the Japanese culture.
2. Ceremonial drinking to celebrate annual events, such as *Obon* or New Year's. This type of drinking also has long historical roots.
3. Drinking after meetings and conventions. The purpose is for relaxation, for a show of friendship, and to facilitate communication. This style of drinking is most prevalent among salaried men and is more of a post-World War II phenomena.
4. Drinking with one's close friends after work. This style has also grown much more popular after World War II.

Each of the styles refer primarily to male drinking, although we have observed and heard of more housewives joining in the banshaku. Since banshaku drinking is done primarily in the home, the chances of some of the problems associated with drinking occurring, such as drunk driving, are slight, and drunkenness remains invisible.

Drinking at ceremonies is more public and could spill out into problem behavior except for the assistance of friends, colleagues, and neighbors. Drinking at weddings and other occasions can also be quite heavy, but social control by friends and extended family in a public setting reduces the probability of problem behavior.

Drinking after a meeting or seminar also involves a collegial network. After a seminar at Meiji University in the summer of 1982, the senior author was taken for drinks *(kon-shin-kai)* and dinner with a selected group of faculty members. It was a time to "express feelings" and the drinking was steady but mild, in that no one appeared to be inebriated.

Hayashi (personal communication, 1982) mentions the wide variety of occasions where drinking occurs. There is the New Year's party *(sin-nen-kai,)* the farewell party *(kanso-kai,)* the welcoming party *(kangei-kai,)* the recreation party *(ian-kai,)* the appreciation of someone's service party *(kangei-kai,)* the encouragement party *(gekirei-kai,)* and the year-end party *(bonen-kai.)* There is also the *setsutai* style of drinking, which is to drink with people in order to obtain favors.

After-work drinking can lead to a high consumption of alcohol. Nakagaki (personal communication, June, 1982) describes a style called *hashigo-nomi*, in which the first step might be drinking at a beer garden, followed by drinking at dinner, then to a favorite bar with friends for a drink, then possibly to another bar until closing time. The changes in drinking sites can shift the drinking style from a formal to a more informal one and can result in much closer friendship ties. The term refers to a stepladder and step by step drinking.

Hayashi also mentions another style of drinking, called *tsukiai*, which refers to almost any type of informal social drinking that serves several purposes. One common *tsukiai* group is that of married men getting together in order to get away from nagging wives. Supervisors may also take subordinates out to drink in order to find out what is on their minds; colleagues may go out to drink to gossip, exchange frustrations, and share information. These drinking sessions may be likened to therapy sessions at which problems are aired and there is a general talking-out of frustrations and concerns.

The major generalization concerning the Japanese drinking style is its intimate relationship to ongoing social life. Drinking serves a variety of purposes: it is the social cement that binds individuals into group life; it serves to ease communication barriers in a tightly stratified society; and it serves as a vehicle to air problems.

However there are changes. Sargent (1973) has written that Japanese drinking is changing from a convivial pattern to a self-oriented style, so that some modern Japanese drinkers may drink to relieve anxiety and tension rather than to enhance social cohesion. Shibata, Masuda, & Aoki, (1978) sees psychological dependence and depression as strong motives for drinking, whereas Chafetz (1964) mentioned personality disturbances and sociocultural factors as reasons.

Availability

Alcohol is readily available in Japan. It is even dispensed in vending machines so that any person wanting a drink has no problem obtaining it. Much of Japanese drinking is done at "favorite bars." Most bars sell much more than liquor; there is a circle of regulars and an ongoing social group and a high degree of interaction among bartenders, hostesses, and customers. Most bars serve male customers, although occasional bars cater to housewives. Bars attempt to create distinctive social environments; customers often have their own bottles and drinking cups and there are bars where customers sing to the accompaniment of pre-recorded background cassettes. Regular customers are taken care by the *mama-sans* even to the point of private taxis so that the hazards of drunk driving are minimized. These services are not inexpensive, although in many instances the tab is picked up by the individual's company. For young people who have to pay their own way, there is the *compa* or pub and "snacks," which are different from the nightclubs and bars because customers are kept apart from the hostesses. These drinking sites often serve as a place to pick up boyfriends or girlfriends. There are also the small *nomiyas* where drinks and snacks are inexpensive. The Japanese style of drinking is generally away from the home, which is also typical of the way Japanese generally entertain.

Treatment

A clear-cut picture of the treatment of alcoholism is difficult to obtain in Japan for a variety of reasons. Holmgren (1976) mentions that the Japanese believe that drinking is a private matter and that the authorities take a *laissez-faire* attitude toward drinking and most other social problems. There is also the impression that many alcoholics are shielded by their families and only go for treatment in the late stages of their illness.

Drinking is often associated with social problems. Hayashi (1978) writes that excessive drinking was involved in 37 of 47 cases that appeared before a divorce, conciliation court and that physical violence by intoxicated husbands frequently occurred. A rise in weekend intoxication caused in part by an increase in leisure time has resulted in higher traffic accident rates. The historically lenient attitude toward drunk drivers has also changed, so that currently there are heavy fines and the suspension of driving privileges.

Japan has its own version of Alcoholics Anonymous (*danshukai.*) A speech reported in the Roche Report (Fukada, 1982), outlined the features of the All Nippon Sobriety Association (ANSA). Patterned after Alcoholics Anonymous (AA), it fights alcoholism by integrating doctors, local government, and clubs in the treatment effort. The association is currently comprised of 293 clubs with 42,000 members throughout Japan.

Fukuda cited a follow-up study of 110 patients treated over a five-year period as an example of the success of the ANSA approach. He reports that 27 of the 56 patients who joined the club and participated in the meetings abstained from drinking and were leading normal lives (42.2%), but of the 54 patients who did not participate in the club, only one (less than 1%) was leading a life of sobriety. Unlike AA, ANSA publishes the names of its members, since in Japan alcoholism is not viewed as immoral behavior.

It is also interesting to note the clinical history of the 110 alcoholic patients studied by Fukuda. Of these patients, 89% had a clinical history of physical or mental disorders and diseases of one kind or another. The picture is that of individuals less integrated into the ongoing Japanese social structure.

THE ORIENTAL FLUSHING REFLEX

There is evidence that Asians are more prone to the flushing reflex than Caucasians. Cohen (1979) writes that a large number of Orientals manifest a facial flush and a temporary reddening of the face, neck, and upper chest with just one drink. Some American Indians and some members of other cultures may also show this hypersensitivity to ethanol.

Research confirms the existence of this reflex. Wolff (1973) compared adult Caucasians with adult Japanese, Taiwanese, and Koreans and found that 83% of the Asians reacted to alcohol with visible flushing, compared to only 2% of the Caucasians. Similar findings were reported for infants. Seto *et al.* (1978), Zeiner and Paredes (1978), Goodwin (1979), and Hanna (1976) also reported similar results.

Ewing, Rouse, and Pelizzari, (1974) concluded that physiological factors were responsible for the sensitivity of Chinese and Japanese to alcohol. Cohen (1979) reports that the precise cause is unknown, but it is assumed that the accumulation of acetaldehyde in the body could account for some of the unpleasant symptoms.

The reaction is similar to that induced by the drug disulfram (Antabuse), which is used to treat alcoholics, since it causes flushing, headache, nausea, sweating, blurred vision, and generally unpleasant symptoms. Therefore, there is speculation that those with the syndrome have a built-in alcohol "Antabuse" reaction that protects them from over-drinking.

This line of reasoning has been developed by some alcohol researchers in Japan. Kawakami reported in a newspaper article (*Kashu Mainichi*, 1982) that with only a small amount of liquor, the Japanese begins to "glow like a beacon" and to become incoherently drunk. This inability to hold liquor is hypothesized as one reason why so few Japanese become alcoholics in a land where there is so much drinking and so many drunks.

Although the evidence demonstrates the existence of an Oriental flushing reflex, there is contradictory evidence concerning its relationship to alcohol consumption. There are undoubtedly many Asians who abstain from drinking because of discomfort or flushing, but Sue, Zane, and Ito (1979) also found a non-significant negative correlation between the reflex and drinking in self-reports.

Furthermore, Asians were less likely than Caucasians to endorse a questionnaire statement that alcohol harms the body. Although Asians and Caucasians have different physiological responses to alcohol consumption, the extent to which this accounts for their different drinking patterns is still difficult to assess.

JAPANESE-AMERICANS

Research studies concerning alcohol consumption among Japanese-Americans are scarce. Available findings suggest that Japanese-Americans (as well as most Asian groups in America) generally drink less than white Americans (Kotani, 1982), that part of the reason may be attributed to the flushing reflex (Cohen, 1979), and that very few Japanese-Americans participate in alcohol treatment programs.

Chu, Fertig, Sumii, and Yefsky, (1978) found that Japanese-American families were permissive toward drinking, but that there were generational differences, with the first-generation immigrant (the Issei)[1] being the least permissive. Japanese-Americans were characterized as having low social pressures to drink and negative sanctions for excessive drinking and as not showing violent or destructive behavior. The heaviest drinkers were middle-aged Nisei males (many of whom grew up in World War II concentration camps and served in the armed forces) and such professionals as lawyers and businessmen, who regarded drinking

---

[1] *Issei*, the first-generation immigrant, born in Japan; *Nisei* or second-generation American-born children of the Issei; and *Sansei* or third-generation children of the Nisei.

as necessary to their work. Gambling was hypothesized as a possible substitute for drinking. There was little evidence of drinking among Issei and Nisei (second-generation) females.

Kotani (1982), in a number of articles in the *Hawaii Herald*, presented a picture of Japanese-American drinking in Hawaii. In a 1979 state substance abuse survey, heavy drinkers (defined as having the equivalent of four beers or more per day) by ethnicity were as follows: Caucasians, 40.6%; Hawaiians, 19.4%; Japanese, 11.4%; Filipinos, 8.8%; and Chinese, 2.2%.

There are apparent similarities in the function of drinking between the Japanese in Japan and those in Hawaii. Kotani cites Bickerton's study of drinking among the Japanese in Hawaii where alcohol is perceived as the social cement that binds individuals together, rather than as a vehicle to satisfy individual needs or to alleviate frustrations. A similar explanation can be found in Japan. Ceremonies are also occasions for drinking and hosts often feel insulted when guests do not drink (Kotani, 1982).

However, in presenting a case study of an alcoholic, Kotani indicates that the overt motives for drinking included alienation and culture clash. The individual in the case finally attended an AA meeting where only one other person of Japanese ancestry was in attendance. Of an estimated 6,157 alcohol abusers among Japanese-Americans in Hawaii, only 56, or 0.9%, were admitted for treatment into the state system. In 1980, Caucasians accounted for 72.5% of all admissions to alcohol treatment programs, whereas the Japanese accounted for only 3.0%. A 1973 survey of clergymen indicated that the Japanese tended to rely on church leaders for advice on alcohol rather than to seek assistance from professional and institutional treatment. Studies on mental health problems (Kitano, 1982) also indicate the use of clergy and other non-professional mental health workers over the psychologist, psychiatrist, and social worker. Clergymen are generally the most familiar and available resource.

Drinking in Los Angeles

Data for Los Angeles were gathered from a randomly selected, representative sample of Japanese surname heads of household in 1980. The 1980 census reported 116,543 persons of Japanese ancestry residing in Los Angeles County, which makes this group the most populous concentration of Japanese on the mainland.

Findings

Alcohol Drinking Patterns

Table 1 compares Japanese-American alcohol drinking patterns in Los Angeles with the drinking patterns of other Californians (Cameron, 1981). Since alcohol consumption varies widely by area, the Cameron California study was more likely to be comparable to the Los Angeles area sample than were studies based on national norms.

The Kitano sample of males differed significantly from the Cameron sample in alcohol consumption. The Japanese-Americans had more ab-

TABLE 1.
Comparison of Japanese-Americans (Kitano) and California (Cameron) Drinkers by Sex and Age and by QFV[a]

			Percentage					
	N		Abstainer		Moderate		Heavy	
Age	Male	Female	Male	Female	Male	Female	Male	Female
29 and under								
Cameron	130	133	4.6	11.3	68.5	67.7	26.9	21.0
Kitano	33	13	30.3	30.8	33.3	53.8	36.4	15.4
			Male:	Significant at p *.001				
			Female:	Non-significant				
30–49								
Cameron	138	156	10.1	10.9	78.3	70.5	11.6	18.6
Kitano	81	16	22.2	68.8	42.0	12.5	35.8	18.8
			Male:	Significant at p *.001				
			Female:	Significant at p *.001				
50 and older								
Cameron	137	153	19.7	18.3	70.8	66.7	9.5	15.0
Kitano	144	31	48.2	74.2	41.2	19.4	10.5	6.5
			Male:	Significant at p *.001				
			Female:	Significant at p *.001				
Total								
Cameron	405	442	11.6	13.6	72.6	68.3	15.8	18.1
Kitano	228	60	36.4	63.3	40.4	25.0	26.3	11.7
			Male:	Significant at p *.001				
			Female:	Significant at p *.001				

[a] QFV, quantity, frequency, and variability.

stainers (36.4 to 11.6%), more heavy drinkers (26.3 to 15.8%), and fewer moderate drinkers (40.4 to 72.6%).

The Japanese-American female sample also differed significantly from the California sample in alcohol consumption. The Kitano sample had more abstainers (63.3 to 13.6%), whereas the Cameron sample was higher in the moderate drinker category (68.3 to 25.0%) as well as the heavy drinker category (18.1 to 11.7%).

Age also made a difference. All comparisons in the age categories showed significant differences between the Kitano and Cameron samples except for the 29-year-old and under female. The largest differences in drinking were in the moderate and abstainer categories; the California sample consistently had close to twice the number of moderate male drinkers than the Kitano Los Angeles Japanese sample; the reverse was true for the abstainer category. The differences for female abstainers were especially large.

FOREIGN-BORN VERSUS AMERICAN-BORN JAPANESE

There appears to be a Japanese style of drinking among those Los Angeles County residents who were born in Japan. Birthplace is a significant factor in alcohol consumption (Table 2). For example, 51.8% of the Japan-born males were heavy drinkers, compared to 17.3% of the American-born Japanese male. Cameron's California-born sample of heavy drinkers was 15.8%.

The findings indicate that the consumption styles in modern Japan have been brought over by the new immigrants. The United States-born Japanese have drinking styles similar to the California sample in terms of heavy drinking (17.3 to 15.8%), but they also have a much higher proportion of abstainers (38 to 11.6%) and a lower proportion of moderate drinkers (44.7 to 72.6%). The differences are statistically significant.

FREQUENCY OF CHURCH ATTENDANCE

Church attendance is significantly related to alcohol consumption (see Table 3). Among our heavy drinkers, only 4% attended church regularly, 17.6% seldom attended, and 29.2% rarely or never attended. Conversely, the picture for the abstainers was the opposite; 71% attended church regularly, 50% seldom attended, and 43.3% rarely or

TABLE 2.
Comparison of Foreign-Born Japanese to U.S.-Born Japanese in Los
Angeles County by Sex and Age and by QFV,[a] 1980

Age	N	Percentage		
		Abstainer	Moderate	Heavy
29 and under (male)				
Foreign-born	13	15.4	23.1	61.5
U.S.-born	26	40.0	40.0	20.0
Cameron	130	4.6	68.5	26.9
		Foreign-, U.S.-born: significant at $p < .05$		
		Kitano, Cameron: significant at $p < .001$		
30–49 (male)				
Foreign-born	28	10.7	14.3	75.0
U.S.-born	53	28.3	56.6	15.1
Cameron	138	10.1	78.3	11.6
		Foreign-, U.S.-born: significant $p < .001$		
		Kitano, Cameron: significant $p < .001$		
50 and older (male)				
Foreign-born	15	66.7	33.3	0
U.S.-born	106	42.5	39.6	17.9
Cameron	137	19.7	70.8	9.5
		Foreign-, U.S.-born: significant at $p < .001$		
		Kitano, Cameron: significant at $p < .001$		
Total				
Foreign-born	56	26.8	21.4	51.8
U.S.-born	179	38.0	44.7	17.3
Cameron	405	11.6	72.6	15.8
Japanese female	60	63.0	25.0	12.0
Male: Foreign- and U.S.-born significant		$p < .001$		
Male: Kitano & Cameron		$p < .001$		
Male: Total		$p < .001$		

[a] QFV, quantity, frequency, and variability.

never attended. The moderate drinkers had similar habits of church attendance.

The ethnic church still plays an important role in the Japanese community. Most of our sample who identified a church aligned themselves with Japanese churches. Although the influence of the church may have diminished from earlier times, churches still serve both religious and social needs to mixed generations of Japanese-Americans. Our data in-

TABLE 3.
Japanese-American Drinking Patterns by Frequency
of Church Attendance

Frequency	N	Percentage		
		Abstainer	Moderate	Heavy
High	24	71	25	4
Low	74	50	32.4	17.6
Rarely–never	178	43.3	27.5	29.2
		Significant at $p < .05$		

dicate that regular churchgoers tend to be abstainers, whereas those without church ties tend to be heavy drinkers.

SOCIAL CONTEXT OF DRINKING

Table 4 presents data on the social context of drinking. Significant differences were found in the serving of liquor at neighborhood functions; the importance of Japanese friends in terms of drinking; and the importance of friends in drinking. The social context of drinking involves others, especially Japanese acquaintances, so that there is an "in-group" drinking style. In one of our conferences for the Japanese community, several Japanese-Americans mentioned that they felt comfortable drinking with Japanese-Americans, but were hesitant about having more than one drink with non-Japanese companions (unless they were very close friends). This drinking style may be one reason why the dominant community perceives the Japanese-American as a non-drinker. It should also be noted that the old dictum, "If you want stop drinking, change your friends," holds true, since there was a significant relationship between drinking and having friends who drank heavily.

DRINKING BY GENERATION

From our information sources, we will provide a summary of the drinking styles among the Japanese in America. The most appropriate stratification is by generation, which reflects the historical pattern of immigration by the Japanese.

TABLE 4.
Percentage of Japanese-American Drinking (QFV)[a] by Social
Context, Los Angeles County, 1980

Social context	N	Percentage		
		Abstainer	Moderate	Heavy
1. Drinks served in neighborhood gatherings				
Frequently	33	12	21	67
Infrequently	73	48	27	25
Significant $p < .001$				
2. Number of work friends who drink				
Frequently	63	30	24	46
Infrequently	88	49	33	19
Significant $p < .01$				
3. Number of Japanese work friends				
Majority	85	39	20	41
Minority	89	44	37	19
Significant $p < .01$				
4. Frequency of drinking when getting together with other Japanese				
Majority	33	21	18	60
Minority	55	53	20	27
Significant $p < .01$				
5. Number of close Japanese friends				
Majority	241	51	62	57
Minority	52	46	19	9
Significant $p < .001$				
6. Alcohol served with close friends				
Majority	132	28	30	42
Minority	163	67	26	7
Significant $p < .001$				
7. Number of close friends who drink a lot				
Majority	58	31	10	59
Minority	194	50	36	14
Significant $p < .001$				

[a] QFV, quantity, frequency, and variability.

THE ORIGINAL ISSEI

The original Issei immigrated to the United States from 1890 and 1924, at which time the national immigration bill prohibited further Japanese immigration.

This group of Issei were products of the Meiji era of Japan (1867–1912), when values such as hard work, thrift, loyalty, and responsibility were highly prized. There is no systematic empirical data concerning the alcohol consumption patterns of this group, although the general notion is that their drinking was confined primarily to such special occasions as weddings, festivals, and the community picnic. As Kitano (1967) writes:

> The most conspicuous group at the picnic are not the young but the old—the Issei men—because many of them are "tipsy." The picnic is one of the few occasions during the year when workaday gravity, sobriety, and decorum are set aside. Now they gather in convivial groups on the grass, pass the whiskey (overtly if they are single, covertly if they are married), and indulge a license for "racy" talk. . . . The Issei brand of loose talk probably sits unfamiliarly at first on the tongues of the younger Nisei men, who have learned the Western variety. But, to the Nisei, it is a great satisfaction to be admitted to the Issei group; at least to be invited to drink with the Issei at the Japanese picnic is something akin to a "rite de passage." (p. 63)

In our social contacts with Japanese community members, the heavy drinking Issei were discussed. Several mentioned editors of Japanese newspapers as heavy drinkers, although these individuals were able to carry on their work without any visible impairment. Issei females were viewed as non-drinkers. There was high agreement that only the rare Issei, because of shame and the lack of other facilities, would have participated in alcohol treatment programs. Thus, Issei families in which one or both parents were alcoholic generally suffered with the problem.

Most Issei are now deceased and the survivors have achieved senior citizen status. Their attitudes and behaviors towards alcohol consumption have no doubt changed over time, but it is probable that they counseled their Nisei, or American-born, male second generation toward moderate and controlled drinking habits and their female children toward abstinence.

THE NEW ISSEI

The new Issei are products of the 1954 Immigration Law, which allowed a quota for Japanese immigrants for the first time, and the 1965 Immigration Law, which set up a number of preferences.

They reflect the current diversity of Japan itself and the newer immigrants (including businessmen temporarily in the United States) may have brought over the "Japanese styles" of drinking discussed in the first section. Our Los Angeles survey did not anticipate the relatively large number of Japan-born respondents, nor their heavy alcohol consumption patterns so we did not ask special questions of this group. However, there are some observations that may help to explain their drinking habits.

One is the availability of places to drink, patterned after the drinking styles of Japan. There are a wide variety of Japanese bars and nightclubs in Los Angeles, Honolulu, San Francisco, and New York, ranging from very small bars with one or two hostesses to large establishments with many hostesses. The ambience is similar to that in the bars in Japan, so that a Japanese from Japan can feel at home with Japanese friends and Japanese-speaking hostesses. The American-born Japanese who does not speak Japanese is often uncomfortable in such places and seldom becomes a regular customer (with the exception of Japanese-Americans in Hawaii).

We discussed the results of our survey with several visiting professors and businessmen from Japan. They fit into our heavy drinking category. A typical drinking pattern included several beers after work and with dinner (*banshaku,*) followed by after-dinner drinking. They would also regularly join colleagues and friends for eating and drinking so that alcohol intake was steady. None of them felt that alcohol was a problem and indicated that the drinking styles reflected patterns learned in Japan. Their wives were abstainers.

It should be noted that there are several Japanese AA groups in Los Angeles. One group is purportedly composed of Japanese from Japan, and the sessions are in Japanese. There are also individual Japanese-Americans scattered among the various AA treatment groups, but statistics concerning their actual number are difficult to obtain.

THE NISEI

Most of the Nisei, or American-born children of the Issei, are currently in the 50 years and older category. According to our survey data (see Table 2), the U.S.-born 50 years and older drinking pattern for males was abstainer, 42.5%; moderate, 39.6%; and heavy, 17.9%.

There are probably as many different drinking styles for the Nisei as there are for any second-generation American group. Chu *et al.* (1978) wrote about a small group of heavy drinking males in the above 50 category who were World War II veterans and who had spent part of their early lives in the wartime concentration camps. There was also regular drinking among certain professional groups, such as lawyers, businessmen, and social workers.

The following is a case of a Nisei who fits into the occasional-drinking part of the spectrum. It represents a fairly "typical" drinking style.

James

A longitudinal analysis of a 60-year-old Nisei professional provides a picture of one drinking style. Prior to World War II, alcoholic beverages were not a regular part of James' growing-up years. The cost, parental disapproval, and the absence of drinking among his Nisei peers combined to limit his drinking. There was some drinking at private parties and certain Nisei groups were known to serve drinks and to drink generously at these affairs, but he was not a member of any of these groups. He occasionally had a beer with white teenagers who drank more than his Nisei friends.

> I was evacuated with the rest of the Japanese during World War II. Liquor was contraband in camp but I heard of some Issei making saké from rice and of others using hospital alcohol. I didn't drink that stuff but after I joined the army, on my furloughs I used to bring in a full suitcase of liquor. The army sentries never used to search the baggage of GI's.

Liquor was a treasured gift for older Nisei and Issei camp residents, since it was rare and difficult to obtain. Although his army years were characterized by high accessibility to alcohol and a "why save money" philosophy, he noted that he did not drink that much and neither did most of his fellow Nisei. Individuals who did drink were generally "quiet" and the number of loud, belligerent drunks among his Nisei peers was minimal. He remembered that some of the GI's from Hawaii behaved more aggressively.

> When I served with the 442nd in Europe, first in Italy and France, I was introduced to good wine. I was surprised how those families over there served wine to their children, even little kids.

But he did not develop a wine habit.

After the war and release from the army, his drinking was with fellow college students and eventually with fellow professionals. Although liquor was now very common, his style was to "sip" liquor, so that even though he would have as many as four or five drinks, they would be stretched over long periods. His drinking style would be difficult to classify under our three categories, since it would be primarily "social" and could vary from "abstainer" to "heavy," depending on the time period. James might not drink for several months (although he has an extensive private bar in his home) and then attend several conferences at which nightly drinking might occur. Ceremonies, meetings, and visits by friends might also call for drinking, so that in another time frame he could be classified as a "heavy" drinker. In his mind, he is not a drinker, since he knows when to stop, his head remains clear, and he is able to navigate without assistance. He never drinks by himself and he does not need a drink to unwind at home, but he also knows that when he schedules lunch with certain friends he will have several drinks.

James attributes his pattern of drinking to the influence of his parents and acculturation. His Issei father was an occasional drinker; his father-in-law was a heavier drinker. Both of their wives (Issei) were abstainers. He notes that among his Sansei children, one is a regular beer drinker (the influence of college), a daughter has developed a taste for wines, whereas the rest of his children are non-drinkers. He did not talk about alcohol with his children, but he also never drank at home.

James's drinking pattern is a relatively common one among Nisei males. Time, place, and situation determine to a great extent the alcohol intake. A heavy ceremonial schedule, weddings and other get-togethers, means a certain amount of drinking. The role of Nisei females at these occasions (assuming they they are invited) appears to be to control the alcohol consumption of their spouses.

John

An interview with a Nisei who was an alcoholic illustrates another style of drinking. John, a Nisei of about the same age as James, grew up in a Los Angeles suburb in a predominantly white neighborhood. He describes his family as typically Japanese and his Issei father as "lighting up and getting red after even one drink." Therefore, his father limited his drinking, although liquor was consumed at weddings, other ceremonies, and New Year's. His father would often feel quite sick after drinking and his mother's role was to tell her husband that it served him right. Drinking was acceptable for males.

> I had a friend . . . he never came home straight. He'd stay out at night drinking with his cronies and then came home. It didn't seem to create any problems in the household outwardly because the wife just kept her mouth shut. I imagine if this occurred in a Caucasian household the old lady would have kicked the old man out.

Japanese-American wives seemed to put up with drinking as long as the husband came home with the paycheck. On social occasions he saw both Issei and Nisei drink, but very seldom did they become sloppily drunk, and on those few occasions when a person drank too much, his friends would laugh and good humoredly drive the person home.

> I think it's part of the Japanese nature that if anything goes wrong we take care of it.

John did not drink in high school, primarily because of fear of his father. He was told that drinking (and other behaviors in terms of dress and hair styles) was for low life. His first drinking was at his high school graduation party.

> I remember drinking vodka and it burning my throat. I didn't want them to think I was a sissy so I drank another. That night I remember feeling that as I drank, all the things that were Japanese in me got dissolved. All the inhibitions went away. I was talking and dancing with the girls. . . . I thought this is great stuff and I kept drinking more and more.

John eventually became sick, blacked out, and awoke the next day with a miserable hangover. His heavy drinking began in the army, where it became a nightly habit. He felt that his drinking was a part of peer group acceptance, which he so desperately needed. His civilian drinking also was strongly based on group acceptance.

> Funny thing about all of this drinking is that I never did like the taste of booze but I like what it did. My particular style of drinking . . . was to get it on as fast as possible. I'd go to the Brown Derby every night at five-thirty. The waitresses would all know me; they knew the first round was three drinks. . . . Once I got those drinks down I'd be able to talk to people and be friendly.

John's drinking was almost always with Caucasians, as he felt he never was a part of the Japanese-American community. Most Nisei were too uptight and would not loosen up. He felt he could blend in with his Caucasian peers with a few drinks so that as he says:

> I can't really say that alcohol was bad for me because it worked for me for a long time. It provided me with what I thought was a good life.

However, he began to realize that while others could stop after a few drinks, he would go in with the same intention and wind up closing the bar.

> In the beginning [I drank] a couple of times a week but as my position in life got better [as an insurance salesman] it was every night [at least a dozen drinks]. I was always a conservative person but drinking allowed me to be bolder and bolder. . . . I would head towards the bar, feeling like a kid going to a candy store. . . . If they had a piano bar there . . . [I] could sing my lungs out . . . more exciting women . . . bartender would know you . . .

waitresses . . . people would send you drinks and you'd send them back. It gave you status, a kind of status you would not experience at home.

John's drinking led to several drunken driving arrests and the breakup of his first marriage (to a Caucasian woman).

In the insurance company world, he learned to become a "class" drinker. A class drinker was

drinking like a gentlemen. You are not in a beer bar or shit kicker bar. You sit there drinking scotch and soda . . . making conversation, making people laugh. Back then I think it was an essential part of breaking into the Caucasian circle. As Japanese we tend to be traditional . . . stiff, cold. . . . I know other Japanese guys who did the same thing. They would go in and get witty and blend in. Hey, it makes you conclude that these Caucasians aren't bad after all. They invite you out on fishing trips and again you drink. After a period of time they think of you as a regular guy . . . in my drinking days my drinking buddies got me into things where there were no other Japanese. These guys got me into the Elks . . . one colleague even invited me to a John Birch meeting.

John felt that he was trying to break the Japanese mold. For example, when talking about a Japanese bar:

I didn't like to go in and see a Japanese waitress ready to pour my drinks or light my cigarette . . . as a young man I was trying to break away from the old world into the new and they [the Japanese] represented the old. That's what I was trying to get out of. The Nisei were always the ones who didn't get sloppy drunk. They were always the ones who were immaculately dressed and they were indeed gentlemen drinkers. . . . They had their own bowling leagues, clubs and organizations where they could socialize. Now the Japanese from Japan that's a different story. They could get drunk and fall apart and that's O.K.

Drinking eventually became quite expensive ($125 or more per week), but it was like telling the world that he was successful.

But then drinking for me was like telling the world that I'm a macho man, super successful. So you're sending drinks to everybody as well. I notice that this is still true with Nisei today. At Horikawa's I'll see men send drinks here and there the whole evening. You're trying to prove to the world that you're a successful and super guy.

I'd drink all around town because I couldn't get this recognition at home. I wouldn't drink at home. But I never missed a day at work. However towards the tail end it did effect my work.

Finally, his new wife (whose father was a heavy drinker) reacted adversely to his drinking. When he got help by joining an AA group (all white), she became involved in Alanon. John is now a "reformed alcoholic" and spends much of his time talking to community groups about his negative experiences with alcohol.

THE SANSEI

There is a relationship among the generations, acculturation, and the Americanization of drinking styles. The Sansei (and the Yonsei, or fourth-generation) are much more American than previous generations of Japanese-Americans. Family controls and the values of the Issei and Nisei are less relevant, so that drinking styles are more apt to reflect the dominant culture. There is more likely to be steady drinking among the Sansei, and our survey data (see Table 2) shows the following pattern. In the 29 and under age range: heavy drinkers, 20%; moderate drinkers, 40%; and abstainers, 40%. In the 30- to 49-year-old age range: heavy drinkers, 15.1%; moderate drinkers, 56.6%; and abstainers, 28.3%.

There are a large number of small groups and cliques among Sansei of the college-age population and drinking patterns reflect this. For example, Sansei groupings at UCLA include such geographical references as Eastside, Westside, Valley, Gardena, and Orange County and are further stratified by high school. Then there are the basketball players (in all Japanese athletic groups); volleyballers; the Omega's, Chi's, and Theta's (all Asian fraternities and sororities); and the ed-psych library group; the Powell library group; and so on . Clothes, cars, dating, smoking, and drinking styles depend on specific group norms. Less predictable patterns come from the unaffiliated ("the stragglers" and the "nons"), whereas those who have integrated into the larger campus community reflect another type of reference group (Okuma, 1980).

Hal

A Sansei in his mid-thirties, Hal describes his current drinking style as moderate to heavy.

> I have several drinks a day and on certain occasions more than that. Most of my friends (Sansei) take drinks as they come. It's no big thing.

Hal did not drink during his teenage years, primarily because his parents did not drink. His father only drank at ceremonies, whereas his Nisei mother was an abstainer. His high school friends (mostly Caucasian) drank on occasion, but Hal did not remember any drinking until his graduation night when he drank until he got sick.

> My drinking started at college. Most of my friends were Sansei and drinking was a regular part of college life. I think that's where a lot of us were socialized into liquor.

After graduation, Hal continued to drink moderately but steadily.

He went with white friends and with Sansei friends for regular drinks, but he mentioned that the former groups were seldom racially integrated except for himself. Liquor has not posed a problem and up to now has not resulted in problem behavior.

Nisei and Sansei do not seem to have the equivalent of the "Japanese bars" the Japanese from Japan have; therefore their drinking styles do not include regular drinking with hostesses who are paid to listen and to provide emotional support after a hard day at work. The most common style is drinking with friends and at ceremonial occasions.

We would like to emphasize that our findings and generalizations pertain primarily to Los Angeles and the Japanese-American drinking in other areas will reflect the ambience of those neighborhoods and communities. Japanese-Americans isolated from their ethnic communities will be especially prone to adopt local drinking styles unless parents make a special effort to control such behaviors.

In summary, we note that there are similarities and differences between and among our different samples. Differences between the Japanese in Japan and the Japanese-American include the fact that Japanese in Japan drink more regularly and more heavily than Japanese born in the United States. The differences may be attributed to availability of alcohol and its integrative and ceremonial function and a more tolerant attitude toward drinking.

The same difference in alcohol consumption was found between the Japan-born Japanese in Los Angeles and the Japanese born in the United States. The Japan-born sample had drinking styles similar to the patterns in Japan; the Japanese-American reflected a more American pattern.

The major difference between the Japanese in Los Angeles County and a California sample was that there were more Japanese in the "abstainer" and "heavy" drinking category, whereas drinking in the California sample was most frequently "moderate."

Similarities between the Japanese in Japan and the Japanese in Los Angeles include the following:

1. Males drink more than females. However, there is a trend toward younger drinking ages and a rise in female drinking.
2. Both groups report the flushing reflex.
3. Drinking is not closely associated with problem behavior. Part of the reason may be the emphasis on drinking with friends and its ceremonial and integrative function.
4. There is clinical evidence that the drinker with personal problems becomes the individual problem drinker.

5. Alcohol treatment programs are new to the Japanese in both countries and are not well attended.

There should be cause for concern regarding the place of alcohol in the Japanese-American community. The high proportion of "heavy" drinkers, especially among the younger age groups and the Japan-born population, has been virtually ignored. Instead, an image of the Japanese as "non-drinkers," or as a group with "non-problems" drinkers, has been common. However, it is time that we look beyond the stereotypes and the lack of Japanese participation in alcohol treatment programs in order to arrive at a more realistic assessment of the problem.

REFERENCES

Cameron, T. *Alcohol and alcohol problems: Public opinion in California, 1974–1980.* Berkeley: University of California School of Public Health, February (unpublished final draft), 1981.

Chafetz, M. E. Consumption of alcohol in the Far and Middle East, *New England Journal of Medicine,* 1964, *271,* 297–301.

Cohen, S. The Oriental syndrome. *Drug Abuse and Alcoholism Newsletter, 1979, 8, (10).*

Chu, I., Fertig, M., Sumii, S., & Yefsky, G. *A comparative study of alcohol drinking practices among Chinese and Japanese in Los Angeles.* Unpublished Master's Thesis, University of California, Los Angeles, 1978.

Erwing, J. A., Rouse, B. A., & Pelizzari, E. D. Alcohol sensitivity and ethnic background. *American Journal of Psychiatry,* 1974, *131,* 206.

Goodwin, D. W. Protective factors in alcoholism. *Drug and Alcohol Dependency,* Lausanne, 1979, *4,* 99.

Hanna, J. M. Ethnic groups, human variation, and alcohol use. In *Cross-culturual approaches to the study of alcohol.* Paris: Mouton, 1976.

Hayashi, S. Alcoholism and marriage: Family court divorce conciliation. *Japanese Journal Studies on Alcohol, 1978, 13.*

Hokubei Mainichi, January 17, 1981, p.1.

Kawakami, I. Doctors believe they know why Japanese are world's worst drunks, glowing beacon. *Kashu Mainichi,* June 8, English section, 1982, 1.

Kitano, H. H. Mental health in the Japanese-American community. In E. Jones & S. Korchin (Eds.), *Minority mental health.* New York: Praeger, 1982, 149–164.

Kitano, H. H. *Japanese-Americans: The evolution of a subculture,* Englewood Cliffs, NJ: Prentice-Hall, 1976.

Kotani, R. AJA's and alcohol abuse. In *The Hawaii Herald,* Vol. 3, #13, #14, July 2, pp. 4, July 16, pp. 7, 1982.

Ohashi, K., & Nishimura, H. *Alcoholic problem [sic] in Japan.* Paper presented at the 9th World Congress of Sociology, Upsala University, Sweden, August 17, 1978.

Okuma, S. *Traditions of the Japanese-American Sansei in modern times* (unpublished term paper). Sociology 124, University of California, Los Angeles, 1980.

Roche Report: Frontiers of Psychiatry, *Alcoholism in Japan: How caring communities contribute to recovery.* May 1, 1982, 11.

Sargent, M. *Alcoholism as a social problem.* University of Queensland Press, Australia, 1973.

Seto, A. *et al.* Biochemical correlates of ethanol-induced flushing in Orientals. *Japanese Journal of Studies of Alcohol,* 1978, *39*, 1.

Shibata, Y., Masuda, T., & Aoki, I. An Investigation of women alcoholic inpatients. *Japanese Journal of Studies on Alcohol,* 1978, *13*, 123–134.

Sue, S., Zane, N., & Ito, J. Reported alcohol drinking patterns among Asian and Caucasian Americans. *Journal of Cross-Cultural Psychology,* 1979, *10*, 41–56.

Wolff, P. H. Vasomotor sensitivity to alcohol in diverse Mongoloid populations. *American Journal of Human Genetics,* 1973, *25*, 193.

Wynder, E., & Hirayama, T. Comparative epidemiology of cancers of the United States and Japan. *Preventive Medicine,* 1977, *6*, 567.

Zeiner, A., & Pareses, A. Differential biological pensitivity to ethanol as a predictor of alcohol abuse. D. Smith (Ed.), *Multi-cultural view of drug abuse,* Cambridge, MA: Schenkman Publishing Co., 1978, 591–599.

19

Alcohol Consumption among Chinese in the United States

STANLEY SUE, HARRY H. L. KITANO,
HERB HATANAKA, and WAI-TSANG YEUNG

INTRODUCTION

According to the 1980 census, the Chinese, with a population of 806,027, are the largest Asian minority in the United States. The rapid growth in the Chinese population is primarily a result of more recent waves of immigration from Hong Kong, Taiwan, the Peoples' Republic of China, and Southeast Asia. These new arrivals have increased the heterogeneity of the Chinese population. In addition to the demographic, social, psychological, and geographic differences among those residing in the United States, we thus have Chinese immigrants from many different parts of Asia. The heterogeneity makes any simple generalization about the Chinese tenuous, yet there is a general notion that in terms of alcohol consumption, they are a non-drinking group. For example, Cahalan (1978) reports that there are

> several cultural groups within the United States with a consistent record of moderation in the use of alcohol—particularly the Jews and Chinese. . . .

STANLEY SUE • Department of Psychology, University of California, Los Angeles, California 90024.   HARRY H. L. KITANO • Alcohol Research Center, University of California, Los Angeles, California 90024.   HERB HATANAKA • Special Service for Groups, 1313 West 8th Street, Los Angeles, California, 90017.   WAI-TSANG YEUNG • School of Social Work, Chinese University of Hong Kong, Shatin, Hong Kong.

We should study these groups to determine exactly how they manage to
maintain their record of moderation even when immersed in a heavy-drinking society. (p. 24)

The prevailing belief that Chinese in the United States consume
alcohol in moderation raises several issues. First, what kinds of empirical data are available with respect to the nature and extent of alcohol
consumption among Chinese in the United States? Second, if the rates
and patterns of alcohol use among Chinese differ from those of other
groups, what are the social, cultural, psychological, and demographic
correlates of consumption? And third, what conceptual models can be
proposed to account for rates of alcohol consumption among Chinese
Americans?

We shall examine these questions in this chapter. The literature on
alcohol consumption is reviewed, and physiological and social/cultural
models are examined. The results of a survey of the drinking patterns
of Chinese and other Asian groups in the Los Angeles area are presented,
along with several case examples of Chinese drinking patterns.

ALCOHOL CONSUMPTION

Two means are used for ascertaining the extent of alcohol use and
abuse among various groups: the treated and the untreated case methods. In the treated case method, the assumption is that members of a
group with alcohol-related problems will come to the attention of mental
health practitioners to a greater extent than members of a group in which
alcohol problems are minimal. Thus, if few Chinese are treated for alcohol abuse or alcoholism as compared to other ethnic groups, the assumption is that Chinese have fewer alcohol-related problems. For example, studies of hospital admission rates in the United States indicate
that a significantly low proportion of Chinese are admitted for alcoholism
treatment (Barnett, 1955; Chu, 1972; Rosenthal, 1970; Wedge & Abe,
1949). Using this method, the findings suggest that the incidence of
alcohol abuse and alcoholism may be relatively small and that Chinese
may be an ethnic or cultural group in which alcohol consumption is low.

The problem in the treated case method is that utilization of services
may be an invalid means of determining disturbance in the population.
For one reason or another, many Chinese who are experiencing drinking
problems may not seek treatment or may not come to the attention of
mental health or alcohol treatment facilities. In fact after reviewing the
literature on help-seeking behavior, Sue and Morishima (1982) concluded that Chinese and other Asian groups in the United States are
less likely than non-Asian groups to seek mental health services. Thus,

the issue whether the rates of alcoholism or alcohol-related disorders among the Chinese are low or high cannot be meaningfully addressed using the treated case method.

The untreated case method involves the assessment of disturbance in a population through the use of interviews, psychological tests, behavioral measures, or surveys. By definition, this method does not rely solely on those who undergo treatment. If appropriate sampling procedures are used, it is possible to obtain a representative sample of the target population, and if valid assessment tools are used, one can estimate the extent of alcohol consumption (or psychopathology) in the population. Several untreated case surveys have been conducted on the Chinese and other Asian-Americans. In a survey of ethnic drinking patterns in Hawaii (Schwitters, Johnson, Wilson, & McClearn, 1982; Wilson, McClearn, & Johnson, 1978), 4% of the Caucasians were abstainers in comparison to 18% of the Chinese. Even among those who reported drinking, the Chinese drank less, and less frequently, than Caucasians. Chu (1972) found that Chinese in San Francisco consumed less alcohol than Caucasians, and in a national survey by Rachal, Williams, Brehm, Cavanaugh, Moore, and Bokerman (1975), Asian adolescents in the United States had a high rate of alcohol abstinence, although among drinkers, consumption was high.

In Chapter 18, this volume, Kitano, Hatanaka, Yeung, and Sue reported on the alcohol drinking patterns of the Japanese in Los Angeles. Although not reported in the chapter, data from other Asian groups (i.e., Chinese, Koreans, and Pilippinos) were also collected. Using various sampling techniques for each of the groups, interviews with approximately 300 heads of households were conducted. A survey by Cameron (1981) on a California sample was used for comparison purposes. The results of the Cameron (1981) and Kitano, Hatanaka, Yeung, and Sue (Ch. 18) studies (see Table 1) indicate that of the four groups, the Chinese were among the highest in the abstainer category (54.7%) and among the lowest heavy drinkers (8.4%). The non-drinking pattern is even more pronounced among Chinese females—the abstainers represented 73.8% and there were no females in the heavy drinker category. Therefore, in comparison to other Asian groups and to a California sample, the Chinese had one of the highest percentages of abstainers and one of the lowest percentages of heavy drinkers. The impression that the Chinese are a relatively non-drinking population is confirmed by the data. Further the very low proportion of heavy drinkers suggest that alcoholism may be a relatively minor problem among the Chinese, although it may also mean that the small proportion of Chinese who drink and have alcohol-related problems are probably ignored.

TABLE 1.
Alcohol Consumption Patterns by Group and Sex[a]

Group N		Percentage		
		Abstainer	Moderate	Heavy
Chinese	298	54.7	36.9	8.4
Male	218	47.7	40.8	11.5
Female	80	73.8	26.3	0
Japanese	288	47.0	37.2	20.8
Male	278	36.4	40.4	26.3
Female	60	63.3	25.0	11.7
Korean	280	66.8	19.6	13.6
Male	155	55.5	20.6	23.9
Female	125	80.8	18.4	0.8
Cameron	847	12.6	70.4	17.0
Male	405	11.6	72.6	15.8
Female	442	13.6	68.3	18.1

[a] Kitano *et al.* surveys.

DEMOGRAPHIC FACTORS AND DRINKING PATTERNS

Table 2 shows the relationship between selected demographic variables and alcohol consumption patterns from the Kitano data. In the total sample, only 35 or 12% were American-born Chinese. Generally, the American-born Chinese were moderate drinkers (57.2%), with none in the heavy drinking category. The foreign-born Chinese had 9.5% among the heavy drinkers, but there were also a high percentage in the abstainer category (56.3%). The results conflict with a previous study by Sue, Zane, and Ito (1979) who found that the American-born Chinese were heavier drinkers than their foregin-born peers. Sue *et al.*, however, sampled college students.

Age was related to alcohol consumption. Older Chinese tended to drink less than younger Chinese. The heavier drinkers were in the 18- to 29-year-old group; moderate drinking was most prevalent in the 30- to 39-year-old group. Neither marital status nor education was related to alcohol drinking patterns. Single and married Chinese had similar percentages of abstainers, moderate drinkers, and heavy drinkers. The proportion of Chinese with some college or more education was high (over 60 %), but education was not related to drinking.

TABLE 2.
Alcohol Consumption Patterns and Demographic
Characteristics for Chinese[a]

| Characteristic | N | Percentage | | |
		Abstainer	Moderate	Heavy
Birth	298			
Foreign	263	56.3	25.0	9.5
American	35	42.8	57.2	0
Age	290			
18–29	77	48.0	39.0	13
30–39	94	55.3	60.6	5.3
40–49	57	60.0	31.2	8.8
50–59	29	48.3	44.7	7.0
60 and over	33	72.7	21.2	6.1
Marital status				
Married	186	54.8	37.1	8.1
Single	100	54.0	38.0	8.0
Div./sep.	9	50	20	20
Widowed	3	66.7	33.3	0
Education	298			
No high school	17	71	29	0
Some high school	38	66	21	13
High school	49	57	35	8
Some college	59	51	37	12
College	121	50	45	5

[a] Kitano et al. surveys.

DISCUSSION

Three general conclusions can be drawn. First, Chinese exhibit low rates of alcohol consumption when compared to Caucasians or other Asian groups. Particularly striking is the high proportion of abstainers or light drinkers. Second, females report far less drinking than males, which is a consistent finding not only for Chinese, but also for all other Asian groups. Third, those in the age category, 18 to 29, consume the most alcohol. With the exception of these findings, other factors in the Kitano et al. study do not seem to bear simple relationships to consumption. Place of birth, marital status, and educational attainment are not related to drinking in any straightforward manner. Nor does analyses of other data concerning the relationship of alcohol consumption and such variables as social environment and family background reveal consistent patterns.

What is clear from this study and a review of other studies is that consumption rates are low among the Chinese and that there is no simple explanation to account for the low rate of drinking. One previously mentioned factor to be emphasized is related to the scattered birthplaces of the Chinese. For example, in the survey, the following picture concerning the birthplace of the respondents emerges: 12% born in the United States; 26%, Taiwan; 41%, mainland China; 14%, Hong Kong; 4%, Vietnam; and the remaining 3%, other countries. Although many investigators have discussed Chinese culture as a traditional influence on Chinese, the degree to which these cultural influences are maintained or modified depends upon the country of origin. Sue and Morishima (1982) note that even for Chinese in the same country (e.g., Chinese in Hawaii versus those on the mainland United States), there are considerable differences in self-concept, identity, attitudes, and values. Heterogeneity can also be found in language (e.g., Mandarin versus Cantonese), place of residence (Chinese enclaves versus white suburban neighborhoods), social class (nationally, Chinese appear to have high proportions of college graduates and those with little education), and degree of acculturation to the United States. Given this diversity, it is not surprising that simple correlates of consumption are difficult to find.

REASONS BEHIND ALCOHOL CONSUMPTION

In Chapter 18, which analyzes the drinking rates of Japanese in the United States, two major perspectives have been used from which to view Japanese consumption: physiological–genetic and social–psychological. The former perspective hypothesizes that ethnic differences exist in reactivity to alcohol. If after the ingestion of alcohol, certain Asian groups tend to have a more adverse reaction than Caucasians, they may drink less or be more likely to avoid consumption in the future. Since Kitano et al. (Ch. 18) have already raised the possibility of physiological differences between Asians and Caucasians, our discussion regarding Chinese is brief. In response to the ingestion of alcohol, Chinese, in comparison with Caucasians, have been found to exhibit more flushing (Ewing, Rouse, & Pellizzari, 1974; Seto, Tricomi, Goodwin, Kolodny, & Sullivan, 1978; Wolff, 1972). These studies suggest that differences in physiological reactions can serve as an explanation for decreased alcohol consumption among Chinese. However, the exploratory power of the physiological approach is limited. For example, in the Kitano et al. survey

(Ch. 18), recent Japanese immigrants to the United States reported heavier drinking than Japanese born in the United States. Furthermore, the different Asian groups—the Chinese, Japanese, and Koreans, presumably all equally vulnerable to the physiological response—showed different alcohol consumption rates. The evidence points to the importance of historical, social, cultural, and psychological factors so that attitudes, the availability of alcohol, the role of the family, and learning processes are as important as the physiological response.

A historical-cultural explanation of moderate alcohol consumption among the Chinese is offered by the anthropologist Hsu (1970). He writes that alcohol-related problems and alcoholism in pre-Communist China were quite rare. Although intoxication did occur, the Chinese tended to use alcohol as a means of intensifying feelings and acting more hospitable and friendly in social functions. Hsu contrasted the general low-keyed behavior of Chinese under the influence of alcohol to the American style of intoxication, which often leads to aggression, uninhibited self-expression, and rowdy behavior. He cited the differences between the two cultures as reasons behind the different behavior: Whereas the Chinese way of life concentrates upon the social situation such that the individual's place and behavior among others are of primary concern, Americans are socialized to emphasize the individual. Assertiveness, independence, individuality, and self-gratification are important American values so that alcohol could serve as an important and consistent factor in enhancing individuality as opposed to the Chinese model that emphasizes social drinking with its concomitant social controls.

Barnett (1955) found that the Chinese in New York's Chinatown were quite permissive over the use of alcohol at social functions, weddings, anniversaries, and other important events that allowed for heavier drinking, especially among adult males. However, drinking outside of specified situations was generally discouraged. The family and the community played an important role in controlling excessive "inappropriate" drinking. Chu (1972) indicates that Chinese values emphasize moderation in alcohol consumption. In his survey of drinking attitudes in the United States, Chinese and Caucasians were similar in their ratings of approval–disapproval of moderate alcohol consumption. However, in the case of drunkenness, the Chinese respondents had higher disapproval ratings than the Caucasians. Since traditional Chinese values stress group harmony, the importance of the social situation, gaining the respect of others, and the opinion of members of the family and the community serve as important means of social and behavioral control. Chinese philosophy, based on Confucianism and Taoism, indicates the importance of drinking in moderation (Wang, 1968; Singer, 1972), and

Hendersen (1976) mentions the repugnance toward public drunkenness in India and China. The famous Chinese poet of the Tang period Li Po mentioned the pleasures of wine, and the existence of wine shops was noted by Marco Polo (Sargent, 1973, pp. 66–67).

Singer (1972) attributes the low amount of alcohol consumption in Hong Kong to four major factors: the lack of drinking-centered institutions, the Chinese philosophy of moderation, the general restriction of drinking to males, and the habit of drinking primarily at mealtimes. He also found that although the Chinese in Hong Kong had a low to moderate prevalence of alcoholism, they had a higher prevalence of addiction to narcotics, especially opium. Singer (1974) indicates that the traditional ideology of the Chinese, which included acquiescence, moderation, propriety, intellectual control, and familial authority, would support the use of drugs over alcohol. Alcohol is more often associated with violence and a lack of restraint, whereas the opiates encourage more introspective behaviors. There were also historical events and the factor of availability as other reasons for the use of narcotics. He also wrote that neither drug addiction nor alcohol was considered a major problem in the People's Republic of China (Singer, 1974).

Chafetz (1964), in an early study of Taiwan, indicates that in 17 years there were no more than 10 cases of reported alcoholism in the Chinese population. Possible explanations included no solitary drinking (since alcohol was primarily taken at parties and meals), strong social sanction against drunkards, and well-defined ways of interacting socially. The strong moral Confucian code, with its clear role prescriptions for age and sex, was also mentioned. However, the influx of the Chinese apparently affected the alcohol consumption patterns of the native Taiwanese. The confusion of sex roles, economic frustration, and the encroachment of Chinese on the native culture were sources of stress that were hypothesized as being related to alcohol consumption.

Lin (1982) found low rates of alcoholism among the 50,000 Chinese in Vancouver, British Columbia. He speculates that drinking takes place primarily with eating. The ingested food modifies the effect of alcohol on the body and where alcohol is only taken with meals, the frequency and hours of drinking are restricted. Furthermore, Lin notes that many Chinese believe in the medicinal value of alcohol, but only if consumed in moderation.

Chu (1972), in a survey of the drinking patterns of the Chinese in San Francisco, reported that although use of alcohol was common, alcoholism was relatively rare. Attitudes toward alcohol indicated very high disapproval of drunken behavior under any circumstance. Observational data indicated that heavy social drinking among single Chinese males did not occur in bars. The Chinese bar clientele was primarily the second- or third-generation college-educated Chinese and middle-aged businessmen. He concluded that there would be a rise in alcohol consumption and alcoholism among the Chinese as a result of changing patterns of paternal authority and the acquisition and learning of American values.

CASE EXAMPLES

In order to illustrate current drinking styles, we present a number of cases based on interviews with selected Chinese respondents in Los Angeles. These cases should be viewed as selected instances from which insight into individual drinking patterns can be obtained.

THE ABSTAINERS

Annie L.

Annie is a 32-year-old, single woman who came to the United States from Hong Kong at the age of 11. Currently working as an assistant bank manager, Annie lives with a girlfriend. Although she dates regularly, Annie considers herself an alcohol abstainer. On the few occasions that Annie tried to drink wine or mixed drinks, she found the taste of alcohol to be aversive. Once when she consumed a glass of wine, she reported face flushing and initial feelings of light-headedness. Soon afterward, she experienced drowsiness and dysphoria. Annie now avoids drinking alcohol beverages, preferring fruit drinks at parties. Some of her dates do drink and Annie is not critical of others who do drink, unless they act foolishly. Her parents, still in Hong Kong, do not drink very much. Occasionally, her father would drink heavily at family gatherings. Her two brothers consumed very little alcohol.

It should be noted that Annie did not object to others' drinking. Her own behavior was one of abstinence, primarily because alcoholic beverages did not taste good and made her feel sick. She did not express any moral convictions against drinking, although she was critical of others who drank and then acted foolishly or misbehaved.

A more succinct statement concerning abstinence came from a graduate student from Hong Kong. "Alcohol tastes bad, it's not good for you, and it costs too much." Other attitudes from non-drinkers came from young Chinese–American females who commented, "It's not lady-like," "It doesn't look good for a woman to drink," and "It makes me feel sick."

INFREQUENT DRINKERS

Ben H.

Ben is a 56-year-old restaurant cook who came to the United States 40 years ago from a small village in southern China. He speaks very little English and lives near Chinatown. Ben does not drink very often. However, about once every two months he becomes thoroughly intoxicated with friends or at celebrations. During these situations, Ben's friends would typically pour him about six ounces of straight bourbon. Ben would emphatically tell his friends that they had poured too much. However, the statement was basically a polite gesture, since Ben and his friends implicitly understood that heavy drinking was permissible, if not encouraged, at these gatherings. After a few glasses, Ben's face would flush and he would act quite jovial. Some of his other friends would also consume large quantities of alcohol in a similar style. Ben's wife did not strongly criticize this type of drinking, telling other wives, "I hope he doesn't drink too much again; oh well, he enjoys himself." Privately she wanted to see him reduce his consumption, but as long as these drinking bouts were infrequent, she could afford to be tolerant.

Ben's drinking is characterized by long periods of abstinence, punctuated by drinking to intoxication at gatherings with friends. This pattern is often described as the loss of control over drinking, once it has begun, rather than an inability to abstain (characterized by daily drinking). However, Ben's drinking has not affected his social or occupational functioning to a marked degree. He does not perceive himself to have a problem with alcohol. In actuality, he has maintained this pattern over many years. Ben also believes that he is different from alcoholics; he does not drink alone, he drinks with friends, and his drinking is always accompanied by food.

HEAVY DRINKERS

Pao L.

Pao is a 53-year-old unemployed immigrant who lives in meager circumstances in a Mexican–American *barrio*. He is unable to speak much English and has never been able to find satisfactory work. The few jobs he found have not lasted, primarily because of his drinking. He would drink to intoxication whenever he could acquire alcohol. Pao has a married sister who would sometimes give him some money or invite him for a meal. However, at present, she wants little to do with him because she feels that he is "no good." Pao would constantly ask for money. If she refused, he would break down and cry, repeating that she felt no pity for him. If this did not work, he might become enraged. Pao knew he was an alcoholic and attributed his problems to his being Chinese in the United States.

Pao has never come to the attention of mental health or alcoholism programs. Once he was detained by the police for being intoxicated, but he was released after he sobered up. Except for minor and unnoticed offenses (e.g., shoplifting) and for his interactions with his sister, Pao has not caused problems for anyone. Pao can be called a "hidden alcoholic," and the chances are that he will not change his ways for the rest of his life.

Alan T.

Alan is a 23-year-old, American-born graduate student in sociology. About three times a week he consumes great amounts of alcohol. He typically drinks six to eight gin and tonics. He considers himself a heavy drinker who can outdrink most other Chinese or Caucasians. He never misses an opportunity to drink with various friends. Despite the amount of time spent drinking with friends and the occasional hangovers, Alan has been able to function successfully in graduate school and to maintain social relationships. He does not see his drinking as problematic except for some concern over drinking and driving. Under criteria adopted by the American Psychiatric Association (1980), Alan is not strictly considered an alcoholic. Although there is a heavy consumption of alcohol, social and occupational functioning has not (yet at least) been significantly affected.

There appears to be a U-shaped curve among many of our Chinese–American acquaintances. A high proportion are abstainers, whereas another high, but smaller proportion are heavy drinkers. It may follow the saying that "Most of them don't drink, but those who do drink will drink you under the table."

Several summary statements can be made concerning Chinese-American alcohol drinking practices. First certain Chinese cultural values stress moderate drinking. Intoxication is permissible in certain social situations, such as weddings, banquets, celebrations, and family association meetings. However, drunkenness to the point of being ill, of openly expressing hostility, or of disrupting social occasions is considered inappropriate. Even among heavy drinkers, drinking seldom occurs alone, and drinking is generally accompanied by food. It is as if alcohol consumption can be viewed as nondeviant if conducted in an approved social situation or if accompanied by food. Many Chinese seem to feel quite inhibited in their drinking if they are home, alone, or in non-sanctioned drinking situations. Chinese may be less likely to have a few cocktails for lunch and martinis before dinner, although exposure to this American model may change this pattern.

Second, although the cultural variable is important, what constitutes Chinese culture in the context of present-day America is difficult to specify. Chinese have emigrated from a variety of nations (and in Los Angeles we have also met Chinese who grew up in Mississippi, the East Coast, and the Caribbean); there are age and generational differences and differential exposure and identification with the American culture. For example, in a study of Chinese and Japanese college students, the amount of alcohol consumed varied directly with the degree of assimilation to American society (Sue *et al.*, 1979), whereas the Kitano study (Ch. 18) indicated that some of the newly arrived Japanese immigrants were already heavy drinkers, so that exposure to the American society may have different results.

Third, besides cultural values, the Chinese are a visible ethnic minority so that strategies of adaptation may influence consumption and other behaviors. It could very well be that as long as drinking continues to occur on special occasions, often out of view of the dominant culture, the Chinese will continue to be perceived as non-drinkers with few problems related to alcohol. This perception ignores the number of heavy drinkers with alcohol problems, the newcomers with already established heavy drinking styles, and the continued influence of the American culture in shaping Chinese-American behavior toward higher consumption.

Finally, there is the physiological–genetic response to alcohol. Aside from identifying the reaction to alcohol (many Chinese report adverse reactions such as sickness, headache, and nausea), there is little knowledge on how and why it affects within-group and between-group (other Asians) alcohol consumption.

REFERENCES

American Psychiatric Association. *Diagnostic and statistical manual of mental disorders.* Washington, DC: American Psychiatric Association, 1980.

Barnett, M. L. Alcoholism in the Cantonese of New York City: An anthropological study. In O. Diethelm (Ed.), *Etiology of chronic alcoholism,* Springfield, IL: Thomas, 1955.

Cahalan, D. Implications of American drinking practices and attitudes for prevention and treatment of alcoholism. In G. A. Marlatt & P. E. Nathan (Eds.), *Behavioral approaches to alcoholism.* New Brunswick, NJ: Rutgers Center of Alcohol Studies, 1978.

Cameron, T. *Alcohol and alcohol problems: Public opinion in California, 1974–1980.* Berkeley: University of California, School of Public Health (unpublished final draft), 1981.

Chafetz, M. Consumption of alcohol in the Far and middle East. *New England Journal of Medicine,* 1964, *271,* 297–301.

Chu, G. Drinking patterns and attitudes of rooming-house Chinese in San Francisco. *Quarterly Journal of Alcohol,* 1972, Suppl. 6, 58–68.

Ewing, J. A., Rouse, B. A., & Pelizzari, E. D. Alcohol sensitivity and ethnic background. *American Journal of Psychiatry,* 1974, *131,* 206–210.

Hanna, J. M. *Alcohol metabolism in four ethnic groups in Hawaii.* Technical Report, Honolulu, 1982.

Henderson, S. The Contributions of Social Psychiatry. Part II. *Medical Journal of Australia,* 1976, 632–719.

Hsu, F. L. K. *American and Chinese.* New York: Doubleday, 1970.

Lin, T. Y. Alcoholism among the Chinese: Further observations of a low-risk population. *Culture, Medicine, and Psychiatry,* 1982, *6,* 109–116.

Rachal, J. V., Williams, J. R., Brehm, M. L., Cavanaugh, B., Moore, R. P., & Bokerman, W. C. *A national study of adolescent drinking behavior, attitudes, and correlates.* Research Triangle Park, NC: Research Triangle Park Center for the Study of Social Behavior, 1975.

Rosenthal, D. *Genetic theory and abnormal behavior.* New York: McGraw-Hill, 1970.

Sargent, M. *Alcoholism as a social problem.* Queensland, Australia: University of Queensland Press, 1973.

Schwitters, S. Y., Johnson, R. C., Wilson, J. R., & McClearn, G. E. Ethnicity and alcohol. *Hawaii Medical Journal,* 1982, *41,* 60–63.

Seto, A., Tricomi, S., Goodwin, D., Kolodny, R., & Sullivan, T. Biochemical correlates of ethanol-induced flushing in Orientals. Quarterly Journal of Studies on Alcohol, 1978, *39,* 1–11.

Singer, K. Drinking patterns and alcoholism in the Chinese. *British Journal of Addiction,* 1972, *67,* 3–14.

Singer, K. The choice of intoxicant among the Chinese. *British Journal of Addiction,* 1974, *69,* 257–268.

Sue, S., & Morishima, J. K. *The mental health of Asian Americans: Contemporary issues in identifying and treating mental problems.* San Francisco: Jossey-Bass, 1982.

Sue, S., Zane, N., & Ito, K. Alcohol drinking patterns among Asian and Caucasian Americans. *Journal of Cross-Cultural Psychology,* 1979, *10,* 41–56.

Wang, R. P. A study of alcoholism in Chinatown. *International Journal of Social Psychiatry,* 1968, *14,* 260–267.

Wedge, B., & Age, S. Racial incidence of mental disease in Hawaii. *Hawaii Medical Journal,* 1949, *8,* 337–368.

Wilson, J. R., McClearn, G. E., & Johnson, R. C. Ethnic variation in use and effects of alcohol. *Drug and Alcohol Dependence,* 1978, *2,* 147–151.

Wolff, P. Ethnic differences in alcohol sensitivity. *Science,* 1972, *125,* 449–451.

20

Hmong Drinking Practices in the United States
The Influence of Migration

JOSEPH WESTERMEYER

INTRODUCTION

A small literature has begun to document those changes in drinking practices that occur in association with acculturation. One subtopic within this general field has been the appearance of problematic drinking and alcoholism among people who did not previously use alcohol and then were conquered by and/or integrated with a group who did use alcohol and had problems with it. These studies have primarily involved American Indian groups in North America (e.g., Carpenter, 1959; Kunitz & Levy, 1974; MacAndrew & Edgerton, 1969; Stull, 1972; Westermeyer, 1972). A second category of studies has originated in societies that used alcohol or some other intoxicant in traditional times; but massive influence by a colonial or conquering power led to the replacement of traditional drinking by new forms. These reports have mostly come from Oceania (e.g., Gluckman, 1974; Hocking, 1970; Ogan, 1966). In a third

JOSEPH WESTERMEYER • Department of Psychiatry, University of Minnesota, Minneapolis, Minnesota 55455. This study was supported in part by the National Institute of Health (Grant No. 1 RO1 MH37632), the Minnesota Medical Foundation, the Center for Urban and Regional Affairs, and the Alcohol Beverage Medical Research Foundation at Johns Hopkins University.

category, Heath (1971) and the Honigmanns (1945) among others have studied drinking practices of ethnic groups in contact with each other, along with the new social functions served by these drinking practices when the two groups get together. For example, intergroup drinking can serve to reduce interracial and interclass tensions while facilitating intergroup communication (Heath, 1971; Honigmann & Honigmann, 1945). And fourth, Karayannis and Kelepouris (1967) have examined the changes of drinking practices over time in association with economic and social change.

Despite these extensive reports, some major questions remain unanswered:

1. What occurs to the drinking patterns of those who migrate to a markedly different sociocultural milieu?
2. Do stable drinking patterns persist or are they notably altered?
3. In what direction do they change?

This report reviews the drinking practices of the Hmong over the period 1965 to 1975 in Laos and the period 1976 to 1983 in the United States. The Hmong drinking patterns had been stable and highly uniform in Asia (Westermeyer, 1971), and drinking always occurred in a social context. Virtually all seasonal celebrations, milestones in the life cycle (e.g., birth, name day, marriage, death), and major crises (involving animal sacrifice) involved drinking. In migrating to the United States, they underwent major shifts in work, family relationships, occupation, social status, and (for most) religious affiliation. Thus, their move to the United States comprises a good test case for examining the questions listed above.

BACKGROUND

The Hmong are quite unlike other Indochinese refugees and quite different from other migrants to the United States. In many respects, they resemble tribal American Indians in their cultural background more than they resemble other migrants from Asia. Most of them now over the age of 10 years were born at home in autonomous, self-sufficient villages numbering 40 to 200 people. Few over the age of 20 are literate even in Lao or their own language, and even fewer over the age of 30 are literate. All were animists until a few decades ago, when American Protestant and French Catholic missionaries arrived in the mountains of Laos. They supported themselves by raising upland (dry) rice, gardening, practicing animal husbandry, and hunting and gathering from their nearby mountain forests. The opium poppy had been raised as a cash crop for generations. Greatly valuing their independence, they join together in large groups only in times of external threat. Over the last 2,000 years, such threats have been common, as the lowland Chinese, Vietnamese, Lao, and Thai have tried to conquer and assimilate this people. Surrounded by powerful lowland neighbors, they have become as adept at political accommodation as they are skilled in warfare. Several million Hmong inhabit a vast mountainous expanse of Asia, ranging from China to Thailand and from Vietnam to Burma. Of the approximately 200,000 Hmong in Laos, about one-half have fled as political refugees. Most have come to the United States, but many are in Thailand, Australia, France, Switzerland, and Guiana.

METHODS OF STUDY

This paper is based on observation–participation, survey–epidemiology, and clinical studies among the Hmong during the period 1965 to 1983. Methods of data collection included the following:

1. Survey of community drinking practices in Laos, 1965–67.
2. Clinical experience among the Hmong as a general physician in Laos, 1965–67, and as a psychiatric physician in the United States, 1977–82.
3. Observation–participation in Hmong homes and social settings in Laos over a total of three years of field work, 1965–75, and in the United States, 1976–83.
4. Survey of Hmong refugees in Minnesota (including their drinking practices) in 1977, 1979, and 1982 to 1985.

The Hmong people studied in Asia lived in remote villages of Sam Neua, Xieng Khouang, and Luang Prabang provinces. By the 1970s, Hmong students and some literate Hmong families were moving into the towns. However, the data here concern primarily the rural Hmong. Data collected in the United States were based primarily on a group of 102 Hmong, aged 16 years and older, who resided in Minnesota during the Fall of 1977. Other information has come from Hmong patients encountered in various clinical situations and from visits to Hmong families and communities in other states (Wisconsin, Illinois, Massachusetts, Connecticut, Colorado, and California).

HMONG DRINKING IN THE UNITED STATES

Their former social imperative to drink in social contexts has weakened considerably in the United States. Their previously highly uniform drinking patterns have given way to much variability. New drinking patterns, as well as specific characteristics of abstainters and drinkers, are presented here.

I have observed a marked decrease in the frequency and amount of alcohol drinking among most Hmong people in the United States. Soft drinks alone are often served at weddings in the United States, whereas alcohol was always served at Hmong weddings in Laos. This is also true for certain other ritual events formerly marked by alcohol use (e.g., old friends visiting, business or political meetings, New Year parties), whereas in the United States coffee, tea, fruit juice, or soft drinks are served.

In addition to my observations, Hmong people themselves report that their drinking in the United States has been reduced. In the year 1982–83 we conducted a community survey of 50 Hmong people who had been in the United States since 1976, between six and seven years. They rated their drinking *vis-à-vis* drinking in Laos as follows:

Current drinking status	N	Percentage
Not drinking	30	60
Drinking less	12	24
Drinking the same	4	8
Drinking more	4	8
Total	50	100

Twenty-three of these subjects previously drank in Laos and now reported no or less drinking in the United States. They were asked why, and their first reponses were as follows:

Explanation	N	Percentage
Don't like any beverage alcohol	2	9
High cost of beverage alcohol in the United States	2	9
Don't like American beverage alcohol	2	9
Other individual reasons (e.g., abstinent because Christians)	7	30
Don't know, cannot say	10	43
Total	23	100

Religious conversion was mentioned by only one of these subjects, but it has probably had an important impact on drinking practices. Few Hmong had been Christians, Catholics, or Buddhists in Laos. Over the last few decades, some literate Hmong had converted—often after attending denominational schools. The numbers of converts increased in the Thai refugee camps and then increased further in the United States. This occurred in association with several factors: greater literacy, sponsorship by churches, more active proselytizing by churches, access to a place for meeting other Hmong (i.e., in churches and church halls on Sundays), access to second-hand clothes and furniture and appliances, access to jobs, and intermarriage with Buddhist Lao or Thai or with Christian or Catholic Hmong. Since much Hmong drinking occurred in the context of animal sacrifice, the abandonment of animal sacrifice by these converts reduced the traditional occasions when drinking was mandated. Most Hmong converts also have joined fundamentalist Christian religions that forbid or oppose drinking.

Among those 20 subjects who did drink in the United States, the occasions for their drinking were as follows:

Drinking occasion	N
New Year	15
Wedding	12
With relatives	12
With friends	11

The frequency of using alcohol among these 20 drinkers was as follows:

Period of time	N
Drinking in last week	8
Drinking in last month	11
Drinking in last year	16

Volume of alcohol consumed was estimated for the last week and the last month by all 20 drinkers. One "drink" equaled 12 oz beer, 4 oz wine, or 1 oz whiskey or other distilled liquor. Their volume of drinking was quite low, in comparison to usual amounts consumed in Laos:

Time period	Number of drinks per unit time		
	Range	Mean	Standard deviation
In last week	0–6	1.4	2.1
In last month	0–8	2.1	2.6

At times, drinking behavior in the United States manifested certain traditional group-control aspects (to be presented later), such as when drinks were served in rounds and each person received the same amount of alcohol at the same time. Generally, however, drinking rates in the United States were set by individuals themselves—a remarkable departure from drinking practices in Hmong villages.

Another new practice was the medicinal use of alcohol. In the 1982–83 survey, I asked 50 Hmong if they were using alcohol for self-treatment for such symptoms as insomnia, pain, or sadness. Seven of the fifty reported doing so. I also encountered Hmong patients who were using alcohol this way. This medicinal use of alcohol had appeared in Asia during a time when the Hmong, forced by war out of the mountains, were living among the Lao and the Thai. The latter traditionally used whiskey–herb concoctions as remedies, and the uprooted Hmong had begun to emulate this practice—albeit in small numbers.

Over the last few years, I have observed a new type of drinking among young Hmong men. Although Hmong people traditionally drink in multigenerational settings, some young men have begun to drink among themselves in the United States. They sometimes drink rapidly and become obviously intoxicated. Individuals set their own rate of drinking. I have observed about 20 of them vomit, pass out, or have to be carried away. Most of these men have been in their twenties. Their raucous and uncoordinated behavior has been a source of much consternation and embarrassment to their spouses and elders (at least when this drinking occurred in multigenerational settings). This occurs among the more literate, educated Hmong men, but it is not restricted to them. Female Hmong patients have told me about being beaten when their men drink in this fashion, although I have never personally observed this kind of family violence. (I do know of several such cases among the Lao, but not the Hmong, in Asia.) The men typically drink with other young men, avoiding their spouses and older men. Such drinking up to this time has occurred infrequently, less than weekly. Between drinking bouts the men maintain abstinence or have only a few beers in a quiet setting. The heavier drinking episodes occur typically at large ceremonial or party-like occasions (although the latter do not always include such drinking).

So far I have not encountered daily heavy drinking among Hmong people, whether they are seen in clinical or nonclinical settings. Self-medication with alcohol in psychiatrically ill Hmong does occur (usually to relieve insomnia associated with major depression).

The following are case examples of these general trends:

A 23-year-old single man had seldom used alcohol up to the time he left Asia at age 16. His use had increased during the last few years while at college. He drank two or three beers per weekend in the company of fellow students. On a few occasions each year, he drank twice that amount, to the point of mild intoxication. There were no problems in association with drinking, and he considered his use to be moderate.

A 32-year-old woman did not drink in Laos, as she had married a Christian man whose religion opposed alcohol. In the United States she and her husband had frequent social and official contacts with Americans, including dinner parties. Both she and her husband tried wine with meals, which they both enjoyed. Over the previous two years she regularly had a glass of wine with dinners at restaurants or dinner parties.

A 44-year-old woman had stopped drinking entirely since arriving in the United States. She had never liked the taste of Hmong corn whiskey in Laos, nor did she care for the scotch or rye whiskey many Hmong drank on ceremonial occasions in the United States. She attended a nearby Christian church on Sundays, primarily to socialize with other Hmong people who gathered to spend most of Sunday together at the church hall. Since the American pastor there preached against drinking, and other Hmong had stopped drinking, she decided to stop also.

A 53-year-old married man had drunk regularly in Laos, often in association with his role as a local militia leader. In Asia, he had a few to several ounces of corn whiskey once or twice a week over a 10-year period. Since coming to the United States, he had two or three beers about once a month, always in a ceremonial setting. He refused to drink at any time that he might drive, since he considered that to be dangerous. Although both he and his wife held jobs and his finances were good, they considered beverage alcohol in the United States to be too expensive. He had no leadership roles in the United States.

In order to appreciate these current patterns of Hmong drinking, it is necessary to consider former Hmong drinking patterns. Traditional drinking in Hmong villages of Laos was highly stereotyped and homogeneous. Later "town" drinking was similar to the Lao pattern of drinking; it persisted and spread in the Hmong refugee camps in Thailand.

TRADITIONAL VILLAGE DRINKING IN LAOS

In Laos, the Hmong produced their own distilled corn whiskey, a strong beverage that was often flammable (indicating 100 proof or 50 percent ethyl alcohol). Each family grew their own corn and distilled their own whiskey, using two iron wok-like pots. The alcohol-containing mash was boiled out of the lower pot. The second pot was suspended above the first one in a wooden frame. The ethanol fumes, upon contacting the cooler pot above, condensed into a fluid, ran along the bottom of the pot, and dropped into a tube of bamboo, whence the fluid was conducted off to the side into a bottle.

Production of alcohol required the use of two materials always in limited supply: grain carbohydrate and firewood. Each of these resources was accumulated by family members, and (except in a few lowland areas) they could not be purchased. Each family made their own alcohol; none could be purchased in traditional Hmong villages. The family made alcohol once or twice a year, so that alcohol was produced usually several months prior to its consumption.

Each adult drank at specific times and in defined social contexts, regardless of personal preference. Teenagers and women typically drank only one-quarter to one-half the amount drunk by the men. Adolescents and females participated in all the rituals and celebrations that involved heavy or moderate drinking, but not in those involving light or symbolic drinking.

A rigid etiquette governed the drinking event. The host, who supplied the whiskey, offered a toast or made a speech appropriate to the setting. He then drained the vessel, a bamboo or small glass container $1\frac{1}{2}$ to 4 oz in size, generally containing only 1 to 2 oz of whiskey per round. Each of the guests then matched the host glass for glass. The drinking vessel was passed around the circle from person to person. Often a female relative of the host filled the drinking vessel and presented it to each guest. Since the number of participants typically varied from several guests to a score or more in a circle, this might require considerable time for a round—especially when some guests offered a toast in response. Good-natured teasing focused on the differences between the participants (e.g., different villages, different clans, and different skills and personal characteristics) and on the amounts of alcohol being imbibed. Drinking events possessed the attributes of a drinking contest, with the host setting a moderate to heavy drinking rate.

Passing up a round also involved its own specific etiquette. The drinker might ask one of the attendants not in the drinking circle—often a woman—to serve as a helper. The attendant might or might not accede to the request. If she refused, the guest would ask another and then another attendant until a helper was obtained. This breaking of the drinking circle, with the extra time it required and the switch from male–male and female–female conversation to a (usually) male–female conversation, became the occasion for much teasing and punning directed at the person who could not keep up the drinking. As one might anticipate, drinkers displayed considerable ingenuity in order not to ask for help (e.g., dribbling some of the alcohol, spitting some of it out on one's sleeve or on the ground). But these maneuvers, if repeated, were eventually called to the group's attention and became the focus for more teasing.

Seemingly opposed to this cultural imperative for mandatory drinking was a norm against loss of control while drinking. Actors in the drinking event were expected to be able to walk without staggering, talk without slurring, and converse in a skilled, even intellectual fashion. Anyone imbibing to the point of impaired speech, vision, or ambulation was expected to slip quietly away on some pretext or another (such as going to the bathroom) so that he or she would not call notice to him- or herself. If a person did become impaired, this became part of the local oral tradition, and subsequent badgering might persist for weeks or even months.

The heaviest drinking—which occurred at Hmong New Year and at weddings—continued from afternoon to late at night, sometimes even into the early morning hours. The celebration proceeded generally outside as well as in the home, with 20 to 40 participants. The social activity included loud joking, playacting, and uproarious mirth. At weddings, efforts were directed at inducing the bride and groom to drink heavily.

With more moderate drinking, participants usually remained seated in the house. Convivial, lively conversation and laughter continued over several hours. Toasting occurred first, with eating later. Such drinking occurred at harvest time, the visit of a friend or relative, and the naming day for an infant.

Light drinking and symbolic drinking were much like moderate drinking with social feasting and conversation. These events—mostly religious in nature—continued over only a few hours during a meal. At funerals alcohol was splashed on the ground under the person being waked or on the grave, thereby symbolically providing alcohol to the deceased.

In the Hmong villages of northern Laos, all alcohol consumption occurred in groups, and only at times and in contexts approved by the community and by tradition. There was no individual use of alcohol as a medicinal, a food, or a recreational intoxicant. Alcohol abuse and alcohol dependence did not occur. Drinking always involved eating and was closely integrated with other aspects of Hmong culture: annual festivities (e.g., New Year, harvest), crises (e.g., healing ceremony for illness), rites of passage in the life cycle (e.g., birth, marriage, death), and major non-kin relationships (e.g., friendship, exchange labor, establishment of a client–patron relationship).

NEW "TOWN" DRINKING

In the 1960s, a few Hmong towns began to appear. At first they numbered only several hundred people. By the mid-1970s, however, the population of two Hmong towns (Long Cheng and Ban Son) was between 10,000 and 20,000 people each. Inhabitants of these towns included dependents of soldiers involved in the 1954–1974 Indochina War, refugees, staff from educational and medical facilities, civil administrators, Christian missionaries, truck farmers, and merchants.[1] In addition, many educated Hmong around the same time began moving into the Lao town of Vientiane, especially the Na Hai Deo district. There they were exposed to many aspects of Lao culture, including Lao drinking practices.

Drinking largely continued unchanged in these towns, and most Hmong drinking occurred in the traditional fashion described earlier. However, several new elements were added. First, alcohol was available for purchase, including Hmong corn whiskey, rice wine made by nearby Lao villagers, rice whiskey made in Lao towns, and beer from Thailand and Japan. Second, many salaried and mercantile Hmong people entered a money economy for the first time and were able to purchase beverage alcohol rather than relying on their own production. Third, although Hmong drinking styles persisted, other drinking styles (to be described later) began to occur also. Fourth, alcohol drinking events became frequent enough to cause alcohol-related problems for some leaders with heavy social obligations (to be described later).

The early appearance of an alcohol-related problem resulted from the admixture of traditional Hmong drinking and new means of transportation available in towns. It is likely that I treated the first case (or at least a very early case) at Sam Thong Hospital in 1966:

> A man was driving home from a family gathering on Sam Thong's first motorcycle. Bounding along footpaths in the middle of the night after several ounces of whiskey, his motorcycle went out of control and he sustained a fractured leg from a classical alcohol-related, single-vehicle accident.

---

[1] These towns included Sam Thong (where the author lived for eight months, 1965–67), Long Cheng, and Ban Son (also called Na Su). Sam Thong—which had a maximum population of about 4,000 people—was attacked and destroyed by the North Vietnamese in the late 1960s, leading to a subsequent migration of its inhabitants to the Ban Son area.

The man was not a chronic drinker, and he had not drunk more than others at the same event. This man, now living in the United States, has had no further alcohol problems over the last 18 years. This was the first of many such problems, with subsequent victims involved in jeep and truck accidents, ordnance accidents (with grenades and small arms), and so forth. In traditional times, a man's companions or his mountain pony would have conveyed him to a safe sleeping place. Thus, traditional non-problematic drinking turned into problematic drinking with technological change.

The case above exemplified a new public health problem superimposed on a traditional drinking pattern. This next case involves an elaboration upon, but a definite change in the traditional drinking pattern, which led to another kind of health problem:

> In his early forties, a Hmong man relocated as a refugee from a remote rural area to the Hmong town of Long Cheng. Intelligent and possessed of good social and administrative skills, he resumed his role as a local leader in Long Cheng. Soon he was a *nai khong*, representing the political interests of several thousand people. In this role it was necessary that he host various visitors and constituents once or twice a week.
>
> He (along with his wife) had always enjoyed drinking and was known for the large volumes he could imbibe. However in his rural home, he would have occasion to drink only every month or two. In Long Cheng not only did he drink more frequently, but his volume also increased from several ounces up to a pint of corn whiskey per evening, and up to a quart when drinking began during the day and continued late into the night. He did not drink on the alternate days when there was no social occasion to drink (even though he often craved alcohol). Over time however, he sought out social occasions at which he might drink three or four times a week. His drinking in this pattern continued over eight years, from 1967 to 1975. At that point he became a refugee in Thailand (1975–77) and then subsequently migrated to the United States. His alcohol intake during this time ceased, as he used all his available resources to support his family.
>
> I first encountered him as a patient in 1978 when he sought help for a severe memory problem. His memory loss consisted of being unable to recall recent events more than a few hours old (e.g., what he ate for breakfast, where he laid his money the night before, a promise made to a friend or relative). Onset of this problem had begun in the mid-1970s, while he was still in Laos. Currently the most disabling aspect of this problem was his complete inability to learn any English, since he forgot all that he studied within a few hours. On physical examination he manifested several neurological signs commonly observed with brain damage from chronic, heavy drinking (i.e., frontal lobe signs including positive palmo-mental and snout reflexes, and central cerebellar signs including positive Romberg sign and scanning speech). A diagnosis was made of chronic organic brain syndrome secondary to earlier alcohol abuse.

LAO TOWNS IN LAOS

The next step out of their early Iron Age life-style for the Hmong occurred in Lao towns. Through the mid-1960s, most Hmong were timid about visiting Lao towns, much less living in lowland Lao towns. They did not like the topography or climate. Local people often made fun of their dress or speech and tried to take advantage of them. New illnesses afflicted them (such as illness from strains of malaria and intestinal parasites to which they were not accustomed), supporting their notion that unfriendly ghosts and spirits lay in wait for them in the lowlands, where their own spirits and charms were impotent.[2]

As early as 1970, a number of Hmong had taken up residence in Lao towns. This had occurred for several reasons. As teenagers many young Hmong adults had gone for education or technical training to Lao towns, and so they did not share their parents' aversions. Greater military pressure from the North Vietnamese had fostered closer civil and military ties between the Hmong and Lao. Consequently, more Hmong were working in Laotian government ministries and institutions. A small number of young Hmong, educated in Lao towns, had become Lao-philes, adopting Lao dress, language, religion, and friends, and even taking Lao spouses. Heavy casualties among Hmong militia and guerilla units led to more integration of Lao and Hmong personnel in various military units.

It was in this setting that Hmong were first introduced to markedly different drinking patterns. Rural Lao drinking strongly resembled traditional Hmong drinking, but Lao drinking in towns—especially by young soldiers and students away from home—was similar to that in Bangkok, Hong Kong, Honolulu, or New York. That is, decisions to drink were more the province of individuals than groups. Alcohol might be used purely for recreation, away from any ritual or feasting, by anyone able to afford it. Although group rather than solitary drinking persisted, the group might consist of non-kin age-mates rather than the traditional group of multigenerational kin.

Despite the appearance of this type of drinking among young Hmong people, 1970–75, alcohol-related problems were relatively infrequent. For one thing, alcohol of any kind was beyond the means of most Hmong people. And despite reversals in the war with the North Vietnamese, morale and hope for the future remained high—fueled by the peace

---

[2] Many Hmong friends stayed in my Vientiane home during this period, but they were generally most uncomfortable and anxious to return to their mountain homes. Younger students tended to tolerate the towns better than their older siblings and parents.

treaty of 1973, as well as by the many educational opportunities and new occupational options available to the Hmong. In addition, opium remained the primary drug of abuse during this time among both urban and rural Hmong people.

The following case of an alcoholic woman in a Lao town demonstrates the exhange of alcohol for opium, and vice versa:

> A 32-year-old Hmong widow, addicted to opium for several years, moved out of the northern mountains following the death of her husband. He had died as a soldier in battle and, as his widow, she was entitled to a small pension. In order to receive it, however, she had to show up personally at the Veterans Affairs Ministry in Vientiane (a Lao town) every month.
>
> Opium was more expensive in Vientiane than it had been in her northern home. The mother of two children, she decided she must overcome her addiction for the financial benefit of her small family. Taking the advice of some Lao neighbors, the woman began drinking an alcohol-based medicinal compound containing certain herbs. Soon she was drinking 4 oz of rice whiskey four times a day (i.e., 16 oz a day). Whenever she attempted to stop drinking, she became physically ill, could not sleep, and was emotionally very despondent.
>
> Aware that she was now addicted to alcohol, she nonetheless persisted in taking it because it was cheaper than opiates. After several months more, however, she began to experience upper abdominal pain, headache, and increasing weakness—symptoms that had not bothered her when she was addicted to opium. At that point she stopped alcohol and resumed her opium use. Subsequently I interviewed her two years later in 1972 during a study of opium den users. She remained addicted to opium.

REFUGEE CAMPS IN THAILAND

I did not conduct any surveys in these settings, so only retrospective and anecdotal data are available. During this time, the ritual use of alcohol largely abated or decreased. In part, this was due to the fact that some drinking occasions were not occurring (e.g., harvest); in part, this was due to lack of resources to make or purchase alcohol. Food was in short supply in the camps, especially during the year 1975–76, so that available carbohydrate was not fermented into alcohol.

Although social use of alcohol decreased, the individual use of alcohol for medicinal purposes appeared. In particular, Thai rice whiskey was used to relieve chronic insomnia and various somatic complaints—symptoms for which a Hmong person might have used opium as a medicinal in Laos. Social drinking among young age-mates was not common due to limited resources.

The following is an example of self-medication beginning in a refugee camp:

> A 32-year-old Hmong man had never abused alcohol or drugs, nor been depressed in Laos. Within several months after arriving in a Thai refugee camp, however, he began to feel constantly sad. He also lost his appetite, had trouble sleeping, and had severe headaches. Family members complained that he was irritable and unpleasant to be around.
>
> He began taking a 16-oz bottle of beer at night to help him fall asleep. [American beer bottles usually contain 12 oz, but Thai beer bottles hold 16 oz.] A nightly pattern of beer drinking developed, increasing from 16 oz of beer to double and triple that amount [i.e., three to four and one-half beers per night by American standards]. This pattern continued in the United States. When I saw him two years later he was drinking five to six 12-oz bottles of beer in the evening.
>
> In addition to persistence of his earlier symptoms, he had begun to think about killing himself. He had alienated his wife and children with his verbal harangues and occasional physical abuse while drunk.
>
> This man responded to treatment for a major depression over several weeks. Subsequently he was able to abandon his nightly drinking habit without difficulty once his insomnia and other symptoms of depression had cleared.

DRINKING AS SYMBOL IN HMONG SOCIETY:
CHANGE AND CONTINUITY

Two groups of Hmong have changed their traditional drinking patterns in a major fashion since arriving in the United States. One of these groups, composed mostly of women and a few older men, have stopped drinking entirely. The other group consists of younger men who have adopted episodic, heavy drinking. Both groups can be seen as embarking upon their new patterns in association with other changes, many of which concern attitudes and values. It is posited that the new drinking patterns symbolize their changing attitudes and values.

The women's new-found abstinence reflects major changes in their self-identity and social role. Although Hmong women continue many traditional activities and roles, their residence in the United States has brought them new family and social status. Their husbands can no longer take a legal second wife—a common fear of some Hmong women in Asia. In the United States a Hmong man who divorces his wife must legally continue to provide for her, whereas he could send her back to her family in Laos. Hmong women in the United States have a claim on their children in the event of divorce, whereas in Asia they had to surrender the children to the husband or his clan. A woman physically abused by her husband in the United States can report him to the police. Cases of Hmong women who have obtained the support of American social workers and attorneys against their husbands are well known in the Hmong community. Since many Hmong women are now employed, they are also economically important to the family's material progress. Drinking ceremonies in Hmong villages had often placed women in a subservient role; only infrequently were they full and equal participants. This subservient role in the drinking context is no longer in keeping with their new legal status and new occupational roles. Thus, avoidance of their traditional role in drinking alcohol is in keeping with the new social role and self-image of many women. Fundamentalist Christianity—which preaches total alcohol abstinence—also reaffirms certain conventions now favored by many Hmong women, such as monogamy and egalitarian relationships between the sexes.

A few Christian Hmong women have abandoned abstinence under another set of circumstances. They are literate, educated women who have frequent contact with American professionals and political leaders. They are regularly exposed to moderate social drinking in American homes and restaurants. Although they rarely drink in Hmong settings, they drink moderately in other settings. Drinking in this "American" fashion also reaffirms egalitarian roles for women, since the women do not serve the men in this setting. Instead, they drink along with the men, behaving as they do.

A few older Hmong men have also become abstinent. Like the women described above, they have had something to gain, or at least nothing to lose, by abstinence. They typically have less status than other men of their age. They are not the clan chiefs and political leaders who would host drinking events. When called upon by the clan chiefs or leaders to help purchase alcohol for such an event, they can withdraw from the obligation by pleading their abstinent orientation.

Young Hmong men who drink heavily and individualistically from time to time also comprise a major departure from past convention. Just as many women and some older men, they have abandoned the strictly ritualized Hmong drinking style. Like the women, they would have been subservient to older men had they continued to drink in the traditional pattern, since the drinking host was typically a mature male. The new drinking pattern frees them from a symbolic domination by elders. They are also considerably more "American" in their dress, behavior, and values than are their elders. Their individually determined drinking style fits with this new view of self. But why have they not chosen abstinence, like the others? (It would save them money and probably some hangovers as well, for example.) I believe that their episodic heavy drinking serves several purposes. It sets them apart from the women, much as the traditional pattern once did, since Hmong women do not drink in this way. It may serve as a means of expressing anger indirectly against their elders, whose wisdom and leadership no longer protect them and who lost the country, thereby forcing them into refugee status. It provides the young men with a social "time out" during which they can behave in quite un-Hmong ways (e.g., staggering about, flirting with or behaving in a hostile manner toward women). These young men often bear heavy responsibilities in their extended families. They earn the primary income, translate for the family, make major decisions, and provide care for parents and children. These responsibilities have occurred a decade or two earlier than they would have in Laos. This earlier-than-anticipated assumption of family burdens may have also stressed and angered these younger men.

DRINKING BELIEFS AND DRINKING PROBLEMS

In traditional Hmong society, alcohol was a valued and desirable thing. Eveyone was expected to drink at certain times. Alcohol was offered, along with food and other valued consumables (e.g., tobacco, tea, opium), to honor visitors and guests. Even at death, alcohol was symbolically given to the dead person, that they might have it along with other good things (e.g., canned fish, a new pair of shoes) to accompany them to their life-after-death. It was a powerful substance governed by groups and by society, not by individuals.

That has changed tremendously over the last several years as the Hmong have come to the United States. Some Hmong now see alcohol as a bad, immoral or even evil substance that should be avoided. Others now perceive it as a highly valued substance, but one that groups or society should not control. Rather they should govern their own use of this valued substance. And still others maintain their old attitudes, as well as their old customs vis-à-vis alcohol drinking. Often these latter remain animists or have converted to Catholicism rather than fundamentalist Protestant sects.

Even before their move as refugees out of Laos and subsequently to the United States, some Hmong were experiencing alcohol-related problems. This occurred in the context of an intact, but changing Hmong culture. One change factor was a new means of transportation. Hmong ponies could carry their intoxicated riders slowly and safely home, whereas the non-intelligent motorized vehicles depended solely on the drivers for direction, speed, and judgment. Hmong did not run their ponies in the middle of the night down a mountain path, but some Hmong did speed motorized vehicles down mountain paths at night despite recent drinking. A second change factor was the move from communities of 50 to 100 people to communities of 200 or 300 to over 10,000 people. Infrequent drinking rituals in the former could become regular drinking bouts in the latter. These two changes can be considered intracultural changes related to technological and social changes.

Other changes can be considered intercultural, with diffusion of drinking practices from Lao, Thai, and American culture to Hmong culture. In particular, these practices involved three new drinking contexts that had been entirely absent among the Hmong. First, age-mates (especially young men) engaged in recreational drinking that involved neither feasting nor ritual nor celebration. Second, alcohol began to be taken as a medicinal for disorders ranging from opium addiction to insomnia, pain, and dysphoria. And third, some Hmong (especially, women) began to ignore the cultural imperative to drink in certain social contexts and insisted on remaining abstinent. The first two intercultural changes had appeared in Asia in association with increased intercultural contact with town-dwelling Lao and Thai peoples who engaged in self-determined recreational drinking and employed alcohol–herb concoctions as medicinals. The third change, abstinence, accompanied religious conversions to Christianity, as well as changes in female work roles and male–female relations (with females expressing greatly increased freedom from cultural imperatives and more independence from male leadership). This had begun in Laos to a limited extent, but expanded considerably in the United States.

Self-treatment with alcohol for psychosomatic disorders, depression, and other emotional maladies began at a time when these problems had greatly increased among the Hmong (Westermeyer, Vang, & Neider, 1983a,b,c). With longer residence in the United States, the prevalence of such symptoms has decreased considerably. Nonetheless, the unfortunate apposition of these two factors (medicinal use and increased symptoms) may have favored the widespread adaptation of this new drinking behavior. This could lead to a problem of alcoholism among the Hmong in the United States.

Some Hmong leaders and many Hmong women are concerned about the recreational drunkenness that now occurs among some young men. Their response to this problem has usually been to plea for abstinence, rather than to insist on a return to traditional drinking—i.e., multigenerational, ritual, socially controlled drinking along with feasting at socially prescribed times. Even if there were a consensus for a return to traditional drinking, the momentous changes in Hmong life-styles around the United States would not readily support a return to former drinking customs.

REPLACEMENT OF OPIUM BY ALCOHOL

The events described above might be expected to set the stage for widespread alcohol dependence among the Hmong in the United States. So far that has not been observed. In fact the overall prevalence of abuse of and dependence on all psychoactive substances among the Hmong in the United States is much less than it was in Asia, where the Hmong have had some of the highest rates of opium dependence ever observed—up to 9 to 12% prevalence of addicts in poppy-producing villages (Westermeyer, 1983). Of particular interest, however, is the fact that their current medicinal and recreational use of alcohol is identical to their traditional use of opium in Asia. Some recreational and medicinal alcohol users would no doubt be consuming opium were they still living in Asia. Many of the Hmong I have treated for psychiatric disorders (especially depression) report the same symptoms they once used to justify opium addiction (Westermeyer, 1971, 1983). Thus, the concern heard among Hmong themselves regarding episodic alcohol abuse and the potential for chronic alcohol dependence seems well justified.

HYPOTHESES FOR FUTURE WORK

This complicated assortment of historical, cutural, economic, and environmental factors suggests that the development of changing drinking patterns is entirely an idiosyncratic process, a matter for historians rather than social scientists. That may well be true. However, the following hypotheses are presented in an attempt to understand the Hmong case, possibly to predict future drinking changes in other groups, and to suggest potential means for reducing alcohol abuse in such populations:

1. Stable, nonproblematic patterns of psychoactive drug or alcohol use are liable to change in the context of rapid and extensive sociocultural change.
2. When an encompassing culture with alcohol abuse and a client culture without alcohol abuse come into contact, the former pattern is likely to prevail over the latter.

3. Elective alcohol abuse and elective abstinence accompany one another; drinking imperatives are antagonistic to this concomitant abuse–abstinence pattern. Widespread abstinence in an ethnic group may be associated with alcohol abuse. Ritualized, socially controlled drinking, with an imperative to drink with the group, is apt to reduce alcohol abuse.

4. Societies in which opium serves as a medicinal–recreational intoxicant will, when deprived of access to opium, substitute other psychoactive substances (such as alcohol) to meet the same purposes.

REFERENCES

Carpenter, E. S. Alcohol in the Iroquois dream quest. *American Journal of Psychiatry*, 1969, *116*, 148–151.

Gluckman, L. K. Alcohol and the Maori in historic perspective. *New Zealand Medical Journal*, 1974, *79*, 553–555.

Heath, D. Peasants, revolution, and drinking: Interethnic drinking patterns in two Bolivian communities. *Human Organization*, 1971, *30*, 179–186.

Hocking, R. B. Problems arising from alcohol in the New Hebrides, *Medical Journal Australia*, 1970, *2*, 908–910.

Honigmann, J., & Honigmann, I. Drinking in an Indian-White community. *Quarterly Journal of Studies on Alcohol*, 1945, *5*, 575–619.

Karayannis, A. D., & Kelepouris, M. B. Impression of the drinking habits and alcohol problems in modern Greece. *British Journal of the Addictions*, 1967, *62*, 71–73.

Kunitz, S. J., & Levy, J. E. Changing ideas of alcohol use among Navaho Indians. *Quarterly Journal of Studies on Alcohol*, 1974, *25*, 243–259.

MacAndrew, C., & Edgerton, R. B. *Drunken comportment: A social explanation*. Chicago: Aldine, 1969.

Ogan, E. Drinking behavior and race relations. *American Anthropologist*, 1966, *68*, 181–188.

Stull, D. D. Victims of modernization: Accident rates and Papago Indian adjustment. *Human Organization*, 1972, *31*, 227–240.

Westermeyer, J. Use of alcohol and opium by the Meo of Laos. *American Journal of Psychiatry*, 1971, *127*, 1019–1023.

Westermeyer, J. Options regarding alcohol use among the Chippewa. *American Journal of Orthopsychiatry*, 1972, *42*, 398–403.

Westermeyer, J. *Poppies, pipes and people: Opium and its use in Laos*. Los Angeles: University of California Press, 1983.

Westermeyer, J., Vang, T. F., & Neider, J. Refugees who do and do not seek psychiatric care: An analysis of premigratory and postmigratory characteristics. *Journal of Nervous and Mental Disease*, 1983a, *171*, 86–91.

Westermeyer, J., Vang, T. F., & Neider, J. Migration and mental health: Association of pre- and post-migration factors with self-rating scales. *Journal of Nervous and Mental Disease*, 1983b, *171*, 92–96.

Westermeyer, J., Vang, T. F., & Neider, J. A comparison of refugees using and not using psychiatric service: An analysis of DSM-III criteria and self-rating scales in cross-cultural context. *Journal of Operational Psychiatry*, 1983c, *14*, 36–41.

VII

Case Studies
Religion and Family

21

Irish-American Catholics in a West Coast Metropolitan Area

JOAN ABLON

INTRODUCTION

Belief systems concerning alcohol usage available to us in emic terms often only tell us certain *values* about drinking. They may indicate whether drinking is positively accepted at all, who is allowed to drink, in what quantity, under what circumstances, and with what results. Much less frequently we learn about the insiders' explanations of *why* they believe they drink. Causal models exist, but they tend to be etically derived. Ordinarily they are imposed *post hoc* by psychologists, psychiatrists, sociologists, or less often, anthropologists. As anthropologists, we aim to make our proposals as informed as possible by the insiders' point of view.

With respect to the Irish, we have a wealth of data from historical sources and folklore dealing with the Irish in Ireland. This material provides ample evidence for the intensive integration of alcohol in Irish culture and its extensive use. In contrast many of the explanatory models of *why* alcohol has been used have been proposed by contemporary scholars. However, the etic models presented to explain drinking pat-

JOAN ABLON • Medical Anthropology Program, University of California, San Francisco, California 94143. The research on which this paper is based was supported in part by grants 2 R01 AA00180, NIAAA, and MH08375, NIMH.

terns of the nineteenth and early twentieth century would not have been very helpful for planning interventions against alcoholism because they tend to emphasize broad economic and social factors impinging on the individual. This is not sufficient for effective intervention, since it is critical to take into account the individual's own perceptions of his drinking practices and explanations for excessive drinking, and then to create a model for intervention on the individual or family scale that can be feasibly implemented. In this chapter, I review some of the better-known etic models that deal with Irish drinking, presenting data from my own work in an attempt to explain some of the reasons for excessive drinking in Irish-American Catholic families in one contemporary middle-class population. The data presented offer clear directives for feasible treatment interventions.

Rarely have the drinking practices of a cultural group been as richly chronicled as those of the Irish. Their reputation for hard drinking is notable in popular lore and well documented in alcohol literature, as well as graphically and tragically illustrated in numerous works of drama (O'Neil, 1939/1957; 1941/1950) and fiction (Farrell, 1932/1976; O'Connor, 1961; Donleavy, 1958; Joyce 1922/1934). In their homeland, the Irish rank among the highest groups for prevalence of alcohol-related problems internationally. Similarly, Irish Catholics in the United States rank near the highest among ethnic groups for heavy intake, loss of control, and the provision of social support for heavy drinking. Indeed having Irish friends is almost as clear an indicator of a person's heavy drinking as being Irish oneself (Cahalan, 1970; Cahalan & Room, 1974). Greeley, McCready, and Theisen (1980) state that "of all American groups, the Irish are least likely to be 'non-abstainers,' the most likely to report drinking twice a week or more, and the most likely to consume three or more drinks of hard liquor at a sitting" (p. 3).

THE SAMPLE AND ITS DRINKING PRACTICES

The Irish sample reported on here is part of a population of Irish, and Irish mixed with German and Italian families in a large middle-class Catholic parish in a metropolitan area of Southern California. The families are second- and third-generation Americans of European stock. They were educated in a "parochial school culture," which has produced remarkably homogeneous individual, marital, and family histories. High school graduation is the normative level of educational attainment. Most husbands are employed in the civil service or by large utility companies. The mean age of informants in the sample is 50 for the husbands and

47 for the wives. Family life is church-oriented and child-centered, and family members busily take part in parochial school, sports, and religious activities. Most of these events are centered around the parish church.

Thirty families who were highly representative of the ethnicity, life-style, and values of their parish population were followed over a four-year period and studied through in-depth interviewing and observation in their homes and at parish events. Although these families were not chosen on the basis of their evidencing alcohol-related problems, many such problems emerged during this study.

The parish population is unequivocably a drinking population, and liberal alcohol usage is an expected part of the good life. Alcohol is served at such church events as fairs and festivals and at receptions or celebrations of parishioners for the sacraments held at church and in homes. Bottles of different hard liquors and wines are common raffle prizes at church events. The flip side of such liberal drinking practices is that alcohol abuse and alcohol-related family problems are common, although hidden from the larger society. Ongoing problematic drinking usually appears in two forms. The first is binge drinking among the men, which begins at a party, wedding, school reunion, or special gathering, or during spontaneous trips to a local bar in the evening, and continues through the night. The second pattern is individual, continuous solitary drinking at home during the evenings after work, and over the weekend. In both instances, drinking patterns are never or are very rarely allowed to interfere with work, no matter what havoc they may wreak upon the family. Family members tend to accept and adjust to alcoholic behavior, which usually takes the form of verbal abuse or sleeping off a binge. Physical violence, although reportedly common in other cultural groups and among the Irish in other areas, is extremely rare and would not be tolerated. Wives often comment that their husbands' behavior is a familiar harking back to their childhood and the behavior of their fathers. Similarly, such behavior was even more common among their husbands' fathers. Parishioners are often aware of alcohol-related problems in the homes of their neighbors and discuss these as matter-of-fact, common, and unfortunate happenings. Massive denial exists in regard to help-seeking by problem drinkers, and to a lesser extent, by their families.

Initially, I classified families' ethnicity according to the ethnicity of husband and classified individuals by father's ethnicity. Later it became evident that informants, when asked, tended to vary in how they defined their own ethnic identity. Although this study reveals the significance of *both* the mother's and father's ethnicity for predicting drinking problems, survey research data customarily do not distinguish the specific ethnicity of individual parents or grandparents. For example, Cahalan and Room (1974, p. 203), one of the most useful compilations available, classified subjects by their responses to the questions, "What country or countries did most of your ancestors come from?" and "Which one do you feel closest to?" if there was more than one country of ancestry. Greeley *et al.* (1980) in choosing sample households for a two-generational study asked potential parent respondents to identify their ethnic background or that of their spouses. Thus, ethnicity was an approximate conception rather than a specific empirically derived one.

Problematic drinking by husbands was most often found among the Irish. In all eight cases in which the husband had an Irish father *and* an Irish mother, he was, or had been in the past, a serious problem drinker. In cases in which the husband's mother was Irish and the father had been at least a heavy drinker, I also found alcohol problems. In fact the presence of an Irish mother was the most consistent factor in the perpetuation of drinking. My own explanatory model for this phenomenon is derived from an analysis of family case history materials that linked massive, culturally patterned controls governing significant areas of the life-course to the husband's drinking problems. These patterned controls were usually conveyed through the Irish mother and later through the wife, often also Irish.

REVIEW OF EXPLANATORY MODELS

The dramatic patterns of heavy drinking among the Irish in Ireland have elicited several explanatory models for Irish drinking, which have been explicated in the literature. These have related the prevalence of hard drinking, most notably among males, to characteristics of the family, social, and economic systems.

Bales (1962) has presented a vividly descriptive analysis of Irish drinking practices, liberally illustrated through excerpts from historical works, folklore, and song:

> Why, liquor of life, do I love you so,
> When in all our encounters you bring me low?
> You're my soul and my treasure, without and within,
> My sister, my cousin, and all my kin! . . .
> Surely you are my wife and brother—
> My only child—my father and mother—
> My outside coat—I have no other—
> (reprinted from Bales, 1962, p. 157)

Bales portrays the use of alcohol as integrated in virtually all aspects of Irish life, pointing out the particular significances of alcohol use for convivial and utilitarian purposes. Drinking served socially as a chief element of conviviality and constituted an important ritual of social equality, as well as a central element in business transactions. Alcohol was also used medicinally for almost all ailments and as an antidote for the damp and the cold. Poverty-stricken Irish who could not dress adequately for the harsh weather braced themselves with alcohol, the "outside coat" referred to above.

Bales notes that because of certain situational and psychological strictures on eating in Irish culture, there is a tendency to substitute drink for food. Famines have contributed to an irregular and often inadequate food supply in Ireland, both limiting food and causing shame about its quantity and quality. Guilt, inculcated in children who were often given food at the expense of their father's going hungry, created a strange ambivalence about eating. Furthermore the Church imposed fasts associated with the sacraments. The social, religious, and psychological strictures on food did not pertain to drinking. Differing social definitions of eating and drinking made possible a "symbolic substitution" of drink, oftentimes alcohol, for food.

Bales also discusses the relationship of alcohol usage to the rigid segregation of the sexes and to sex roles. He further posits the function of male peer group drinking as suppression of forbidden aggression and sexuality, an explanation Arensberg (1937) had earlier suggested. In fact the teetotaler was suspect because he might be chasing women and getting them in trouble. Bales (1962) notes that alcohol was perceived as the balm to "drown" out all emotional problems. Arensberg reported to Bales "any emotional difficulty of young men is best treated by advising them to 'drink it off.' Drowning one's sorrows becomes the expected means of relief, much as prayer among the women and young children, or among the older people" (p. 169). Bales noted that generations of children in Ireland grew up with models of heavy-drinking adults—both men and women—around them. Thus a circularity of behavior was assured.

Despite its numerous untoward consequences, drunkenness was not condemned unless the drunkard was married and his drinking threatened the family's cash resources or tenure on the land. Arensberg communicated to Bales that drunkenness was laughable and applauded. The drunkard was treated with care and affection, with a special connotation of sympathy, love, and pity. He was sometimes envied by the sober (Bales, 1962, p. 170). Up to the present day, drunkenness is not condemned in Ireland. Walsh and Walsh (1981), in a summary statement on contemporary Irish drinking, note "the level of acceptance of excessive drinking and even drunkenness is very high in Ireland. This permissiveness also extends to road vehicle use. The drunk person is often seen as an object of amusement rather than criticism" (p. 120).

Stivers (1976, 1978) similarly has presented incisive analyses relating economic and social factors to drinking patterns of the Irish *and* Irish–Americans. The traditional family "stem" system of inheritance in Ireland allowed only one son—usually the eldest—to inherit the farm, and often as late as his middle years, since the parents customarily chose to keep the farm until they could no longer comfortably manage it. The eldest son could not marry until he was a landholder, and younger siblings were even more disadvantaged. They remained landless or had to emigrate. Stivers maintains that hard drinking among men became a basis for status in the bachelor group, membership in which constituted the redeeming masculine identity for landless, single males.

Stivers traces the history of Irish immigration to this country and notes the severe early discrimination against the Irish. He posits (Stivers, 1976) that hard drinking in the United States was disembodied from its cultural context in Ireland and became important among Irish immigrants as a national identifying trait, both to Irishman and to their fellow Americans: "In Ireland, drink was largely a sign of male identity; in America it was a symbol of *Irish* identity" (p. 129). Stivers (1978) has explained the more recent patterns of Irish hard drinking in this century, a period of relative affluence and consumerism, as a newly sacred mode of "consuming one's ethnic identity and heritage." Alcohol became the item of choice for the consumer-oriented Irish-American. "Thus, at first one drank heavily because one was Irish while now one was Irish because one drank heavily" (p. 130). In contrast to Stiver's position that drinking is a "sacred" search for an elusive Irish-American identity, my data explore specific aspects of Irish-American family life and suggest that a combination of restrictive characteristics of life-style continue to offer a contemporary variant of the traditional and historic motivations for heavy drinking and alcoholism.

A totally different approach has been presented by O'Carroll (1979). In a provocative unpublished study of the relationship of religion and ethnicity to drinking behaviors of Northern European immigrants to the United States, he suggests that in the case of Catholics and particularly Irish Catholics, religious conflict and fostered dependency needs are causal elements of heavy drinking. Leaving aside the economic factors often posited as causally critical, O'Carroll suggests that the conflict caused by the relationship of an authoritarian religious institution and its psychologically dependent constituency initiates a routinized cycle of rebellion (abusive drinking) and resentment (confession, forgiveness, and reincorporation into group life) that is easily transferable from religious to secular domains. The author hypothesizes that it is functionally imperative for Catholic ecclesiastics to both maintain institutional authority and to enforce prohibitive norms surrounding premarital sex by concurrently tolerating deviant drinking practices and relaxing regulations of alcohol consumption by its constituents in order to discharge the psychosexual tensions inherent in the relationship.

The relationship of hard drinking to sexual activity—or lack of activity—and marriage has been discussed by a number of researchers. Irish marriages are fewer and occur later than in any Western country. Attitudes toward sex and marriage have historically been problematic. Both Bales (1962) and Stivers (1976) discussed the rigid segregation of the sexes and sex roles. In their studies they noted that because of few and late marriages and rigidly enforced celibacy, drinking and the reliance on male companionship were regarded *by the Irish themselves* as a safety valve for sexual tensions.

Historically, relationships between the sexes have been problematic. Irish mothers and wives have been repeatedly pictured as domineering matriarchs who rule their husbands and children. Relationships between spouses, while involving a great deal of dependency, are often characterized by withdrawal and lack of emotional expression and gratification (Scheper-Hughes, 1979; Messenger, 1969). Greeley (1972) has presented a sharp commentary on the affective characteristics of the Irish-American family that provides a very congenial basis for the analysis I present below of the family life of my Irish Catholic informants. He (Greeley, 1972) states that Irish-Americans have "practically no capacity to give themselves in intimate relationships. Emotions are kept under control by internal guilt feelings and external ridicule" (pp. 104–105). Tenderness is not tolerated. Greeley (1972) presents a dim view of the sexual relations among Irish-Americans:

> One has the impression—obviously from the secure but biased perspective of the celibate—that many, if not most, Irish-Americans get rather little enjoyment out of sex and are not very skillful in the art of lovemaking. This may stem from the fact that the Irish are generally not very good at demonstrating tenderness or affection for those whom they love. The Irish male, particularly in his cups, may spin out romantic poetry extolling the beauty of this true love, but he becomes awkward and tongue-tied in their presence and clumsy, if not rough, in his attempts at intimacy. She, on the other hand, finds it hard to resist the temptation to become stiff, if not frigid, in the face of his advances, however much warmth she may feel. For her especially, sexual relationships are a matter of duty, and if she fails in her "duty" to her husband she will have to report it the next time she goes to confession. Some Irish women, with obvious pride, will boast that they have never once refused the "duty" to their husbands, even though in twenty years of mar-

riage they have not got one single bit of pleasure out of it. A sex encounter
between a twosome like that is not likely to be pleasurable. (p. 114)

Greeley states that the Irish often tend to drink for reassurance and
escape from their "intolerable" psychological burdens and their need to
repress sexuality and aggressiveness. He describes the domination of
the Irish-American mother who rules her family by her strong will or
by subtly manipulating the sympathies and guilts of her husband and
children. Greeley (1972) notes that "many if not most of the alcoholic
Irishmen I know come from families where the mother rules the roost
and have married women who are very much like their mothers" (p.
135).

McCready, Greeley, and Thiesen (1983) have detailed a socializa-
tion-personality model which, they posit, accounts for the differences
in drinking behavior among various ethnic groups. The authors conclude
that

> the Irish are much more likely to have drinking problems than the Jews, for
> example, not because of social-class differences, and surely not because of
> genetic differences, but because the Irish have parents, spouses, and friends
> who drink more and because the family structure of the Irish is more likely
> to produce men with low feelings of efficacy. (p. 313)

> The differences in alcohol intake—as distinguished from differences in al-
> cohol problems—among the four groups can be accounted for by the influ-
> ence of mother, father, spouse, drinking environment, and Jewish religious
> tradition without any need to appeal to personality or family structure, drink-
> ing problems in the family background, or genetic influences. The Irish drink
> more than Jewish adults, for example, because their fathers drink more,
> because their mothers drink more, because they are in a heavier drinking
> environment, and because religious Jews have a special attitude toward al-
> cohol. As far as quantity of alcohol intake per year is concerned, in other
> words, one need not appeal to personality or family structure to explain the
> differences in drinking behavior. (p. 324)

My own research suggests that the continuance of restrictive and
frustrating features of Irish family life and spousal relationships, and a
repetitive role modeling process by which children continue to be exposed to domineering mothers, and hard-drinking fathers, uncles, and
male adults, as also suggested by McCready, et al. (1983) are chiefly
responsible for the maintenance of hard drinking and the frequent development of severe alcohol-related problems among men.

Analyses of Irish drinking have tended to focus only on male drinking patterns, although Bales (1962) quoted historical sources that com-

mented on the widespread drinking of both sexes. Room's (1968) study of deaths attributed to alcoholism in major ethnic groups in U. S. cities, 1889–90, suggests that greater emphasis may have been placed on the Irish male bachelor group and male drinking in general as opposed to female drinking, than is warranted. His figures demonstrate that although Irish males, and particularly Irish single males, had a very high rate of deaths from alcoholism, Irish women, and particularly Irish married women, exhibited even higher rates proportionately. Irish men were two to four times as likely to become alcoholics as American or German men of the same status, whereas Irish women were eight to ten times as likely as American or German women of comparable status (p. 108). Although most wives in my sample drank amply, I found only one instance of acknowledged alcohol abuse or "alcoholism" among the 30 women in my sample families, and I observed only two other alcoholic women in the total parish during four years of observation.

PATTERNS OF SOCIAL CONTROL AND ALCOHOL ABUSE

In my sample of Irish families, many historical characteristics appeared to be remarkably persistent, despite the seeming acculturation and homogenization of the Irish in American society. Greeley (1981) has commented on the persistence of cultural traits among Irish-Americans.

> What do I mean by becoming American and still being Irish: I mean that they have become the most successful educational, occupational, and economic gentile ethnic group in America while at the same time maintaining cultural patterns whose origins can be traced to their past and which do not seem to be eroding with education, generation in America, movement to the suburbs and, in some cases, intermarriage. Such patterns can be found in their family structure, their religious behavior, their political attitudes and participation style, their reaction to death, their attitudes toward family size and the role of women, their occupational choices, and their use and abuse of alcohol. (p. 9)

The existence of persistent, culturally patterned social controls encompassing almost all features of family life in my sample appear to be correlated with the recurrence of severe alcohol-related problems. Although these social controls have contributed to the development and maintenance of strong stable families in a geographic area and a time of notorious pervasive family instability, they also serve to lock both men and women into painfully restricted family roles and to stifle personal and family interaction, creativity, and freedom of choice in major areas of life and family careers. Irish Catholics who exhibited the greatest amount of control in all features of life also exhibited more problem drinking than did the German and Italian Catholics in the sample. The model presented here is developed in detail in Ablon (1980).

These traditional patterns of control are maintained in large part through the unchanging nature of primary relationships over many years. These relationships are made up of ties to extended kin and friends going back to childhood and parochial school days. Frequent social activities with these persons reinforce traditional beliefs and patterns and buffer new ideas from the rapidly changing world around them. Few or no new significant relationships have entered the lives of my informants since young adulthood.

The major areas of control in these families are most vividly illustrated through a number of themes that were reported to be injurious to family life by precluding or diminishing communication and growth of positive expressive relationships. These themes were most prevalent in families with alcohol problems. Some of the themes are (1) the domineering nature of informants' mothers as contrasted to that of fathers who were absent working or in their rooms quietly drinking; (2) strong religious sanctions surrounding premarital sexual activity, birth control, and related sexual issues; (3) lack of preparation for the suddenness of large families with heavy emotional and economic demands; (4) a busy social life centering around extended family and childhood friends, which left little room for intimate expressive relationships to develop between spouses; (5) fathers who traditionally work for civil service or large bureaucracies, which give them little control over their time or home schedules, and who were often absent during their children's waking hours; and (6) a strong masculine peer group composed of workmates, schoolmates from parochial school, and churchmates, all of whom are also often Irish and heavy drinkers.

Among these themes, those that deal specifically with sexual relationships and their broad implications appear to be the most significant for stifling spousal affection and communication. For example, informants have expressed the view that both early and contemporary heavy drinking patterns have been caused by sexual frustration that developed during premarital and later marital abstinence and the continual worry over the realities of raising and supporting large families. For example, one informant said:

> Well, being devout Catholics we really couldn't have any sexual relations outside of marriage. . . . Most of us, most of the people I knew, got married so they could have sexual relations. I knew my husband was a virgin and B (an alcoholic friend) too. That is why I think there was a lot of drinking

among these young men—they couldn't have sexual relations because that was so disapproved.

Drinking wasn't disapproved?

No, that is just it. The disapproval from the families and the Church went to the sex, and it didn't go to the drinking.

And another reported:

R and I dated five years before we married. We played it straight and were both virgins. After we married, we stuck strictly to the rhythm method— that meant for more than half of the month we couldn't sleep together. . . . Then when I nursed my four children, I did not have periods for seven months each time, so we abstained from sex. It seemed all right then; maybe we didn't have such a sex drive. But looking back, maybe it bothered him more than he said.

Sexual problems are imbedded in a generic *void* of communication between spouses. Spouses talk about logistic matters of family life, but there has been little development of expressive relationships. Early taboos and the ambivalence about sexual issues and church strictures surrounding sexual matters and birth control have particularly confounded communication of emotional and sexual needs and feelings. It is significant that although wives often perceive the biggest problem to be excessive drinking, husbands state that the biggest problem is sex— particularly the unwillingness of wives to have sex. Most Irish women find little physical pleasure in sex, and relationships often lack romantic feelings, intimacy, or sharing of personal thoughts or desires. Refusal to have relations is one way—and a highly significant one—in which these women can express their underlying resentments over years of excessive drinking. Even though women know that "the giving" of sex often helps their relationships, patterned resentments and willful begrudgings usually win out, and they refuse sexual overtures, often for years on end. An apparent function of husbands' drinking appears to be the dulling of their sexual needs and frustrations, a function posited by many researchers to serve both single and married men in Ireland in historic and present times.

In this cultural milieu in which divorce is not a thinkable alternative, I suggest that excessive drinking on the part of husbands enmeshed in this life pattern of massive social control, generally passed on and enforced by wives and mothers, is a culturally familiar and ambivalently condoned mode of registering their complaints without actually endangering their marriages. Women register *their* complaints through refusal to have sex and through their domination of their spouse and children. Thus, the drinking patterns of Irish men may best be understood through an examination of the behavior of family members as they relate through a complex of social and cultural expectations. This behavior is part of a self-perpetuating cultural paradigm that continues to produce families beset with alcohol problems. These problems, because of the drama surrounding them, often mask the many other marital problems that exist.

CLINICAL RELEVANCE AND INTERVENTION

The population described here tends both to deny excessive drinking as a source of problems and to avoid professional mental health or alcohol-related services at all costs. Hard drinking is integrated into family life, and even abusive drinking is tolerated for decades without serious help-seeking. Yet the obvious reality of pain and chaos in which both spouses often live, and the frequently serious acting-out on the part of children in many families, clearly indicates that treatment sources should identify acceptable intervention strategies for this population. Based on the cultural materials presented above, I suggest the following directions such interventions might take.

Recent studies of alcoholic families utilize family systems theory and view the total family unit as the necessary functional context for understanding the individual family member's drinking patterns. This conceptual thrust, led chiefly by Steinglass and his colleagues, posits that the major perpetuator of drinking patterns is the overriding need for the maintenance of the *status quo* or homeostasis that has developed within the family system. In 1971, Steinglass, Weiner, and Mendelson outlined a number of functions that alcohol fulfills for the total family unit and individual family members. The drinking behavior of one member, state the authors, may serve as a symptom or expression of stresses created by conflicts within the system. Thus, the drinking may serve as an escape valve for such stress, or drinking may constitute an integral part of one of the working programs within the system. In this form it may serve the unconscious needs of family members in its effect on such areas as role differentiation or the distribution of power or "might express culturally learned attitudes toward alcohol" (p. 406).

The strong homeostatic theme that appears to perpetuate heavy drinking in these Irish families clearly fulfills *all* these functions for the total unit and individual family members. Thus, here the realistic unit for intervention would be the total family, if possible, or the spousal dyad at a minimum. A primary task would appear to be the development of basic and trusting communication patterns between spouses, difficult as this might be, relatively late in their life and marriage cycle. Within the context of a more explicitly loving and trusting relationship, sexual dysfunction as a major issue can be broached. As a "naturally occurring" program in the Church environment, the Catholic Marriage Encounter has proven of great help to those few families I know who have experienced it. This program focuses on the establishment and maintenance of candid spousal communication. It is family-oriented, non-confrontative, and culturally acceptable. Thus, it seems to be tailor-made for the Irish-American population.

Secondly, prevention strategies should be developed in consultation with the perpetuators of this family cultural system. For example, attempts at primary prevention must start with outreach to and acceptance by those organizations having particular meaning for the families. Those suggested by the research include the parish church, parochial schools and their parents' clubs, civil service associations, ethnic and nationality societies, and alumni groups of various kinds. Educational programs for adults including values clarification, childrearing practices, and realistic information on sex and alcohol use and abuse should be introduced.

Preventive programs also should be planned for children. Family life education courses in the parochial schools and public schools should clearly discuss issues of family role behavior and family problem-solving, so that children begin to recognize dysfunctional interaction patterns in their homes and identify negative modeling behavior of their parents. Children grow up, as did *their* parents, my informants, in households in which their mothers are often dominant in almost every sphere and their fathers, away at work or when home, leave childrearing and other major aspects of family life to their wives. The heavy drinking of husbands is not lost on these children, just as Bales (1962) observed that heavy drinking of both parents was not lost on Irish children in Ireland through the generations. In parochial schools, Irish youths particularly tend to exhibit drinking problems more than do other young people. One mother observed bitterly, "Yes, the *flower* of Irish youth, sons of all the politicians, policemen, and firemen, and they drink just like their fathers."

Far too rarely have clinicians utilized either emically or etically derived cultural materials for the planning of services for alcohol-related problems. The historical and cultural data presented here for the Irish provide an example of the potential of such materials to assist in the planning for and implementation of effective services for individuals and families.

CLOSING COMMENTS AND CAVEATS

This discussion has focused on the excessive drinking of husband-fathers within the context of family life in one population of Irish-Americans. Epidemiological data indicate that historically Irish men and women, single and married, have exhibited both high rates of excessive drinking and alcohol-related problems in their own country and in the United States. The family has been used in this study as the context within which to examine the drinking of Irish-American married men. Other contexts would also be useful for exploring the drinking patterns of single and married men or single or married women in this or other Irish populations.

The analytic framework adopted by the researcher predetermines in large part his or her focus, and thus what he or she may see or not see and often the explanatory model that is proposed from the study. For example, most anthropologists who have looked at alcoholism on this continent in the past have been concerned with American Indians. Furthermore, they have ordinarily studied Indian drinking practices as influenced by the sociocultural or psychosocial contexts of culture change, acculturation stresses, economic pressures, or peer group support. Rarely have the dynamics of the *family unit* of identified Indian drinkers been examined. The anthropologist tends to place the blame, so to speak, for the widespread alcohol-related problems shared by a great many American Indians on these larger extrafamilial cultural or economic issues.

The arbitrary nature of the issue of focus was made clear to me in the course of a seminar I presented on the dynamics of Irish Catholic family behavior and how I perceived the role of these in perpetuating the drinking patterns of Irish husbands. A number of American Indians who participated in the seminar, all with social service or case work training, later approached me to discuss the immediate parallels they had perceived in thinking about drinking problems within the family contexts of their own respective tribes. Strangely enough, as an anthropologist who over many years has worked with relocated American Indians, many of whom have had severe drinking problems, I had never *thought about* alcohol problems as being seriously affected by family dynamics. Because my studies had focused on acculturation and social organization and not alcohol behavior *per se*, I had perceived excessive alcohol use almost purely as a by-product of these larger forces. My own research framework and focus had distracted me from a consideration of the significance of family interaction as a factor in the patterning of alcoholic behavior among American Indians. Thus, I caution here that the researcher's conceptual framework and focus may similarly obscure significant factors in the understanding of drinking behavior.

REFERENCES

Ablon, J. The significance of cultural patterning for the "alcoholic family." *Family Process,* 1980, *19,* 127–144.

Arensberg, C. M. *The Irish countryman.* New York: Macmillan, 1937.

Bales, R. F. Attitudes toward drinking in the Irish culture. In D. J. Pittman & C. R. Snyder (Eds.), *Society, culture and drinking patterns.* New York: Wiley, 1962.

Cahalan, D. *Problem drinkers: A national survey.* San Francisco: Jossey-Bass, 1970.

Cahalan, D., & Room, R. *Problem drinking among American men.* Monographs of the Rutgers Center of Alcohol Studies, 7. New Brunswick, NJ: Publications Center of Alcohol Studies, 1974.

Donleavy, J. P. *Ginger man.* New York: Astor-Honor, 1958.

Farrell, J. T. *Studs Lonigan.* New York: Avon, 1976. Originally published in 1932.

Greeley, A. *That most distressful nation.* Chicago: Quadrangle Books, 1972.

Greeley, A. *The Irish Americans: The rise to money and power.* New York: Harper & Row, 1981.

Greeley, A., McCready W., & Theisen, G. *Ethnic drinking subcultures.* New York: Praeger, J. F. Bergin, 1980.

Joyce, J. *Ulysses.* New York: Random House, 1934. Originally published in 1922.

McCready, W., Greeley, A., & Theisen, G. Ethnicity and nationality in alcoholism. In B. Kissin & H. Begleiter (Eds.), *The pathogenesis of alcoholism: Psychosocial factors.* New York: Plenum Press, 1983.

Messenger, J. C. *Inis Beag.* New York: Holt, Rinehart & Winston, 1969.

O'Carroll, M. D. *The relationship of religion and ethnicity to drinking behavior: A study of northern European immigrants in the United States.* Doctoral dissertation, School of Public Health, University of California, Berkeley, 1979.

O'Connor, E. *The edge of sadness.* Boston: Little, Brown & Co., 1961.

O'Neill, E. *The iceman cometh.* New York: Random House, 1957. Originally published in 1939.

O'Neill, E. *Long day's journey into night.* New Haven: Yale University, 1950. Originally published in 1941.

Room, R. Cultural contingencies of alcoholism: Variations between and within nineteenth-century urban ethnic groups in alcohol-related death rates. *Journal of Health and Social Behavior,* 1968, *9,* 99–113.

Scheper-Hughes, N. *Saints, scholars and schizophrenics: Mental illness in rural Ireland.* Berkeley: University of California Press, 1979.

Steinglass, P., Weiner, S., & Mendelson, J. H. A systems approach to alcoholism: A model and its clinical application. *Archives of General Psychiatry,* 1971, *24,* 401–408.

Stivers, R. *The hair of the dog.* University Park, PA: Pennsylvania State University, 1976.

Stivers, R. Irish ethnicity and alcohol use. *Medical Anthropology,* 1978, *2,* 121–135.

Walsh, G., & Walsh, B. Drowning the shamrock: Alcohol and drinking in Ireland in the post-war period. In E. Single, P. Morgan, and J. de Lint (Eds.), *Alcohol, society and the state, 2, The social history of control policy in seven countries.* Toronto, Canada: Addiction Research Foundation, 1981.

22

Alcohol Writ Accountable
The Episcopal Diocese of Washington, D. C.

LINDA A. BENNETT

Churches in American society provide a rich cultural context within which alcohol-related beliefs are espoused and practiced. They also serve as organizations through which alcohol problems can be identified, counseling given, and treatment referrals made. The historical circumstances of the settling of North America and its original colonies led to a highly pluralistic religious environment with an overarching Protestant–Catholic–Jewish identity. In the seventeenth and eighteenth centuries, especially, America held great promise for religious freedom for many emigres. Over the centuries Protestant denominations, in particular, have evidenced considerable splintering into distinct theological and organizational domains. Witness the proliferation, for example, of new congregations in the Baptist Church and the organizational complexity of the Lutheran and Methodist church bodies. A half-hour drive through most urban neighborhoods, small towns, and rural settlements is likely to take us past an interesting mix of different churches.

Beliefs about alcohol and its position in church life are no exception. Denominational stances range from total proscription against drinking al-

LINDA A. BENNETT • Department of Psychiatry and Behavior Sciences, George Washington University Medical Center, Washington, D. C. 20037. This work was supported in part by a grant from the National Institute on Alcohol Abuse and Alcoholism, No. R01 AA04784.

coholic beverages (such as found in fundamentalist denominations) through precise prescription of wine in religious ritual and allowance of alcoholic beverages served socially in designated areas of the church (such as in the Episcopal and Serbian Orthodox churches). A middle ground—in which alcohol is forbidden on church property, but not explicitly sanctioned in the private lives of congregants—is relatively common in such denominations as the Presbyterian Church. Furthermore, we find considerable variability within specific denominations regarding alcohol use strictures, depending on cleric and lay perspectives and on parish traditions. Consequently alcohol patterns can differ widely from congregation to congregation, even in the same geographical area.

This study was undertaken to explore alcohol belief patterns among different church bodies of a single denomination located in one geographical area. I chose the Episcopal Church, in general, and the Diocese of Washington, in particular, for several reasons. First, alcohol is a natural component in the ritual and social life of the Church. In Episcopal parishes, wine is an intrinsic part of the Holy Communion (Eucharist) ritual. Generally there are also no strict rules against the social use of alcohol, either on or off church premises. Second, individual parishes have considerable flexibility around which they can establish guidelines on their own alcohol-related issues. This permits an examination of differences among parishes in (1) alcohol use and (2) response to alcohol problems. Third, the Diocese of Washington has demonstrated a real concern in addressing alcohol-related problems. In this diocese, as in many others in the Episcopal Church, a Diocesan Commission on Healing Alcohol and other Substances (CHAOS) has been created to recommend guidelines and voice concerns about prevention, intervention, and treatment of alcoholism. Thus, the Diocese is taking a stand on these alcohol-related issues.

On a national level, too, the Diocese of Washington has been instrumental in founding the National Episcopal Coalition on Alcohol, which was formally organized in Washington in 1982 and which held its inaugural meeting in May 1983. The Coalition is "a nation-wide network of Episcopal laity and clergy, dioceses and provinces, seminaries, schools and agencies, with a common commitment to join together in addressing the problem of alcoholism and other such chemical dependencies" (1983 brochure).

The Coalition is an outgrowth of a resolution passed in 1979 at the General Convention, in which the illness of alcoholism was defined as a legitimate province of the Church. The Convention resolved to request that each Diocese form a Committee on Alcoholism to, among other things, address "issues of education, prevention, intervention and treatment."

More specifically, it requested (Johnson, 1980) the following of the Diocesan committee:

1. Provide a written procedure for treatment of clergy and Diocesan employees and members of their family who suffer from the illness of alcohol.
2. Include in its policy a statement covering the use of alcoholic beverages at church functions and/or on church property, with particular emphasis on the provision of non-alcoholic choices.

Organizational movements such as these are especially meaningful when considered within the cultural context of this denomination. The Episcopal Church has a reputation for being liberal regarding the role of alcohol in parish life. Reportedly alcohol problems are relatively common among Episcopal clergy, as well as among parishioners, and the formation of Commissions on Alcohol and the National Coalition reflects, in part, a concern with this reality. Jokes such as "whenever you see four Episcopalians, you see a fifth" and designations such as "whiskypalians" were frequently noted as signs of a heavy-drinking notoriety among Episcopalians. Thus, the Episcopal Church offers a good opportunity to explore some of the differences and similarities in Church policy and customary practice across a group of parishes at a time when alcohol addiction has become a publicly acknowledged problem within the Church. The Diocese of Washington is in the forefront of regional and national movements to call attention to alcoholism as a critical concern.

Material for this study is based upon interviews conducted with rectors in 12 parishes, with other clergy in the Diocese, and with couples participating in a study of family ritual and alcoholism. Interviews with clergy covered (1) sociocultural features of the parish; (2) the ritual of serving wine in the Eucharist; (3) customs and policy around serving alcoholic beverages in the social life of the parish; (4) extent of recognized alcohol problems among parishioners; (5) strategies drawn upon for alcoholism identification, counseling, and referral; and (6) the influence of diocesan-level recommendations on alcohol awareness within the parish. From these data, a general profile of the diocese can be drawn, with distinctive inter-parish patterns noted.

THE EPISCOPAL CHURCH IN THE UNITED STATES

One of 20 autonomous national branches of the Anglican Communion, the Protestant Episcopal Church in the United States of America traces its history back to the turn of the sixteenth century with the arrival of British colonists in the New World. During this period, church services were conducted by Sir Francis Drake in California, Sir John Hopkins in Florida, and the Reverend Robert Hunt in Virginia. By the time the Revolutionary War had broken out, Episcopal parishes had been established in all the American colonies, though concentrated in seaboard towns and cities. The Church's role in the religious life of the founding fathers is seen in the fact that a majority of the signers of the Declaration of Independence and the Constitution were Episcopalian (Bernardin, 1983, p. 7).

During the same period (1789), a General Convention of the Church drafted its own Constitution, creating the first official spin-off of the Anglican Communion (Emrich, 1973, p. 4). Expansion of Church membership commenced in the nineteenth century and continued through the middle of the twentieth. Concentrations of Episcopalians remain in the Northeast, in urban centers throughout the country, and along the Atlantic Coast. Other independent branches of the Anglican Communion have been founded in areas colonized by the British—for example, Canada, Australia, South Africa, Kenya, and the West Indies—as well as in other parts of Asia, Africa, and South America and the Pacific Islands. Forty-seven million members belong to the Anglican Communion and are "bound together by a common heritage, common worship, and a deep spiritual allegiance to the Mother Church. The Archbishop of Canterbury is symbolic head of the Anglican Communion, but exercises authority only in his own see" (Emrich, 1973, p. 3).

The Episcopal Church combines "Protestant congregationalism within a Catholic structure" (Gray & Gray, 1974, p. 3), warranting the label of "Bridge Church." Governed by bishops (a Catholic feature), the Episcopal Church is ruled democratically (a Protestant trait). Those Anglican churches favoring more Catholic characteristics are often referred to as "high church (or Anglo-Catholic)," and those more inclined toward Protestant features are usually referred to as "low church." Two of the twelve parishes in the study are Anglo-Catholic. The rectors in these churches are not married, whereas those in the more Protestant parishes are. In addition, one Anglo-Catholic priest clarified how the theology and ceremony in his parish is distinctive:

> Our theology tends to be rather more Catholic. The liturgy is so different. It is very rich in terms of motion, action, and noise. There is lots of singing and chants, and we have incense, bells, vestments, lights being carried, and processions. It is distinctive from many more mainstream Anglican parishes.

The intermingling of Catholic and Protestant features is also seen in the governmental process of the church. In a General Convention held every three years, cleric and lay delegates representing each Diocese make up the House of Deputies, whereas the bishops constitute the House of Bishops. The Convention serves as the final church authority, and all acts must pass through both houses. The Constitution and Canons set the basic rules for church organization. Although these Canons have direct roots in the Church of England and in the Catholic Church, the use of Canon Law in the Episcopal Church is relatively limited. Custom, instead, informs and directs much of church life. The Episcopal Church is not inclined to dictate to its members what they must do, but instead it is oriented toward assisting them in making their own decisions responsibly (Emrich, 1973, pp. 7–8).

In structure, the diocese is the central organizational unit. A diocese, in turn, is made up of parishes or congregations located in its geographical domain. Under this arrangement and the rules of Canon Law, parishes are under the administrative direction of their Diocese. However, they have control over such matters as budget and selection of rectors (priest-in-charge) and operate relatively autonomously. The rector oversees the worship, educational program, and community services of the parish, in conjunction with the vestry (elected lay leaders) (Emrich, 1973, pp. 10–11).

The structural nature of the Episcopal Church has considerable relevance to the manner and extent to which national and diocesan positions on alcohol-related issues are transmitted to individual parishes. With the governmental foundation of the Church as congregational as it is and with the church community so inclined toward custom rather than Canon Law, the opportunity for wide variation in parish approaches to alcohol policy is substantial. The cultural makeup of the parish, the personal perspective of the rector, the realities of the neighborhood, the social background of the parishioners, and parish tradition are only some of the mediating influences between national and diocesan recommendations and the realization of these ideas at the local parish level.

The study indicates that in this Diocese, at least, individual parishes have far and away the greatest say over what the parish does when it comes to alcohol use patterns, alcohol awareness, and intervention to solve alcohol problems. Although the Diocese is not an insignificant force in setting the tone among its constituent parishes about concerns over alcohol problems and although diocesan resources are available and sometimes drawn upon to address problems when they arise, in the end it is the individual parish that handles the situation as it sees fit.

Accordingly, one rector explained:

> Regardless of how important it may be that alcoholism has become a social problem and, therefore, a political issue in the Church, the outright ban of alcohol wouldn't work because it would be like taking an individual problem and raising it to a law.

Furthermore, since communicants tend not to look askance at moderate drinking, it would not be realistic for national or diocesan church bodies to legislate curtailed drinking. "We are a drinking bunch, along with the Roman Catholics," another rector asserted. "We believe in the virtue of moderation, not abstinence." However, as is clear from this study, rectors—as well as parish members—vary enormously in the extent to which they think that alcohol consumption might be controlled within the context of church events. Because the organizational and governmental structure is democratic, variations in local "policy" around alcohol use are bound to persist.

SOCIOCULTURAL PROFILE OF THE WASHINGTON PARISHES

The Episcopal Diocese of Washington encompasses all parishes within the District of Columbia as well as those located in several Maryland counties. The two primary Maryland counties are Montgomery—situated north and west of the city—and Prince George—north, south, and east of the city. The Diocese is made up of six regions, five of which were sampled for this study.

The 12 parishes involved in this project are quite distinct sociocultural entities. Many of their sociocultural characteristics are relevant to current alcohol use patterns and alcohol awareness. To begin with, the churches span more than two centuries in age. Although one (St. Paul's Church, Rock Creek Parish) was founded in 1712 and is the oldest church in Washington, most were established in the mid- to late-nineteenth century. In respect to parish size, there are two ways to count membership: by the total number of individuals or by "pledging units" (households). Based upon the number of individual communicants, the parishes range from five having less than 250; three having between 250 and 500; two having 500 to 1,000; and two having 1,000 members or more.

Five of the twelve have experienced recent expansion or revival of the congregations, whereas the others have remained stable or declined in membership. The length of tenure of the rectors (the head clergy) is similarly variable, averaging ten years and ranging from two to twenty-four. The number of clergy in each church ranges from one to four. Though women can be, and are being, ordained as priests in the Episcopal Church, the particular rectors interviewed are all men. One who is a woman did take part in one interview.

Reflecting to a great extent geographical location and church tradition, these parishes also evidence considerable variation when it comes to the ratio between black and white congregants. Seven are almost exclusively white, with no more than 10 to 15% black parishioners on their rolls. The remaining five vary from 25% black (two); 50% black (two), and 80% black (one). The latter three are in predominantly black neighborhoods. In most of these parishes, at least one-half of the black families are foreign-born, from African countries and the West Indies. The shift from white to predominantly or majority black has occurred over the past 20 years and has resulted from demographic changes in the Washington area, as well as from a large influx of Africans and West Indians to the metropolitan region.

Although the majority of the parishes (seven) draw their communicants from throughout the Washington area, five are characterized as neighborhood parishes by their rectors. The bulk of the membership of these churches consist mainly of families, rather than unattached individuals. Most of the parishes are made up of congregants who are, on the average, highly educated and working in professional and administrative managerial positions, often with the federal government. Although each parish naturally evidences some variability when it comes to socioeconomic composition and household structure, many are quite homogeneous and can be described as having a certain élan. One, for instance, is described as "a place for questioning people," and another similar, but considerably larger and more neighborhood-based church espouses "a willingness to try things and to take a chance on failure." This attitude extends to their alcohol policies.

In the Episcopal Church, a sizable portion of current members are converts from the Roman Catholic Church and other Protestant denominations. This generalization holds for the clergy, as well. Only five of the rectors estimated that at least 40% of their members were raised Episcopalian. Five others gave the percentage at 25% or less. One interesting distinction among the various religious affiliations that congregants were raised in is seen in the backgrounds of American-born and African-born blacks. The Africans were usually raised Episcopalian; the Americans, Baptist or Methodist.

Cultural contrasts in religious upbringing, in combination with the "bridge" nature of the Church, contribute considerably to parish differences around alcohol-related beliefs and practices. Such diversity, however, is far more evident in the social life of parishioners—on and off church premises—than in the liturgical traditions. The only liturgical act in which alcohol is an intrinsic component is Holy Communion or the Eucharist, and the use of wine is prescribed under orders of the Rubrics in the Book of Common Prayer.

THE HOLY COMMUNION RITUAL

The sacrament of the Eucharist—also known as Holy Communion or the Lord's Supper—is central to Episcopal Church theology (Gray & Gray, 1974, p. 50). Today the Eucharist is administered to both communicants and unconfirmed people who have been baptized, including children. Although the Eucharist varies from very simple and unadorned services in some parishes to quite elaborate in others, it always consists of two material elements—bread and wine—and a prescribed canon of words: "This is my body" (the bread) and "This is my blood" (the wine). Bread and wine had been associated with religious ceremony well before the advent of Christianity and were the most common "foods" in the Middle East. It was natural that they were adopted in the Christian Eucharist, which directly derives from the Jewish Passover meal (Gent & Sturges, 1982, pp. 33–34).

Many varieties of bread, including paper-thin wafers, are currently served in Holy Communion (Gent & Sturges, 1982, p. 33). Parishioners sometimes provide homemade bread for the sacrament. To take and consume the bread alone may suffice for Holy Communion. One rector explained that "there is a tradition in the church that to receive Communion of one kind—bread or wine—is in a spiritual sense to receive it all."

In modern Christianity—especially in American Protestantism—the symbol of wine for the blood of Christ has become a topic of some dispute in regard to its incorporation in the Eucharist. For one thing wine has both negative and positive connotations as a substance, depending on cultural context and personal experience. Anthropologist Gillian Feeley-Harnik (1981) in her book *The Lord's Table: Eucharist and Passover in Early Christianity* documents ambivalence going back to the early Christians over the consumption of wine:

> On the one hand it was a gift from God, a sign of richness and plenty. . . . On the other hand, there was danger in its abuse. . . . Wine represented both God's cup of salvation and his deadly cup of wrath. It was symbolic of the Jews, faithful and faithless, but also of heathen gentiles; drunkenness was associated with lack of understanding and idolatry. (p. 156)

In the Episcopal Church wine symbolizes "the cup of salvation" and not the "deadly cup of wrath," within the context of the Holy Communion. It is available to all parishioners, including children. Strictly speaking, to take the wafer or the bread ("the host") is sufficient, but most people attending a Eucharist service will also drink the wine. A common cup is either handed to each parishioner individually by a member of the clergy or by a licensed lay chalice bearer, or it is passed from person to person under the supervision of church clerics or chalice bearers. By and large parishes do not pay much attention to the type of wine being served. For one thing, very little wine is normally drunk in the course of the Eucharist.

Some parishioners choose not to sip from the cup, for various reasons. First, they can abstain, or pass their lips over the cup without actually drinking. Or, they can dip the wafer or bread into the cup, which is called "intinction." This obviates, ostensibly, sanitation problems. All in all, wine is an intrinsic part of the Eucharist, but actual drinking it is not so inflexible that worshipers cannot adapt to its presence in Holy Communion.

Two types of worship services predominate in the Episcopal Church: the Morning (or Evening) Prayer Service and the Eucharist. In the former, Holy Communion is not served; instead, Anglican chants are sung, prayers given, scripture read, and a sermon delivered. Although 20 years ago the traditional pattern was to hold the Eucharist once or twice a month, today a majority of services are the Eucharist. With the provisions of the New Book of Common Prayer (1979), the increasing popularity of the Eucharist became institutionalized.

The ambivalence that Feeley-Harnik (1981) reports for wine consumption during the Eucharist among early Christians does not seem to hold for these parishes. During the period when there was an increase in the demand for and holding of Holy Communion, awareness of alcohol-related problems heightened. The lack of any contradiction between these two developments may be due to a firm belief in the spiritual nature of the ritual of the Lord's Supper. Comments from some priests suggest this: "When I say the blood of Christ, that's what I mean, and I think that's the way it is perceived. So it doesn't seem like alcohol. It's not a drink." Thus, the physical quality of the wine is superseded by the spiritual nature of the act.

The significance of this ritual and its performance in a precise way is reflected in one parishioner's account:

> You realize that our central sacrament has drinking in it. We pass a chalice, and don't have fake wine. We have real wine. And we don't drink the sweet stuff. And we don't have fake bread either. We have real bread that people bake. And the cup is administered by the members of the congregation to each other. The minister passes it to the first person in the line, and they pass it around. A lay person can be a minister for the wine. So you're not controlled in what you take, although obviously you can't really swill it down and not be noticed.

THE ROLE OF ALCOHOL IN SOCIAL FUNCTIONS

In addition to serving wine as part of the Eucharist, alcoholic beverages—most often table wine or other kinds of wine such as sherry—are sometimes provided during social events held in the church. One dictum regarding the serving of alcohol socially is universally observed in these churches: If you provide an alcoholic beverage, an *equally attractive alternative* (EAA) must also be available and readily visible. Oversights of the EAA rule were reputed as having been quickly noted and rectified. Associated with this rule is the stipulation that people should not be pressured to drink. Coffee is usually available, but drinks like Perrier water, cranberry juice, and cider (special drinks) are also served. These stipulations point up a preference on the part of the Church to retain alcoholic beverages within its social life, while at the same time assuring non-drinkers that partaking of such beverages is not a necessary aspect of social inclusion. Choice is the key word.

A few other rules have been established regarding the serving of alcohol socially. No church in the study permitted hard liquor (usually referred to as "spirits") at church functions. Most were quite restrictive in fact about which events could legitimately have alcoholic beverages available. These sanctions on the provision of alcoholic beverages are by and large recent creations. Every rector could identify the specific events when alcohol is permitted to be served. Most frequently mentioned are vestry conferences, stewardship dinners, receptions, annual parish lunches, potluck dinners (usually those without children in attendance), Mardi Gras dances, Christmas dinners, and Lenten dinners, as well as annual parish planning conferences (usually held off church premises).

Quite disparate views were expressed regarding serving alcoholic beverages. One rector flatly stated that he would prefer to eliminate them altogether at social gatherings. "To have sherry at a reception is totally unnecessary. It is civilized to have a glass of sherry; it is the second and third glass that is not civilized." In contrast, another parish houses a bar in its basement, which is open at specified times, usually following evening meetings, and which serves non-spirit alcoholic beverages. Another parish is considering a modified version of this idea, a once-weekly evening get-together in which adult study and discussion would be followed by a relaxed social period for which alcohol would be available. Even considering the widely divergent views expressed by parish priests regarding the social provision of alcoholic beverages as part of church life, most churches exert considerable control over the extent of alcohol made available during parish functions. For the most part, the interviewed rectors exhibit at least a moderate level of awareness and concern about alcohol-related problems in the parish.

From these accounts, it appears that moderation in drinking does not always pertain to social gatherings of parishioners in private homes, nor to parish events held off church premises. Although every interviewed priest commented on the fact that social events outside the church are likely to have more alcoholic beverages consumed than social functions in the church, most also pointed out that today—in comparison with a decade or two ago—parishioners are practicing a fair amount of moderation in their drinking. A substantial shift from hard liquor to wine has occurred, along with people seemingly being more careful about how much they drink.

One particular parish is worth describing in regard to its moderate drinking tradition. This parish has a particularly large group of African-born blacks, who often invite the rector and his wife to their homes for social occasions. Alcohol consumption always characterizes these get-togethers. Wine and exotic warm beers—in the cultural style of the hosts—are brought out on a tray, placed on the coffee table, and made available throughout the evening. Although alcoholic drinks are ever-present, drinking during these social gatherings always remains very light.

Some parishes are highly social in that parishioners regularly gather together in each other's homes, outside of church-related functions, and establish close friendships among the church group. Of the twelve parishes studied, four can be defined as highly social. "Neighborhood" parishes, these churches have a core group of parishioners living in close proximity in the parish. All are family-oriented, with a serious concern about their children's religious education. As part of this concern, they pay rather close attention to the drinking practices and contexts of alcohol use in their social life.

Rectors similarly vary widely in terms of how much they socialize with parishioners. Some feel strongly that it is better to spend relatively little time at such gatherings, whereas others are active participants in the ongoing social life of the parish. They tend to think that the drinking that occurs when they are present at these social gatherings is typical and that no special behavior is being exhibited on their account. One rector who is a recovering alcoholic and who has not drunk any alcohol in years, continues to serve alcohol in his home to members of his parish. He is comfortable doing so, and posits that "alcohol is a pleasant thing for people who can handle it." He does not call attention to the fact that he is abstinent, and characterizes his parish as very congenial for recovering alcoholics.

This leads to the issue of drinking habits among the rectors themselves. Of the twelve interviewed, three abstain from alcohol, two for reasons of alcohol problems. Two others drink very moderately. Six are moderate drinkers, and one drinks heavily. In addition, one rector is head of a parish in which the previous rector had been an alcoholic. Although the personal predilections of rectors toward drinking alcoholic beverages and the prior experience of parishes with alcohol problems can certainly influence each parish's alcohol policy, none loudly proselytize a particular viewpoint about drinking to parishioners.

One particular social–liturgical event described several times as the most common site of heavy drinking is the annual parish planning conference. Apparently this generalization does not hold for all parishes, but drinking at these conferences does tend to be more extensive than at other gatherings. Usually held at retreat centers outside Washington over a weekend, these meetings provide an opportunity for relaxed social interaction. They have also, in some churches in the Diocese, become the target of internal dispute over parish alcohol policy. In one parish "people would load up their cars with tons of booze, and stay up into the wee hours of the morning drinking while they were there in this very convivial atmosphere."

As alcohol awareness has increased, heavy drinking at annual conferences has been challenged by some parishioners and rectors. However, in one parish, when the rector suggested that they change the location of a conference, the group balked because no liquor was permitted in that particular center. "And I realized that a big part of the culture of this event is that everyone brings a bottle, and after the Friday night session there is a party." Attempts to prohibit alcohol at these conferences have been met with strong resistance. People enjoy the conviviality and see this aspect of the weekend as part of the attraction to attend the meeting. Although little prohibition of alcohol has resulted from these disputes, the discussion has led to a reexamination of drinking generally at parish functions.

EXTENT OF ALCOHOL PROBLEMS

Alcoholism is not a parish-wide issue of central importance in this group of churches. Instead, it is a distressing undercurrent that occasionally surfaces with traumatic force. One rector observes that, as throughout Washington generally, there is too much drinking among his parishioners. He is surprised to hear so little talk about alcohol-related problems. Although in his church alcoholism *per se* is rarely reported or in evidence, most people do drink and a number drink quite heavily.

In fact, in one-half the parishes studied, alcoholism or heavy drinking reportedly seldom appear among parishioners: "Our people are simply moderate drinkers." Five of these six churches have the largest numbers of blacks, foreign-born, and blue-collar members, and the sixth has a substantial elderly population. None have memberships larger than 350. They also tend to be among the relatively "unsocial" parishes in that the social life of parishioners does not revolve primarily around the church itself. By and large, they are not neighborhood parishes. Thus, the picture that emerges is that those churches not conforming to more stereotypic mainstream Episcopalian characteristics—white, professional, middle and upper class—tend to be lighter drinking parishes and apparently evidence fewer alcohol-related problems.

In contrast, the other six parishes—in which the rectors note that alcohol problems have emerged with some regularity—are made up of members who, as a group, have higher status occupations and are better off financially. These parishes tend also to be socially active and involved in community outreach programs. They are knowledgeable about the use of health resources, including counseling and psychotherapy. Alcoholism is ordinarily viewed as one aspect, among many, in the general well-being of the parish: "Alcoholism has touched this parish, as in most churches. We don't have falling-down drunks but we do have problems."

In one very socially active and large parish, the rector noted that alcohol is a problem that keeps emerging. However, it comes up in a special way: "Lots of people come to me and want me to stop their wife, husband, uncle whomever from drinking." This feature—the initial identification of an alcoholic by a family member—was commented on by all rectors who saw alcoholism as a problem in their parish. Being the child of an alcoholic, too, has become a major concern and one that parishioners increasingly take note of.

Additionally, parents in parishes with many adolescents are worried about their children's drinking, but these teenagers usually "refuse to talk about it." More than a few rectors pointed out that adolescents are adamantly opposed to organizing educational programs in the Church around alcohol and drug issues. Whether these adolescents are in fact "into alcohol in a big way" is not clear. In one parish the rector found the teenagers to be a relatively straight group and conservative about alcohol and drugs. In another, where drinking is not seen as a problem among the youth, the parents do worry about the use of other drugs. In short, no consistent picture emerges for adolescent alcohol use. Adolescents were rarely among those specific individuals cited as having alcohol problems.

All parishes have recovering alcoholics as members. Self-identified alcoholics tend to comment on their status to their rector some months after sobriety has been achieved. When alcohol problems come to the attention of a rector while the drinking is current, ordinarily it is a family member or a friend who asks for assistance or advice. An occasional within-church identification of a person who is drinking too heavily is made. By far the most frequently recalled instances of alcohol problems involve members of the clergy and the vestry (elected lay leaders). According to one rector, he has encountered stress due to alcoholism among parishioners in vestry meetings in all his parishes. Since the vestry meets regularly and is highly visible in church functions, the presence of alcohol problems among them is more likely to be obvious.

Similarly, when a rector or other member of the clergy is alcoholic, it can become an extremely difficult matter for the parish and the Diocese to deal with—a "very painful process." Most rectors cited cases where the rector is a recovering alcoholic, although great discretion was used in these discussions. A few intimations about probable alcoholics among the clergy were made, but most references were to those rectors who had stopped drinking and openly acknowledged their problem.

PARISH RESPONSE TO ALCOHOLISM

Rectors prefer not to spearhead confrontations over drinking with parishioners. The first step is usually taken by a family member, a friend, or a concerned fellow-parishioner, who calls the problem to the attention of the rector. All rectors do have a set of strategies they are comfortable drawing upon in responding to requests for help. And almost everyone has a certain resource—a person or an organization—in which they have special confidence. Referral sources are ordinarily church-based services, members of the clergy, or parishioners.

In the Diocese of Washington, two rectors are widely acknowledged to be experts in the field and especially helpful in planning and conducting interventions. They were regularly cited by the interviewed rectors as people they personally trust. One was very helpful in conceptualizing this study and instrumental in getting the project underway, and the other participated as an interviewee. Both have had direct experience with alcoholism. Several members of the clergy in the Diocese of Washington have studied at the Yale, now Rutgers, School of Alcohol Studies and many have worked as counselors in clinical settings in which alcoholism has been a frequent problem among clients. Thus, special training and clinical experience in the alcohol field is not unusual among rectors in the Diocese.

In addition, many parishes have members who are health specialists—psychologists, psychiatrists, counselors, social workers—especially able in the alcohol field. Individuals such as these are involved in the Alcohol Network, one parish's answer to how best to respond to alcohol problems. A component of the pastoral ministry, the Alcohol Network is a "tremendous resource" for the rector. "The clergy can do pastoral counseling, but we don't have the combined experiences of these people." Established a few years ago, the Alcohol Network has maintained its "network" form in order to offer the flexibility necessary to keep a viable organization intact and to respond promptly to problems as they arise.

> They call themselves a network because they want to be known to each other and to the clergy, but they don't want to be an organization that has lots of meetings and gets bogged down in a heavy load.

The Alcohol Network sponsors alcohol-related education programs; maintains a card file of resources available locally and nationally; keeps an updated listing of Alcoholics Anonymous (AA) meetings; and assists in the occasional intervention that occurs in the parish.

"Intervention" as a topic arose several times in our discussions. A relatively new approach in alcoholism identification and confrontation, the intervention concept is somewhat radical in view of more traditional strategies:

> It works in particularly difficult cases, and it works counter to that theory of alcoholism which was very prevalent until a few years ago. Somebody had to want help for himself. A life or death situation. Intervention involves the understanding that it is not necessary for everyone to come to this point, and that a person confronted by a united group—family, employer, church, any part of that person's life that is important and necessary—is given no choice except losing everything.

No rectors undertake interventions lightly. It is a lot of work. It is always painful to go through. And it sometimes backfires. In order for it to work, virtually all the significant people in the alcoholic's life must (1) agree that there is an alcohol problem and (2) be willing to confront the alcoholic. A common scenario is a situation in which a friend or a family member feels strongly that the person in question is alcoholic, but in which a critical person in the alcoholic's life—a spouse most often—denies the problem. Similarly, it is sometimes impossible to obtain the assistance of the person's employer. In some cases, when an intervention has ensued without the involvement of the spouse, the alcoholic responds positively to the intervention and takes steps toward sobriety or moderation in drinking without the support of the spouse. Inevitably these situations can cause major rifts between the non-conforming spouse and the church. For these reasons, parish priests are hesitant to undertake an intervention.

Most, however, were personally familiar with them. Eight of the interviewed rectors had participated in interventions. Two had been involved in cases regarding another rector's drinking. The overall consensus seemed to be that although there is a place for interventions—especially when the stakes are especially high—they should be practiced only under limited circumstances and should be carefully planned.

At the other end of the spectrum is the position that it is best to wait for the alcoholic to ask for help directly: "The person who is drinking too much and doesn't take the initiative to talk to me about it, I doubt whether what I do or say will work very well. I operate on the theory of respect for the individual to make his own decision," one rector explained.

Another rector is very committed to a "spiritual program" of recovery from alcoholism. Although he has participated in interventions, he prefers to focus his counseling directly on the alcoholic and on the spiritual aspects of the disease:

> I compartmentalize the treatment into physical, mental, and spiritual. In AA we always say that it's an illness of the body, mind, and soul. And there are some body doctors and some mind doctors and some soul doctors.

And you're interested in the soul part?

> Yes, both in terms of understanding the illness and the recovery process. AA is a spiritual program of recovery, which means that the most successful program in alcoholism is a spiritual program. So what I try to do is to prepare them for this spiritual recovery program by rebuilding their spiritual values.

Alcoholics Anonymous is an integral component of most of the rectors' approaches to alcoholism treatment. The majority noted AA and Al-Anon as referral resources they routinely draw upon in the course work with alcoholics and their families. Most churches provide space for several AA and Al-Anon meetings. The normal tradition is for parishioners to select AA or Al-Anon groups that meet in other churches:

> Most all the alcoholic groups that meet here are non-parishional. There will be a few parishioners who come to the AA meetings, but not very many.

In some instances the rector prefers to stress the use of AA instead of the church in alcoholism treatment referrals:

> I've seen such good results in AA, and have come to see AA as taking the place of the church in this regard or supplementing it, and I'm very comfortable with what AA represents and those in AA who amplify the church as an adjunct.

The usefulness of AA, according to interviewed rectors, extends into ethnic communities. From one parish in which several AA, Al-Anon, and Alateen meetings are held, which has a large black constituency, the rector commented:

> I'd always been told by people that AA is a white middle-class phenomenon, but it sure isn't. These are big meetings, sometimes 150 people. And it is preponderantly black, but there are a lot of white people who come too.

Another major resource drawn upon in referring parishioners for alcoholism treatment is the Pastoral Counseling and Consultation Centers

(PC & CC) of Greater Washington. Two rectors regularly use it for support in their work with alcoholics:

> Some people, of course, don't recognize that they've got a drinking problem, but will recognize that they have an emotion or stress problem. If that's the approach they're taking, that's fine and we send them often to PC & CC as a starting place.

Using this type of strategy, it is possible to refer people to sources outside those of the parish itself, while keeping them within the religious community.

A final approach that rectors specified is making a referral to a particular person—an experienced priest or a professional in the field—while also retaining involvement with the parishioner:

> When it comes to the technical expertise involved in alcoholism work, I would rather work with someone else. Also I don't usually refer people out. I ask them if I can get in touch with a therapist while staying in touch with the parishioner, with all three of us working together. I have a stake in the process, and the parish does too.

These different mechanisms for working with alcoholics—whether they are self-acknowledged or other-identified—point up some of the underlying differences in approaches in dealing with the problem. Some rectors tend to rely mainly upon knowledgeable people and services outside the parish itself, without retaining direct involvement in the treatment course following referral. They might prefer to direct parishioners to AA or Al-Anon and/or the PC & CC, concluding that the track record and professional competence of these supporters can provide more appropriate assistance to parishioners than those available within the parish.

In contrast, certain rectors attempt to keep the problem at home as much as possible and to have the parish serve as an active force in the recovery process. Parish resources—such as pastoral care groups, the Alcohol Network, health professionals, and recovering alcoholics—take primary responsibility for initiating and sometimes maintaining treatment. "Outside" specialists might be asked to help, but the rector and/or other parishioners are a mainstay of the treatment process. It would be unusual, however, for a rector not to recommend that a recovering alcoholic and family members also attend AA and Al-Anon. In parishes in which internal mechanisms are regularly used, interventions might be more readily considered, with members of the church community taking charge. In large parishes that are organized into pastoral care groups, the sophistication and level of consciousness about alcohol-related problems can be quite high, as highlighted by one rector's remarks:

> I think there has been enough discussion here around the sensitivity and the plain
> good sense of the congregation when it comes to drinking that if in a given crowd
> someone is in bad shape, someone will take the initiative to do something about
> it. I would trust this congregation more than the average group off the street.

Whatever a rector's propensity toward addressing alcohol problems, the subject is handled with great discretion and tact. This attribute is congruent with the cultural style of the Church. Prudence prevails in broaching sensitive issues. Rectors would never barge ahead and publicly identify a parishioner as an alcoholic without considerable forethought and groundwork.

"Quiet observation" is adopted in most parishes. One rector uses such an approach:

> There's one man that I have a strong suspicion about, and I think he's an alcoholic. He comes to church and has a job, and I've never seen him falling down drunk with any regularity. That's what I watch for, and eventually one day I will probably discover that he is drinking out of control. But until that day, I can't say anything. I am not prepared.

DIOCESAN RECOMMENDATIONS AND PARISH ALCOHOL PRACTICES

Several years ago a Task Force on Alcohol and Other Substance Abuse was established by the Diocese of Washington; it was recently elevated to Commission status. Six clergy and six lay members sit on the Commission. Nationally the Commission is in the forefront of a movement to heighten "consciousness of alcoholism, and has had a big impact" on progress made. It was pointed out by a few priests, however, that the Washington Commission has not been as effective as it might be in its work locally, when compared with those of neighboring Dioceses of Maryland and Virginia.

In short the overall impact of the Washington Commission on local parish drinking practices is perceived by some rectors to be less than it might be. On the one hand, a wish was expressed that the Commission become a stronger force. On the other hand, most parish priests indicated a preference to be left to their own devices in deciding specifically how to handle alcohol issues, as these statements indicate:

> I am glad we have a commission, but when we are in real trouble, we use someone we personally have trust in. But I appreciate having them, to suggest ways in which a parish can help.

> I don't care what the Diocesan policy is, in fact. Our own policy in this parish is to offer choice. So that's what we will be doing.

Although all rectors were aware of the Commission's work, their knowledge about its current activities—and even the present members— was limited. Commission guidelines about alcohol have not been as clearly articulated as they should be in order to have an impact, according to these accounts. The message that does not come across from the group is somewhat ambiguous:

> It is okay to drink and okay *not* to drink. And it is appropriate for people to have a choice made available to them. Alcohol is something to be wary of. However, if you go to the big conferences—not the small church meetings—you see a lot of liquor around. It is just part of life.

The picture that emerges is one of mixed messages and not sufficiently firm backing of any "policies" the Commission does espouse.

To look at this from another perspective, it might be that the Commission is sufficiently in tune with the times when it comes to already-existing alcohol awareness and that its particular impact is felt more subtly than it would have been 20 years ago. By the time the task force was formed, many parishes had already developed their own ways of dealing with alcohol problems. From my general reading of the information from the interviews, in order for the voice of a diocesan commission of any sort to be felt with major force among the constituent parishes, a concerned and sustained effort would have to be made, one that was openly and strongly backed by church authority.

These disclaimers aside, the Commission on Healing Alcohol and Other Substances is credited with some very concrete contributions to a broadening awareness of alcohol problems and the development of programs. First, the EEA provision is an outgrowth of the Commission's work. This rule is universally followed in this sample. Second, the Commission has sponsored workshops on alcoholism at national and diocesan meetings, including one on intervention that was positively cited by one of the rectors most skeptical about the Commission's influence. Third, it has provided promotional materials and educational information on alcoholism in conference displays. Fourth, members attending conferences have publicized their efforts by wearing buttons saying that it is okay not to drink. And finally, the work of certain individuals on the Commission over the years is frequently mentioned as a substantial contribution to the advances made in alcohol awareness and treatment.

In summing up, one rector attributes increased sensitivity about alcoholism to several factors:

> The work of the Commission, cultural forces, and the leadership of a few individuals. In the Diocese today there is a greater awareness and general willingness to examine parish practices. And there is more concern about how clergy are having difficulties with alcohol and how alcoholics are treated.

SOCIOCULTURAL PATTERNS AND ALCOHOLISM PREVALENCE

How do the sociocultural characteristics of the parishes correspond to the prevalence of alcohol problems reported for these parishes? According to information from the interviews, two basic profiles emerge: Three parishes with the highest reported incidence of alcohol problems contrast distinctly with four parishes having the lowest frequency of alcoholism among parishioners. The other five churches fall mid-range in terms of prevalence of alcohol problems, and their sociocultural patterns are quite variable.

The four parishes with the lowest incidence of alcoholism are among the smallest in membership: all have fewer than 500 parishioners. Additionally, they have experienced either a decline in membership or remained the same over the past few decades. Socioeconomically, they constitute the entire group of middle- and lower-class parishes, having the greatest number of blue-collar families in the sample. They are among the five parishes having the highest number of black congregants.

These four low-alcoholism parishes are also among the least socially active churches. Two are moderately social; the other two have a very light social agenda of church events. Similarly, the parishioners place very little emphasis on drinking at church functions. Alcoholic beverages are rarely served outside of the Eucharist. Although parishioners evidence a very low level of awareness or concern about alcoholism in the parish—not surprising since there are so few problems—two of the four rectors expressed considerable interest in alcoholism as an issue for the Church in general. Compared with other parishes, these four churches are passive when it comes to alcohol problems. When alcoholism does occur, the rectors tend to refer parishioners to resources outside the parish. Three of these churches do hold AA meetings on their premises. None of these churches are involved in interventions on a regular basis.

The three high-alcoholism prevalence churches present quite a different picture. They are among the largest parishes in the group studied, all having recently expanded in size. These parish communities are composed predominantly of well-educated, upper-middle-class and upper-class professionals. None has more than 15% black parishioners. Compared with the low alcoholism parishes, these three churches tend to have more parishioners living in their immediate neighborhoods.

Very social parishes, these churches reportedly exhibit at least a moderate amount of drinking at social occasions in the church. All three rectors are moderate drinkers. The level of awareness among parishioners about alcoholism as a real or potential problem is the highest by the entire group of churches studied. The rectors are similarly involved in alcohol-related issues in their pastoral work. When alcohol problems emerge in these parishes, the response of the rector and the parishioners tends to be active. Interventions are more frequently carried out, in comparison with all other churches. Resources within the parish are usually drawn upon, sometimes in combination with outside assistance.

Although this study is too limited, in terms of sample size, to allow an interpretation of these contrasting profiles without qualification, the patterns are so consistent that it is worth offering a tentative reading on the relationship between sociocultural features and alcoholism prevalence. Those parishes that conform least to a stereotypic Episcopalian portrait—white, middle and upper class, well educated—seemingly evidence the least amount of alcohol problems. Those that best fit such a stereotype report the highest prevalence of problems.

What might explain these differences? First, the three highest alcoholism-prevalent churches are also parishes that invite "questioning people," communicants who are actively engaged in political, social, and intellectual issues of the day in both their professional and personal lives. These parishes offer a stimulating context for people who want to grapple with such complex problems as alcoholism. It may be in fact that some incipient alcoholics and their families are attracted to these parishes because of a predilection toward confronting such problems.

It is more difficult to explain why the four parishes at the other end of the alcoholism spectrum evidence relatively few problems. It is possible that because these parishes are not socially oriented that the rector is simply less aware of the drinking problems that do exist because there are fewer opportunities to observe parishioners on drinking occasions.

However, the difference may be attributable in part to the religious background of these parishioners before they became Episcopalians. American-born blacks, for example, often convert from Baptist or Methodist churches, denominations with abstinence as the predominant drinking model. African blacks, although usually Anglican in upbringing, reportedly have lived in a very moderate drinking milieu. Although I do not have the data to support this supposition, it is possible that the white parishioners in these churches, most of whom are blue collar or lower middle class occupationally, frequently convert to the Episcopal Church from a fundamentalist Protestant background that does not include drinking as a part of normal social life. Thus far, these four parishes seem not to have moved very far in the direction of moderate social drinking as a regular part of life. Perhaps this is one factor protecting them from alcoholism.

ALCOHOL WRIT ACCOUNTABLE, EPISCOPAL STYLE

"The cup of salvation" in the Eucharist, alcohol is not a feared or unfamiliar commodity in the Episcopal Church. Parishioners are not highly ambivalent about the role of alcohol in Holy Communion; wine does symbolize the blood of Christ. Instead, the virtue of moderation prevails in both ritual and social contexts. Although not a required component of either liturgical or social events, drinking in a "civilized way" is desirable. A normal part of social life, alcohol is to be served and consumed in a moderate fashion. Little overt social control is exercised over drinking at events deemed to be appropriate drinking occasions.

Instead, alternatives or choices are provided. Beverages other than alcohol are made available. In deciding how to select from among available beverages and how much to drink, individual parishioners are left to their own devices. Though not pressured into drinking and not deterred from selecting non-alcoholic beverages, the individual is given great autonomy and leeway in choosing among the alternatives. The constraints against drinking are very subtle. To be less subtle would be rude.

Any initial attempts to moderate abusive drinking are considered the individual's privilege and responsibility. As in the diocese–parish relationship, the individual—like the parish—has the first say in deciding whether, when, and how to moderate or to abstain from drinking. At this point, alcoholism remains a struggle for the individual, not the church community. Privacy is highly valued in this cultural milieu, especially in sensitive matters. It is not that incipient problems are not seen. They are quietly observed, with the hope perhaps that they will eventually disappear with no overt action required.

Earlier it was stated that "alcoholism is not a parish-wide issue of central importance in this group of churches," according to interviewed rectors. However, alcoholism has increasingly come to the fore as a problem that parishes can and should address. It is not to be ignored, but neither should it be pursued just for the sake of pursuit. No "band wagon" mentality exists in this religious community. But parishes are far from apathetic about alcohol problems. The mood is one of "tempered concern." To heed the appearance of alcoholism among parishioners and to establish ways to respond to the needs of these parishioners is seen as an appropriate province of the Church.

As fitting the cultural context of Episcopalianism, discretion in responding to identified alcohol problems is an essential quality. This is seen in (1) the level of professionalism involved in efforts around alcoholism; (2) the recognition of the complexity of the problem; and (3) the care taken in identification, intervention, and treatment of alcoholism. Many diocesan clergy are far from novices in the alcohol field.

Beginning with the post-World War II era, Episcopalian priests began attending educational programs at the Yale, and more recently Rutgers, School of Alcohol Studies. In more recent years organizational structures—task forces, commissions, and the National Coalition—were formed to coordinate and lead efforts to educate parishioners about alcoholism, to identify alcohol problems, and to intervene and promote treatment. Furthermore, there is a high regard for the professional expertise of lay parishioners available for intervention and sustaining aftercare. In short, there is a substantial appreciation for the professional preparation and expertise necessary to work in the alcohol field.

With this also goes the recognition that alcoholism is an illness and not only a biomedical disease. These rectors understand that it pervades the entire social system of the alcoholic, as well as his or her internal physical and psychological well-being. They see it as a complex phenomenon, one that cannot be dealt with simply. At the same time, they operate under the assumption that it is possible to influence patterns of abusive drinking. It is this optimistic perspective that, in part, encourages the universal application of the EAA ruling. However, once alcoholism is clearly acknowledged by the alcoholic, a family member, a close friend, fellow parishioners, and/or the rector, it is necessary to respond to it in an intelligent and professional manner. The underlying ethic is that "if you are going to get personally involved, it must be done right."

When this point is reached, discretion continues to prevail in whatever steps are taken. "When the stakes are high enough"—as when a priest is drinking alcoholically—an intervention might be warranted. A shift occurs at this time from the perception that it is a problem of the individual to the recognition that it is a problem shared by the wider social group. A sense of "church familism" goes into effect. An integrated treatment approach is usually recommended. Although AA is a central component, in most of these parishes other forms of counseling are also usually recommended. Finally, the church community ideally provides a social support system, one in which the recovering alcoholic not only feels comfortable worshipping and socializing, but also one on which he or she can draw for professional assistance.

To summarize, the Episcopal Diocese of Washington is not a context within which you would expect to encounter public furor over alcoholism as a social and political problem. It is unlikely, for example, that sermons would be delivered on the evils of drinking. "Preachiness" is not within the customary Episcopalian genre. Programmatically, too, relatively few events focus on alcoholism. All in all, little public attention is brought to the problem. Instead a measured manner of addressing individual problems is effected. One rector, prominent in the alcohol field, does not promote alcohol-related programs, preferring "to keep a kind of low profile both here and in the community as well" in his case-by-case work with alcoholics. This is the preferred model for dealing with alcohol problems throughout the Diocese.

REFERENCES

Bernardin, J. B. *An introduction to the Episcopal Church,* rev. Ed. Wilton, CT: Morehouse-Barlow, Co., 1983.

Emrich, R. S. *The Episcopalians.* Royal Oak, MI: Cathedral Publishers, 1973.

Feeley-Harnik, G. *The Lord's table: Eucharist and Passover in early Christianity.* Philadelphia: University of Pennsylvania Press, 1981.

Gent, B., & Sturges, B. *The alter guild book.* Wilton, CT: Morehouse-Barlow, Co., 1982.

Gray, W., & Gray, B. *The Episcopal Church welcomes you: An introduction to its history, worship, and mission.* New York: The Seabury Press, 1974.

Johnson, V. E. *Alcoholism and the Church: A call to action.* New York: The Church Pension Fund, 1980.

23

Middle-Class Protestants
Alcohol and the Family

GENEVIEVE M. AMES

INTRODUCTION

Few people would deny that excessive, habitual drinking of one family member has a damaging effect on the whole family unit. There is a general consensus among professionals in the field that alcoholism contributes to family stress and instability and that wives, husbands, and children of alcoholics have relatively high rates of physical, emotional, and psychosomatic illnesses (Straus, 1971).

Until recently, however, these and other indicators of alcohol-related family problems have not been given adequate attention by the medical and research communities. Primarily, this omission was related to the prevailing medical model of alcoholism, which views alcoholism as a progressive, biologically based disease. Since the "disease" process is typically described in terms of the individual, the inclusion of whole families for research and clinical treatment has seemed unnecessary and irrelevant to the traditional establishment in the field of alcoholism.

GENEVIEVE M. AMES • Prevention Research Center, Pacific Institute for Research and Evaluation, Berkeley, California 94704. Preparation of this chapter was supported in part by grants 5 ROI AA00180, National Institute on Alcohol Abuse and Alcoholism and AA06282-02, a National Alcohol Research Center grant from the National Institute on Alcohol Abuse and Alcoholism to the Prevention Research Center, Pacific Institute for Research and Evaluation.

It is only within the last decade that professionals in the field have slowly come to recognize the need to consider alcoholic problems in the context of family life. Increasing awareness of the importance of family factors for the treatment process is clearly represented in the growing number of clinicians who apply family therapy techniques to alcoholic clients (Steinglass, 1976) and—from another treatment perspective—in the growth of such self-help groups as Al-Anon, Alateen, and Alafam, organizations that consider the needs of the families of alcoholics. Unfortunately the enthusiastic demand and support for new treatment techniques has far outpaced workable family-oriented conceptual models of alcoholism (Steinglass, 1980). Attempts to build conceptual models of family interaction and alcoholism to date have emerged from the varying perspectives of stress theory (Jackson, 1954), transactional analysis (Steiner, 1971), behavioral learning theory (Paolino & McCrady, 1977), longitudinal family development (Steinglass, 1980), and family systems (Ewing & Fox, 1968; Steinglass, Weiner, & Mendelson, 1971a,b; Bowen, 1974). New developments in the family systems approach to alcoholism (see Steinglass, 1979), wherein the family is viewed as a system with its own structure, interactional behavioral patterns, and homeostatic mechanisms, have made a significant contribution in the literature dealing with alcoholic treatment and family therapy.

These models for alcoholic family research and intervention have focused primarily on the interaction between the family and the problem drinker. It is proposed here that it is equally important to consider the interaction between the family and the broader cultural environment. As demonstrated in this chapter, and in Chapter 21 by Ablon, the family experience of alcoholism varies with the ethnoreligious background, social class, and sex of the alcoholic family member. Both chapters demonstrate that deep-seated values and beliefs relating to alcohol use and drinking behavior in large part determine family members' adaptation to the disruptive behavior of an alcoholic parent and the degree to which a family is willing to participate in alcohol treatment or prevention programs. Although the findings of both studies suggest subtle differences in the experience of either maternal or paternal drinking problems, here, too, the cultural component emerges as a significant factor in whole-family responses to alcoholism. Simply intervening with the family system, without also addressing fundamental problems of its sociocultural context, will for the most part have a minimal effect. In the case of treatment for alcoholism, and as demonstrated in this study, such negligence may violate basic beliefs of the patient and his or her family, thereby creating a conflict and an ever-widening rift between health professional and client family.

In this chapter, religious and cultural prescriptions for alcohol use and drinking behavior are explained as such issues affect the alcoholic experience (for both the afflicted person and family unit) in middle-class Protestant families. The line of evidence I present in support of a cultural perspective came from a historical review of American beliefs about alcoholism (see Chapter 3 of this volume) and upon my research experiences in the homes of alcoholic families.

THE RESEARCH SAMPLE AND METHOD

The research materials reported on here are drawn from two separate field studies of problem drinking and family life. In both studies the sample populations were predominantly white middle-class Protestants residing in several neighboring and affluent suburban areas in Northern California. Both studies represent a microscopic and naturalistic approach to family research, meaning that a relatively small number of families were studied intensively over an extended period of time and that the primary domain of research was the natural habitat of the family home. Sixteen families were visited, interviewed, and observed in their homes, for periods ranging from six months to two years. In all the families, the mothers were seeking treatment from county or private mental health services, and in each case, the mother had been identified by such services as an alcoholic. The duration of the mothers' drinking problems ranged from one to eleven years.

In the first study, in-depth life histories of the mothers were recorded during a series of two- to three-hour interviews; additionally whole-family interaction was observed in the course of conducting the research in the family home over a six-month period. The focus of inquiry was on (1) Protestant upbringing and socialization to alcohol use; (2) the development of the mother's drinking problem; (3) the treatment experience; and (4) the family's response to alcoholic symptoms and drinking behavior. These data provided insights into the meaning and experience of female alcoholism in the context of middle-class Protestant culture (Ames, 1977) and served as a preliminary data base for the second and more comprehensive study of maternal alcoholism and family life.

The data for the second study (supported by grant 5 ROI AA00180 from the National Institute on Alcohol Abuse and Alcoholism) were collected during two years of intensive study in the homes of eight additional families. My association with these families spanned all seasons of the year and various times of the night and day. In addition to the in-home observations, the data included personal life histories and verbal accounts of everyday family routines obtained from the parents and most of the adult children. These data provided a composite picture of each family's perspective on the mother's drinking problem from its inception and on family and community response to her alcohol-related behavior. Also these materials helped determine whether the family's existing structure and behavioral patterns evolved with the drinking problem, or were predispositions to it, and whether radical changes in family rules and values had taken place.

When asked about their ethnoreligious background, most of the adult participants in the two studies referred to themselves as "WASPs," the common acronym for "White Anglo-Saxon Protestants." In fact the majority were of English or English-German stock, and some claimed ancestry in the earliest generations of American settlers. All the families were either actively involved with their church or professed allegiance to the beliefs and guiding principles of an established Protestant denomination. Six of the families were Episcopalian, five Presbyterian, four Methodist, and one Congregationalist.

The families in both studies were "intact," in the traditional sense that the parents were married to one another and living in the same household together with their own or adopted minor children. All the husbands and all but two of the wives had attended college. In each case the father was the principal provider, and the mother was theoretically positioned in the role "housewife." Although five of the wives managed to hold full- or part-time jobs, they often were unable to report to work during drinking periods or when suffering from alcohol-related illnesses. The husbands were employed in various corporate or professional positions or owned their own businesses. Most of the families had an affluent life-style and, by established socioeconomic definitions for the American middle-class way of life (see Nye and Berardo, 1973), the families ranged from middle to upper-middle class.

In seeking to understand why and how alcoholic families are affected by cultural factors, the following areas of inquiry are addressed:

1. *The Origins of (Protestant) Beliefs about Alcoholism.*[1] Are heavy drinking patterns considered to be a normative or a stigmatizing condition in the family? What is the history of that belief system, and how

---

[1] Depending on the varying personal and ethnoreligious perspectives from among the multicultural groupings in American society, the terms *alcoholism* and *alcoholic* have different meanings for different people. Differing cultural beliefs about the nature of alcoholism are discussed throughout this book, but for overall purposes of defining it as a health problem, I rely here on Plaut's (1967) definitions of alcoholism and problem drinking: "Alcoholism is defined as a condition in which an individual lost control over his

is it reinforced in the cultural domain of family life and other areas of routine social interaction?

2. *Definitions and Beliefs about Alcoholism.* From the viewpoint of the family, what is the nature of alcoholism in terms of both scientific definitions and traditional ethnoreligious beliefs?

3. *The Struggle for Cultural Conformity.* Why, in many cases, does an otherwise normal and caring family allow concerns for the safety and health of a parent or other family member afflicted with chronic alcoholism to be overruled by cultural pressures to conformity?

4. *The Maintenance of Protective Boundaries.* How do families keep up a "front" for culturally prescribed notions of normality, when a family member has a serious drinking problem and repeatedly behaves in a "culturally deviant" manner?

Of course every family may be idiosyncratic in its structural and behavioral adjustments to alcoholism, and intrapsychic and other needs of individual family members, as well as cultural pressures, may affect alcoholic family dynamics. In a more comprehensive analysis of the materials reported on here (Ames, 1982), I have analyzed the alcoholic family system in relation to both cultural and psychological factors. For the more focused purposes of this chapter, the emphasis is on behavioral commonalities that can be identified and explained in the context of a particular cultural environment. These commonalities and the underlying pressures that uphold them are also examined in terms of their implications for alcohol treatment and prevention programs.

THE ORIGINS OF PROTESTANT BELIEFS ABOUT ALCOHOL

The problem of defining beliefs and practices relating to alcohol among Protestant populations in the United States is complicated by the differing views held by various churches. Although the American branches of some large church groups of Europe, such as the Lutherans and Episcopalians, ordinarily have not opposed moderate drinking, other religious groups, such as the Baptists, Methodists, Presbyterians, Congregationalists, and members of smaller and fundamentalist groups,

---

alcohol intake in the sense that he is consistently unable to refrain from drinking or to stop drinking before getting intoxicated" (p. 39); "problem drinking is a repetitive use of beverage alcohol causing physical, phychological or social harm to the drinker or to others" (p. 37). The element of loss of control differentiates between these two definitions, but in terms of the effect of excessive alcohol consumption on individual and family health, the overall consequences of excessive and repetitive drinking overlap.

have a history of strongly opposing alcohol use and drunkenness as sinful (Oates, 1966).

The Methodist creed on alcohol use has had a strong influence on the positions taken by other American Protestant churches. As late as 1964, the "General Rules" of the Methodist Church forbade the sale, purchase, or consumption of alcoholic beverages. Abstinence from alcohol, along with restrictions on gambling, smoking, and dancing, are preferred predispositions in Methodist congregations for church membership and "salvation from sin" (Bucke, Moore, & Pierce, 1964, p. 44). The rule on alcohol use remained unchanged from the time it was first put forth by John Wesley in 1739 (p. 44) until recently, when rules on alcohol use were cautiously relaxed to allow for occasional drinking, although abstinence is still recommended (Bucke, Holt, & Proctor, 1976, p. 93). For over 200 years, such edicts of the Methodist church and similar ones by other ascetic-Protestant churches have continued to be reinforced by religious teachings that drinking alcohol and drunkenness represent sinful behavior and are therefore not to be allowed or accepted under any circumstances (Oates, 1966).

In addition to Wesley's original statements, the positions on alcohol taken by the Methodists and several other fundamentalist religions can be traced back to religious revivals that swept through the United States in the latter part of the eighteenth century, just before the Temperance Era.[2] As noted in Chapter 3, it was during the Temperance Era, and with strong religious sanction, that frequent and heavy drinking (alcoholism) was transformed from a normal state of habituation—albeit unhealthy and problematic—into immoral, depraved, or mentally deranged behavior. According to Rorabaugh's historical analysis of this period (1979), the original impetus for this thinking came from early-eighteenth-century Methodist and Presbyterian "preachers" who used alcohol as a central theme in their religious revivalists efforts; ministers of many denominations followed their lead in preaching the anti-alcohol theme.

> The consumer of alcohol was portrayed as a man of depravity and wickedness, and this idea was supported both by the presence of rowdies at camp meetings and by the emergence of a religious doctrine that demanded abstinence. Although most denominations had long condemned public drunkenness as sinful, it was revivalistic Methodists who most vigorously opposed alcohol. After 1790 the Methodist church adopted rules that imposed

---

[2] Gusfield (1963), a leading historian on this period, places the Temperance Era in a 100-year span from its earliest development by the Federalists in 1820 to the passing of Prohibition legislation in 1920.

strict limitations on the use of distilled spirits. In 1816 the quadrennial general conference barred ministers from distilling or selling liquor; in 1828 it praised the temperance movement; and in 1832 it urged total abstinence from all intoxicants. A similar rise in opposition occurred among Presbyterians. In 1812 their official body ordered ministers to preach against intoxication. In 1827 it pledged the church to support the temperance movement, in 1829 expressed regret that members continued to distill, retail, or consume distilled spirits, and in 1835 recommended teetotalism. (pp. 208–209)

When considering the strength of the molding force behind Protestant beliefs about alcohol, it is important to note (Room, 1982) that out of these religious movements just before and during the Temperance Era (wherein anti-alcohol sentiment was the predominant theme), the "Methodists, American Baptists, fundamentalists, and gospel denominations—and, eventually, at a second remove, the Salvation Army and Seventh Day Adventists—arose" (p. 449).

The revivalist-type meetings and evangelical-style sermons in which abstinence from alcohol use was presented as the road to salvation appealed to the ministry and lay populations alike, at a time when millions of Americans were searching for a moral crusade and a revivalistic cause (Rorabaugh, 1979). But in addition to filling the need for a religious movement, the campaign against alcohol was of benefit to an expansionary industrialism. Rorabaugh (1979) argues that the essential source of the movement's dynamism and lasting power was its accordance with two central impulses of the era, an appetite for material gain and a fervent desire for religious salvation.

The prosperity that the market revolution had brought to the United States turned many Americans to materialism. Americans preached equality but they worshipped success, by which they meant wealth. This pursuit of wealth ran parallel to a rising interest in evangelical religion. . . . The Temperance Movement, in my judgement, was a balance wheel which made it possible for these two principal and often conflicting elements of the national ethos to work together. (p. 202)

In addition to the "loss of grace," temperance reformers declared that those who used alcohol squandered capital, dissipated and destroyed wealth for selfish, nonproductive ends, and deterred opportunities for saving and investing money (Rorabaugh, 1979). This line of argument, coupled with religious proclamations that abstinence was a component for salvation from sin, persuaded the Protestant middle class to accept, and thereafter to maintain, moralistic perceptions of alcohol use, drunkenness, and chronic alcoholism.

From the mid-nineteenth century on, abstinence and a religious-oriented life-style became the touchstones of middle-class respectability and symbols of elevation to that status level. According to Gusfield (1963), temperance and soon abstinence became a status symbol separating the Protestant middle class from the lower working class, most of whom at that time were Catholic Irish and German immigrants. From a similar viewpoint, Lemert (1951) proposes that the strong negative reactions against alcohol use or—as is the case with most Protestant groups today—to anything other than moderate drinking is related to loss of control. Such symptomatic behavior is not compatible with the dominating middle-class Protestant ethic in most regions of American society. Lemert (1951) comments on the philosophical roots of that incompatibility:

> The Protestant attitudes toward drunkenness took their substance from the general Calvinistic condemnation of frivolity and the extollation of frugality, thrift, and industry as religious virtues. Drunkenness among the American Puritans was abhorred along with sexual shenanigans it precipitated, chiefly because it diverted human beings from the earnest task of making a living and capital accumulation, and also because it interfered with parental instruction of children in lessons of work and religion. (p. 355)

The loss of self-control on the part of the drinker is still seen as a deviation from these characteristics of the old Protestant ethic. Character weakness, as represented by a drinking problem, is a vivid distinction in a culture that attributes "morality, success, and respectibility to the power of the disciplined will" (Lemert, 1951, p. 356).

The reasons for and purposes of the intensity of Protestant participation in the Temperance Movement are still under debate. However, the lasting effect of that historical event as a controlling factor for alcohol beliefs and practices among the Protestant middle class has persisted to recent generations. As late as the mid-1940s, all major Protestant churches, including the Episcopalian and the Lutheran, were still espousing official church and public pronouncements on varying degrees of opposition to liquor traffic (Landis, 1946). Since World War II, most Protestant churches have, in the main, moved toward abandoning positions of legalistic, negative judgments, trying to adopt a stand characterized by a concern not only for church principles, but also for persons afflicted with alcoholism (Conley & Sorensen, 1971). Although there is an almost universal recognition among Protestant churches of the need to plan for future church involvement with alcohol prevention programs, there is still considerable controversy regarding the means for achieving such programs. When a Cooperative Commission of Churches (an organization committed to change in church alcohol policies) recommended an acceptance of *both* alcohol use and abstinence, those churches that were historically committed to the tradition of abstinence found it impossible to cooperate in alcohol programs that contradicted their own basic edicts. As long as major churches are intensely divided over the issue of abstinence, it is doubtful that the churches themselves can accomplish a united and significant effect in changing the moralistic climate around alcohol use in the United States (Conley & Sorensen, 1971).

Paralleling recent efforts by Protestant churches to moderate the lasting and powerful effects of the Temperance Era are efforts by the scientific community to deal with the issue. Over the past 60 years, there have been concentrated efforts to redefine alcoholism in varying theoretical frameworks of "disease," personality problems, or learned behavior. Notwithstanding the efforts of both church and science, three surveys over a 20-year period suggest that alcoholism remains a stigmatizing condition in the minds and hearts of many American people. Seventy-five percent of attitude samples from various parts of the United States have persisted in defining alcoholism as a sign of moral weakness (Cumming & Cumming, 1957; Mulford & Miller, 1964; Orcutt, 1976). That this large body of opinion about chronic alcoholism is still significantly prevalent in today's world suggests that it is deeply rooted in the religious and cultural fabric of the dominant middle-class Protestant population. It also suggests that middle-class Protestant Americans, much like other more ethnically homogeneous subcultures, share historical commonalities in their alcohol-related beliefs and practices. These commonalities are variably expressed and perpetuated in religious and other institutional teachings, as well as in popular novels, plays and films with alcohol-related plots or themes.

INTERGENERATIONAL TRANSFERENCE OF BELIEFS AND PRACTICES

The larger generalized taboo on drunkenness and alcoholism obviously varies among and within various Protestant denominations in the more open society of the United States. However, in spite of religious differences, the families in the study reported here demonstrated pronounced consistencies in their unbending and moralistic responses to alcoholism. These consistencies often reflected the parents' early life experiences in homes that adhered strongly to Protestant edicts on alcohol use. As almost all the parents grew up in practicing Protestant families,[3] they were socialized, perhaps more so than people from most

---

[3] The one exception was a husband raised as a Catholic who converted to his wife's Presbyterian religion.

other American subcultures (as illustrated in many of the chapters here), to negative values for alcohol use in general and heavy drinking in particular. Although moderate drinking was tolerated in some of the families of origin, in others abstinence was encouraged, it not demanded. In every case the parents were taught that alcoholism represented a moral rather than a medical debility. The following statements typify the parents' reports of early-life socialization to alcohol: The first represents the more moderate Episcopal view, the second the less tolerant position of Methodists:

> There was no drinking in my home when I was a child. My mother's idea of a drink at a cocktail party was one glass of sherry; then she'd switch to tomato juice for the remainder of the evening. I never saw my father drink. They had alcohol around the house for guests, but their attitude towards people who drank more than one drink was decidedly negative.

> My grandparents were so against drinking and smoking that they were the type—if you were walking down the street and they saw someone smoking or if they saw pictures of people drinking, they would verbally object. I was raised in my grandparents' home, and their moral code was mine. Everyone in my family knew that drinking was a "bad" thing to do.

In their early life experiences, parents recalled that alcohol was rarely if ever used as a social lubricant for family or church-sponsored parties or at such special celebrations as weddings, holidays, and birthdays. In fact most of the husbands and virtually all the wives reported that as teenagers they experienced minimal social pressure to drink and, with few exceptions, abstained from alcohol altogether during their high school years. Upon moving out of their more protected familial and religious environments into the more worldly environments of college, military service, or workplace, they had their first experiences with alcohol use and social drinking. After marriage, husbands developed moderate drinking habits, mostly in response to changing social or professional pressures to use alcohol. Their wives, most of whom entered the "housewife and mother" role, used alcohol only occasionally at special social events with their husbands, or they had an occasional cocktail before dinner. Drinking alcohol in bar environments or at parties at which there was heavy drinking was the exception rather than the rule for both husbands and wives. Even when earlier instilled values around abstinence or extremely moderate drinking were more relaxed, beliefs about heavy drinking and drunkenness as inappropriate or immoral behavior remained firm.

THE PROBLEM OF DEFINITIONS: A SELF-DEFEATING DILEMMA

Although drinking patterns changed across generations from family of orientation to family of procreation, beliefs and attitudes about alcoholism instilled during childhood did not change. Both the identified women alcoholics in this study and their families accepted more or less "on faith" that alcoholic behavior was socially and morally deviant behavior. Although in the course of their treatment experiences the mothers were introduced to other explanatory models (primarily the disease model), other family members were generally unimpressed by or uninterested in biomedical or psychosocial explanations for chronic drinking problems.

The absence of an alternative, culturally acceptable explanation for alcoholism creates an enormous amount of confusion and frustration for Protestant families having an alcoholic member. Repeatedly, during the several years I frequented the homes of such families, the painful, ongoing struggle with definitions for an otherwise culturally taboo problem was evident. Husbands and children alike were reluctant to describe verbally the erratic behavior of the alcoholic mother in terms of alcoholic-related language. The mother's frequent absences from the family circle when she secluded herself during drinking periods and the family's accounts of alcohol-related incidents were more often than not explained as an undefined "sickness" or "problem." In turn her "sickness" was viewed as a symptom of one or more of her various nervous and physical ailments or of some recent stressful incident, but never as being symptomatic of alcoholism as a sickness. Although family members were often humiliated and angered by the unpredictable nature of the mother's drinking behavior, it was never referred to with such words as "drunkenness" or "alcoholic." Within their ethnoreligious tradition, "alcoholism" as a concept is more generally equated with drunkenness and lack of willpower, and "weak-willed drunkard," is an unacceptable label for one's own wife or mother. This conceptual dilemma proved to be self-defeating in terms of early intervention or treatment, from the earliest warning signs of a developing problem to the chronicity stage of alcoholism.

MANAGING A TABOO ILLNESS IN A FAMILIAL ENVIRONMENT

The nature of moralistic definitions of alcoholism and the necessity for concealment impose additional constraints on both the problem drinker and the family. The alcoholic in these situations was forced into a par-

ticularly difficult situation in that she found it necessary to hide both
the drinking and the fact that she had a serious and debilitating drinking
problem. The cultural context in which the alcoholic condition developed
and was maintained disallowed any discussion or open display of its
physical and behavioral symptoms. By admitting to "alcoholism" or
alcohol-related problems, the mother would, in effect, be acknowledging
immoral if not forbidden behavior, and in the process repudiate her
principal life role as wife and mother. Both the family and the mother
managed unspoken, if not unconscious, adjustments that ensured that
such an unthinkable action—that of discrediting her role status as mother
and key family member—would never occur.

For their part, the women alcoholics took steps to avoid disclosure
from the earliest signs of problem drinking. They became seriously concerned about being discovered from the point where they themselves
came to the shocking realization that they not only drank greater amounts
than ever before, but also that their behavior while drinking was becoming recognized by family and friends as aberrant and thereby unacceptable.

In every case, the mother could trace the onset of increased drinking
practices to specific precipitating incidents. These included an unex-
pected death of a family member or friend, being uprooted by a business
transfer of the husband, or a health-related trauma, such as a non-
elective hysterectomy or a spontaneous abortion. In some instances
drinking became a patterned response to an ongoing stressful condition
of their personal lives, such as marital unhappiness, latent sexual dysfunction, a chronically ill child, or in a number of cases, undefined
loneliness and role frustration. The time lapse between the discovery
that alcohol could serve medicinal or relaxation purposes and the development of chronic habituation to its use was relatively short—overall
from one to eight months. For some, as in the following example, the
transition was almost immediate:

> I was emotionally ill after the death of our daughter. I started using alcohol
> as a medicine, and I think I became an alcoholic at the same time I began
> using it. I almost immediately crossed the line into some kind of compulsive
> drinking.

Whereas before the women had been either quasi-abstainers or extremely moderate social drinkers, upon realizing they were addicted to
alcohol they no longer drank in the presence of others, and drinking
became a solitary and, for most of them, an unpleasant experience. In
other words, in a matter of months, drinking was no longer connected
with "good times" or sociability, but instead with fear and uncertainty:

Some people in the AA group at the hospital looked back and said—"we have to remember the good old days when we could drink and have fun." For me, alcohol was never fun. Drinking set me up for something that I could not deal with.

The solitary drinking by the women in this study took place in two ways: either by consuming greater amounts of alcohol over a relatively short period of time, always in seclusion, or by drinking small amounts of alcohol over a longer period of time, usually during the hours when the husband and adult children were away from the home or asleep. As examples of the first type, one woman drank up to 16 oz of vodka between 5 and 7 A.M. on the days that she drank, and another drank an equal amount of brandy between 2 and 4 P.M. in the afternoon. The mother would be passed out (asleep or unconscious) in the first case before the family arose, and in the second before the children and husband returned home from school and work. Although the mother's unconscious condition and the reason for it were obvious, the family accepted and adjusted to this routine, albeit with frustration, in preference to bearing witness to the actual drinking and subsequent drinking behavior. The second type of drinker was represented by several women who started drinking in the morning and sipped alcohol throughout the day, breaking the pattern with long afternoon naps and an early bedtime. Others began drinking after the family was asleep and continued until well after midnight, usually under the pretense of doing housework or special family-related projects. Here too, although the mother's role performance was greatly diminished, family members seldom challenged her drinking and recovery arrangements.

The behavioral and physical effects of the drinking were rarely witnessed by other family members because the mother would remove herself from the family circle, usually to the master bedroom or some other place long enough to recover. One mother went so far as to check herself into a nearby motel for periodic heavy drinking and recovery periods, always calling to inform her husband of her whereabouts. Others admitted themselves to private detoxification centers for a few days' recovery after a particularly heavy drinking period.

In order to avoid disclosing a culturally stigmatizing problem, families went along with the mother's planned absences from the family circle, adapting traditional family rules, roles, and division of labor accordingly. After a period of time, the alcoholic became the expected, if not the routine, way of family life. In one family in which the mother often spent four days out of seven in her bedroom for a cyclic drinking and recovery period, the 18-year-old daughter matter-of-factly stated that neither she nor any other family member had ever seen, nor had any desire to see, their mother drink. She did however explain the immediate cue: "When I look in, I can tell for sure she is having one of her 'deals' because she always lays down on the bed and she rolls herself up in the bedspread, and that's a 'for sure' sign." On any given day that this particular woman was in her bedroom drinking, several family members or the whole family might be in the house also, going about their normal routines or carrying on with social activities outside the home, leaving the mother alone, seemingly unconcerned about her health and safety.

THE STRUGGLE TO MAINTAIN A PRETENSE OF NORMALITY

The major problems facing middle-class Protestant families in their struggle against the stigmatizing effects of alcoholism are not *only* related to religious prescriptions. Bowing to another kind of cultural pressure, that of immediate and external demands to middle-class conformity, families express great concern about and exert considerable energy toward keeping up a pretense of normal family life and maintaining a *status quo* in the face of the disruptive and unpredictable nature of alcoholic behavior. The ongoing reality of an alcoholic family member who repeatedly displays erratic behavior and breakdown in role performance, and the possibility that these discrepancies might be discovered, places otherwise securely imbedded middle-class families in danger of being ostracized or "selected out" as "abnormal" and thereby unfit members of their church, community organizations, neighborhood, or larger cultural environment. In spite of all their efforts to avoid disclosure, there is some point at which the impending possibility of being socially ostracized becomes a reality at the family level. One family was shocked to realize that this could occur when a letter arrived from the president of the volunteer auxiliary group for the community hospital requesting the mother's resignation as a participating member because of her "drinking problem." Although this woman had never drunk while on duty, her drinking behavior had been observed as heavy and inappropriate at a social gathering. The mother expressed her reaction to being labeled as "unfit" for community work:

> To this day, I am embarrassed and hurt by that incident with my hospital auxiliary. I had worked with them for years. Now I never want to see any of those ladies again. I thought I was a normal drinker to all outward appearances up to that point. After that, I had to be so careful about my drinking.

In another case when both the husband and wife regularly visited their minister for counseling about her drinking, the minister announced from the pulpit one Sunday morning that a particular family in the congregation needed prayers to assist them in their struggle to bring a family member "out of darkness into the light." Although he did not mention the family name, even in anonymity they were pointed out as a morally, if not socially, deviant family. In another case, the principal of a private school arrived unannounced at one family home to "check up" on one of their children who had been absent for a week due to illness. When he discovered that the mother had been drinking, he telephoned an alarmed and embarrassed father at his place of business to suggest that his child might be better off in a foster home.

Incidents such as these provoke an ever-growing sense of fear and paranoia among all family members. Given that the families of this particular middle-class Protestant population were socialized according to religiously oriented moralistic attitudes toward alcohol use and affected by cultural pressures to conformity, one could assume that the alcoholic behavior of the mother would weaken if not quickly dissolve the family unit. In fact the opposite was true. In their efforts to protect the mother and in their adaptive measures to maintain a pretense of middle-class normality, these families created a more cohesive, if unhealthy, family unit. In order to "survive" in their natural cultural environment and to appear as productive, participating units within that milieu, the families were forced into certain makeshift behavioral and structural adjustments. Traditional family rules and organization had to be first tentatively altered and, over time, permanently changed to a new way of going about family routines to accommodate the breakdown in the mother's role function.

Family adjustments to external pressures of conformity were manifested in differing behavioral patterns. Identification of these patterns and the ways in which such family practices develop into an "alcoholic family system" will not be discussed here (see Ames, 1982). However, it is important to note that all the families shared the stigma-related consequences of alcoholism to the degree that it took the form of a "family illness." In their efforts to avoid loss of self-esteem, both for the mother and for all other family members, the family organized what Goffman (1963) refers to as a "protective capsule" or what Knupfer (1964) identifies as "cultural protection" around the stigmatizing problem. This "capsule" afforded periodic separation from the broader cultural environment and protection from outside intervention in their own adaptive and "alcoholic" way of life. Unfortunately, it also supported the continuance of the mother's drinking problem, further impaired her health, and lessened the chances of successful intervention and treatment.

In a study of Catholic-Irish American families wherein fathers are frequent and heavy drinkers, a similar pattern of family support for developing drinking patterns was documented (Ablon, 1980), but for radically different reasons. Whereas families in the Protestant population adopted a "protective" system based on moral explanations of alcohol use, the Irish Catholic population accepted the father's drinking bouts as familiar and expected culturally and even religiously condoned behavior. Although the culturally based responses of Ames's Protestant sample and Ablon's Catholic sample were quite distinct, both types of response facilitated the progression of the affected parent's problem.

The Maintenance of Protective Boundaries

Families' efforts to "keep the world out" of the protective and adaptive capsule of alcoholic family life were actualized in the development and maintenance of physical, social, and emotional boundaries.

Each family, as a unit, set up and maintained boundaries around the natural habitat of the family home. Concerted efforts at "closing out" the outside world were manifested symbolically by distinctive physical and social expressions of separateness. As an example of physical boundaries among all the families, curtains facing the street side of the homes were closed day and night, an anomalous feature among the open, convivial appearances of the other suburban homes in this sunny California environment. Well-kept lawns, trimmed shrubbery, and other expressions of suburban affluence were in most cases absent; the spacious front and rear yards were neglected, several homes were in need of exterior painting and repair, and some were obvious eyesores. Although many of the homes were situated on relatively safe streets or cul-de-sacs, with little traffic, the children's play areas were restricted to the family property line and in some cases to the house and the fenced backyard. Social mobility of teenagers in all but one of the families with older children was guarded and controlled with a seemingly paranoid and unnatural intensity. Children were never allowed to have playmates or friends in their own home; entertainment or receiving of visitors was a rare occurrence. One family periodically held meetings for church-related groups in the home; however in this situation, when meetings took place, the mother had to be "confined" to her room. Mothers rarely left the house when they were drinking, and during their non-drinking periods limited their trips outside the home to church services, shopping, or medical appointments. In one of her rare appearances in her own front yard, the mother in one family was shaken by a neighbor's comment, "I heard you had died six months ago." The neighbor and others had not seen her in many months, and they assumed that the rundown condition of the house and yard was an expression of the husband's grief over her death.

A side effect of the boundary maintenance was the development of listless, uncreative, rebellious, and sometimes dull personalities among the younger children. Many of the children demonstrated social and academic inadequacies that in turn created emotional boundaries of difference and separateness. However, those children who were socialized to maternal drinking at an early age and those who encountered it after age 12 evidenced very different degrees of emotional disturbance and social or academic inadequacy. This in turn set them apart from the children in healthier families in their classrooms and neighborhoods.

Boundary development and maintenance emerged as a purposeful manifestation of family protection and distinctiveness. Boundaries keep the outside world out and maintain the protective, alcoholic way of life. However boundaries do not restrict certain family members from interacting with outside social institutions. Like an ethnic group, a family can maintain its distinctiveness by means of internal boundaries around its private domain and yet allow for continued external interdependence and interactive processes (Barth, 1969). In every case without exception, husbands established their own private social relations across the family boundaries, exclusive of their wives. Such liberties were taken or permitted by the situation of a guilt-ridden and alcoholic wife. One woman's late-evening drinking pattern gave her husband a convenient reason to be absent from the home nights and weekends to sell insurance, and these opportunities enhanced his earning power and furthered his career. Another husband maintained separate friendships for fishing and hunting trips and allocated his vacation time to this purpose rather than to family outings. Another maintained close ties to his boyhood friends in another city, and in fact committed the family to relocating near them without seriously considering his wife's feelings. Her guilt afforded him total decision-making power. In another case the husband made payments on an expensive speed boat and mobile home, both utilized for his leisure activities, by drawing from pay his wife received for her care of two severely retarded foster children. As he commented, "this was a convenient job for a woman who has problems which hold her at home anyway."

In sum, in response to cultural pressures against alcoholism, families build and maintain protective boundaries separating their private and public social worlds. These boundaries can be viewed as survival techniques or as temporary delays from the threatening possibility of being "selected out" of the socio-ecological reality of their traditional and immediate cultural environment. Faced with the choice of disbanding as a family unit or organizing and sustaining a family culture that incorporates a culturally deviant member, the families studied here chose the latter. They chose to protect their mother and themselves against exposure and the subsequent negative consequences of chronic alcoholism. In other words, if the condition of maternal alcoholism and breakdown in role behavior were to be discovered, the mother would be socially labeled an immoral woman and unfit mother. The family, by association, would take on the stigmatizing consequences of alcoholism. Rather than externally expose the problem and thus risk loss of status or prerogatives for the family unit, they hide it, deny it, and, in the process, reinforce both the drinking problem and the maintenance of the stigma. The whole family suffers the consequences of these culturally regulated adaptations to alcoholism. By cultural prescription, alcoholism becomes a family illness.

TREATMENT EXPERIENCES

Over a period of months or years, all the women alcoholics of this study attempted to get treatment or counseling from among several, or in some cases all, of the following resources: the alcohol treatment center of the county mental health services; private detoxification and rehabilitation centers; the psychiatric unit of a large county hospital; private hospitals for treatment of alcoholism; Alcoholics Anonymous (AA); and ministers and church-related counselors. Additionally some of the women entered community hospitals for emergency treatment for alcohol-related injuries incurred during drinking periods as a result of automobile accidents, falls, serious cuts, burns, and other household mishaps. Five entered hospitals for a single or repeated suicide attempts.

It would appear, in some cases at least, that cultural prescriptions against chronic alcoholism *and* the symptomatic breakdown in maternal role behavior remain rigid and unforgiving, even in the recovery stage of the illness. Although all made progress with one or another treatment modality on a short-term basis, overall the women were unable to maintain abstinence or controlled drinking through professional treatment alone. In a follow-up contact with the families of the first study (Ames, 1977), six of the eight women reported that through combined participation in AA and ongoing or periodic therapy, they were able to maintain abstinence and recovery. It is interesting to note that, of these six, four of the couples divorced and two separated on a quasi-permanent basis after the mother had achieved abstinence. In every case in which there were minor children, the husband won custody. In recounting the legal proceedings by which this occurred, the women reported that, to their dismay, extended family, friends, school personnel, and in one case, a church minister, supported the husband in his determination to win primary custody of the children.

At the termination of the second study (Ames, 1982), two of the eight women were maintaining abstinence and recovery as a result of a month-long stay in a private hospital for alcohol treatment and continuing counseling and periodic participation in AA. The remaining six women were still unsuccessfully seeking help from various treatment resources.

For the most part, family participation in treatment processes was minimal and only under duress. At the request of alcohol counselors or therapists, some husbands agreed to joint therapy sessions, while others outright refused to participate. After attending a few token sessions, husbands lost interest in cotherapy, for a variety of reasons, not the least of which was the belief that they shared no responsibility for their wives' drinking problems. As one man stated after a particularly painful cotherapy session, "All they do is drag up problem from our past, and I find that depressing. It's *her* problem; let her deal with it." Another complained that the therapists at a county mental health center asked all the wrong questions. "I want to deal with my wife's problem, and all they want to talk about is our personal life." In most cases both husbands and wives were threatened when questioned about their sex lives or other more intimate topics of their personal lives and refused to discuss these matters in the context of conjoint or group therapy settings. Husbands were decidedly uninterested in, and in fact expressed an aversion to, AA, and those wives who did attend AA meetings did so with little or no encouragement from their spouses.

Children did not participate in the treatment process. Parents discouraged or outright rejected specific requests by therapists to use a family-systems therapy approach. In the more religious households, where church and Bible study were routine activities of family life, children attempted to heal their mothers through spiritual means. A teenage son often prayed at his mother's bedside when she was asleep and recovering after a heavy drinking period. He also arranged for a counselor from a church youth organization to visit his mother regularly in the home in the hope that a more spiritually oriented approach would deter the drinking. During repeated stays in hospitals and detoxification centers, two teenage daughters faithfully brought their mother a Bible and other religious literature for "healing and repentance" purposes.

Husbands also attempted to seek help from spiritual, church-related sources. Meetings and counseling sessions were arranged with ministers, either in the church rectory or in the family's home. From the husband's perspective, this was the most successful, and certainly preferred, course of action, primarily because the alcohol problem was addressed as an individual affliction. Wives however found these sessions embarrassing, especially when the more radical elements of alcoholic symptoms and behavior were brought up in the counseling process. Discussion of such things as the breakdown in maternal role behavior, the mother's physical appearance and her condition when drinking, and the nature of her secret and shameful drinking patterns, when addressed in the context of spiritual counseling, implicitly reinforced the alcoholic's role as family "sinner" and malefactor. In view of the fact that husbands were present at these sessions and in full control of the reporting on the wife's drinking behavior, the context of the counseling triad took on the form of accuser, sinner, and judge, although this was not necessarily the intent of the ministers.

To a large degree a family's general lack of interest in professional treatment and continuance of various levels of spiritually centered treatment symbolizes a rejection of current scientific or medical explanations for alcoholism and the maintenance of a unified belief that alcohol problems are moral problems. The conflict of belief systems between the professional and spiritual domains was openly manifested in an encounter between a family and an attending physician–therapist team at a hospital following the mother's third suicide attempt. In the course of conversation with the whole family present, the medical team explained that the mother had a "disease" and that she was "a very sick woman." This incident occurred after many months of attempts by the mother to obtain treatment and at a point when, having despaired of effective professional help, she conceded to her family's implicitly communicated message that she was more morally than physically ill. She responded to the diagnosis of disease with, "I'm not sick; I'm a sinner." The therapist returned, "Sorry folks, the Bible hasn't done it for you." In the verbal interaction that followed, the physician and the counselor responded firmly and negatively to the family's moral explanations for alcoholism. They repeatedly warned that the mother would most likely die without "proper" medical attention and that "neither the Bible nor prayers nor Christian counseling were going to save her." To save face, and in deference to the medical profession, the family withdrew from the debate, feigning agreement with the medical diagnosis. After that incident, they permanently withdrew from all secular and othewise professional treatment for alcoholism.

On Treatment and the Alcoholic Family System

As most experienced alcohol-treatment personnel know, unless the identified problem drinker is unusually resistant to the influence of whole-family dynamics, she or he will not be able to recover in an alcoholic family environment. Successful treatment and rehabilitation of alcoholism in terms of both abstinence and family unity health and happiness in most cases cannot occur without the cooperation of the whole family. The alcoholic family system, structured and maintained as a protective capsule for both the drinker and the family and serving as a system that incorporates alcohol as a homeostatic mechanism for unit stability, must undergo dramatic changes if it is to incorporate a suddenly abstaining alcoholic member. Set patterns of adaptive role behavior and family rules and habits, once developed and maintained to accommodate the alcoholic behavior and protected over long periods of time by family boundaries, will not change without a concerted effort by the family to understand them and restructure family life. Clinician researchers in the field of family-systems therapy have pointed out the importance of understanding how the drinking behavior is serving an adaptive function (Davis, Berenson, Steinglass, & Davis, 1974):

> In each individual or family that presents with an alcohol problem, it is important to ascertain how the drinking behavior is serving an adaptive function. The maladaptive aspects are readily apparent and can usually be recited quite easily by doctor, patient, and family members. Usually, in spite of the agreement by all of how terrible drinking is, the drinking pattern continues with a concomitant increase of feelings of frustration on everybody's part. Care must be taken to avoid this trap and to concentrate during the history-taking and clinical observation on what is adaptive about the drinking. We believe that in this way more useful information can be gathered and a better therapeutic alliance can be established.
>
> Once the adaptive consequences of drinking have been ascertained, therapy may be structured around helping the patient to manifest the adaptive behavior while sober instead of only during drinking and to learn effective alternative behaviors. (pp. 209–210)

In order to determine initially how the drinking behavior serves an adaptive purpose, the therapist must first understand family interaction during both drinking and non-drinking periods. Since therapist–client interaction is usually limited to a clinical setting, "hidden" functions or meaning of the alcoholic experience in the family home are usually not considered. The descriptive and analytical materials presented in this chapter, as derived from in-home, naturalistic family studies, may provide therapists and other health professionals with an alternative approach to information to accomplish the task of therapeutic intervention at the family level.

ON TREATMENT AND BELIEF SYSTEMS

The medical establishment and the alcoholic patients they treat are often operating with conflicting explanatory models for alcoholism. Alcoholics Anonymous is recognized as a treatment modality that utilizes a particular belief system about alcoholism. Whether a treatment establishment is using the disease theory, personality theory, genetic tendency theories, or whatever other theory of etiology without proper knowledge of the *patient's* "theory" on alcoholism, the result is always the same. As was shown in the case studies, an identified problem drinker can be taken out of his or her family or other social community, educated about whatever etiological theories a treatment resource adheres to, and, in terms of achieving temporary abstinence, perhaps be "cured." However, when that person is sent back into a cultural community that has not had the benefit or experience of such education, or a community that is operating within a differing belief system about alcohol use and drinking behavior, he or she does not get the emotional support needed to sustain abstinence. The family, friends, and relatives may listen to the "medical" stance on alcoholism, even "mouth" the words about such "new" findings as genetic tendencies and disease, and then in the end still opt for an ethnic or religious explanation of alcoholism over a scientific one. Deep-seated cultural beliefs about alcoholism, such as Protestant middle-class beliefs, cannot change in midstream just because of the "accident" of an alcoholic family member. For example, the families in this study resisted treatment through AA primarily because of its name and the extended stigma they might absorb by association. As illustrated in the earlier example, when the therapist and physician team countered a family's religiously oriented approach to alcoholism with, "The Bible hasn't done it for you folks!" the family, in deference to the medical profession, did not argue its case. They simply withdrew from all professional alcohol treatment services. This ethnocentric and naive approach of these treatment professionals threw up a permanent ideological wall between the family and secular treatment services.

Successful treatment and rehabilitation of alcoholics and their families may in most cases require health professionals to (1) recognize the existence of belief conflicts; (2) define the differences between client and professional healer; and (3) take sensible, thoughtful steps toward a mutual understanding of such differences. I am suggesting that the healer as well as the patient must be enlightened. A major key to success in therapy is working within the client's belief system, if possible, by gently incorporating it into whatever treatment approach is utilized.

I would further suggest that professionals in alcohol treatment utilize spiritual counseling as a complement to established medical and therapeutic approaches. Practitioners in any area of the United States could make a point of locating what they consider good religious resources for this purpose. Many ministers and priests have good training for working with alcohol problems, and some are well trained in alcohol and family therapy. Clergymen who also have recognized professional training as therapists could be kept on file as a credible resource for treatment or perhaps encouraged to participate in the treatment process as cotherapists. Families who are otherwise resistant to professional alcohol treatment services (as was the case with most of the families in this study) might be more willing to cooperate when whole-family participation in the treatment process is strongly recommended by a minister.

Although such plans require time and effort on the part of the professional healer, in view of the poor success rate of professional alcoholic treatment modes, as witnessed in this population of families, it might well be worth the effort.

REFERENCES

Ablon, J. The significance of cultural patterning for the alcoholic family. *Family Process,* 1980, *19,* 127–144.

Ames, G. *A description of women alcoholics' behavior as affected by American sociocultural attitudes.* Master's thesis, San Francisco State University, 1977.

Ames, G. *Maternal alcoholism and family life: A cultural model for research and intervention.* Unpublished doctoral dissertation: University of California Medical Center, San Francisco, 1982.

Barth, F. Introduction. In F. Barth (Ed.), *Ethnic groups and boundaries: The social organization of culture difference.* Boston: Little, Brown, 1969.

Bowen, M. Alcoholism as viewed through family systems theory and family psychotherapy. *Annals of the New York Academy of Sciences,* 1974, *233,* 115–122.

Bucke, E. S., Holt, J. B., & Procter, J. E. (Eds.), *The book of disciplines of the United Methodist church.* Nashville, TN: The United Methodist Publishing House, 1976.

Bucke, E. S., Moore, L., & Pierce, L. (Eds.), *Doctrines and disciplines of the Methodist church.* Nashville, TN: The Methodist Publishing House, 1964.

Conley, P. C., & Sorensen, A. *The staggering steeple: The story of alcoholism and the churches.* Philadelphia: Pilgrim Press, 1971.

Cumming, E., & Cumming, J. *Closed ranks: An experiment in mental health education.* Cambridge, MA: Harvard University Press, 1957.

Davis, D., Berenson, D., Steinglass, P., & Davis, S. The adaptive consequences of drinking. *Psychiatry*, 1974, 37, 209–215.

Ewing, J. A., & Fox, R. E. Family therapy and alcoholism. In J. H. Masserman (Ed.), *Current psychiatric therapies*, Vol. 8. New York: Grune & Stratton, 1968.

Goffman, E. *Stigma: Notes on the management of spoiled identity*. Englewood Cliffs, NJ: Prentice Hall, 1963.

Gusfield, J. R. *Symbolic crusade: Status politics and the American temperance movement*. Urbana, IL: University of Illinois Press, 1963.

Jackson, J. K. The adjustment of the family to the crisis of alcoholism. *Quarterly Journal of Studies on Alcohol*, 1954, 15, 562–586.

Knupfer, G. Female drinking patterns. In *North American Association of Alcoholism Programs, selected papers presented at the 15th annual meeting*. Washington, DC: NAAAP, 1964.

Landis, B. Y. A survey of official church statements on alcoholic beverages. *Quarterly Journal of Studies on Alcohol*, 1946, 6, 515–539.

Lemert, E. M., *Social pathology: A systematic approach to the theory of sociopathic behavior*. New York: McGraw-Hill, 1951.

Mulford, H., & Miller, D. Measuring public acceptance of the alcoholic as a sick person. *Quarterly Journal of Studies on Alcohol*, 1954, 24, 314–323.

Nye, F. I., & Berardo, F. M. *The family: Its structure and interaction*. New York: MacMillan, 1973.

Oates, W. E. *Alcohol in and out of the church*. Nashville, TN: Broadman, 1966.

Orcutt, J. D. Ideological variations in the structure of deviant types: A multivariate comparison of alcoholism and heroin addiction. *Social Forces*, 1976, 55, 419–437.

Paolino, T. J., & McCrady, B. S. *The alcoholic marriage: Alternative perspectives*. New York: Grune & Statton, 1977.

Plaut, T. F. A. *Alcohol problems: A report to the nation*. New York: Oxford University Press, 1967.

Room, R. Alcohol as an issue in Papua New Guinea: A view from the outside. In M. Marshall (Ed.), *Through a glass darkly: Beer and modernization in Papua New Guinea*. Boroko Papua, New Guinea: Institute of Applied Social and Economic Research, 1982.

Rorabaugh, W. J. *The alcoholic republic: An American tradition*. New York: Oxford University Press, 1979.

Steiner, C. M. *Games alcoholics play*. New York: Grove Press, 1971.

Steinglass, P. Experimenting with family treatment approaches to alcoholism 1950–1975: A review. *Family Process*, 1976, 16, 97–123.

Steinglass, P. Family therapy with alcoholics: A review. In E. Kaufman & P. Kaufmann (Eds.), *Family therapy of drug and alcohol abuse*. New York: The Gardner Press, Inc., 1979.

Steinglass, P. A life history model of the alcoholic family. *Family Process*, 1980, 12, 211–226.

Steinglass, P., Weiner, S., & Mendelson, J. Interactional issues as determinants of alcoholism. *American Journal of Psychiatry*, 1971, 128, 55–60. (a)

Steinglass, P., Weiner, S., & Mendelson, J. A systems approach to alcoholism: A model and its clinical implications. *Archives of General Psychiatry*, 1971, 24, 401–408. (b)

Straus, R. Alcohol and alcoholism. In R. Nesbit (Ed.), *Contemporary social problems*. New York: Harcourt Brace, 1971.

VIII
Synthesis and Application

24

American Experiences with Alcohol
Commonalities and Contrasts

DWIGHT B. HEATH

INTRODUCTION

This book offers unusual strengths in several fields: It is a collection that sheds new light on important aspects of American culture; it incorporates a variety of fresh approaches that anthropologists have only recently brought to bear on the study of alcohol use and its outcomes; and it gives us insights into the several meanings of alcoholism that have rarely been articulated, despite the enormous volume of literature that exists on the subject. The contributions all appear here for the first time, and they are presented in styles that are clear and generally concise and well organized, without being monotonous or formulaic.

This volume represents a pioneering attempt to look at the diverse patterns of belief and action that relate to alcoholic beverages in a modern nation-state and to do so in a way that emphasizes viewpoints of various segments of the population. It is a tribute to the editors that they were able to elicit original work from so many well-qualified contributors, in a way that provides an interesting sampling of the varieties of American

DWIGHT B. HEATH • Department of Anthropology, Brown University, Providence, Rhode Island 02912.

experiences with alcohol. By avoiding the temptation to prescribe a standard format for the contributions, the editors gave each author an opportunity to stress those data, methods, theories, interpretations, and conclusions that seemed most salient to the experience of each given population. As a result, the chapters do not fit together in a way that adds up to an overall portrait of the meanings and uses of alcohol in the contemporary United States—but neither is the book a patchwork of unrelated vignettes.

In one sense this book would constitute a valuable contribution if it did nothing more than document in rich detail some of the diverse American experiences with alcohol that had not previously been chronicled. The case studies do that effectively—but they go far beyond that novel, but relatively simple accomplishment. Another valuable contribution that they make is to offer us some insights, by way of commonalities and contrast among the case studies, into some of the variables that appear to be important in shaping the interrelations between alcohol and human behavior in general. That importance is sometimes from the point of view of scientific analysts, of various disciplines and from various cultural backgrounds, who have a broad overview of the extant literature; sometimes the importance is from the point of view of members of a given population, who know and care little about drinking in any other context, but for whom certain aspects of their own way of life have special salience.

The anthropological enterprise is often marked by different kinds of dialectic. On the one hand there is a fundamental concern for what is universal in the human experience—an attempt to highlight uniform, constant, or at least analogous elements that recur among diverse populations. A contrasting fundamental concern is the celebration of diversity in the human experience—an attempt to highlight what is distinctive, unique, or at least variant among populations. On the other hand another dialectic is operative. Much of our work is dedicated to the detailed communication of data, meticulous descriptions of human activities and the products thereof, in ways that ideally make them amenable to alternative interpretation by others who may never have witnessed them; the other side of that coin is a widespread conviction, at least within the living memory of any practicing anthropologist, that serious efforts should be made to relate any body of data to at least some set of concepts or theories that might allow interpretation of meanings beyond positivistic description and classification. Another of the customary dialectics that pervades anthropological work is the emphasis on socially shared patterns of belief and behavior, and recognition that the range of variation around any norm is important, to the point that differences among individuals may sometimes become the bases of new norms. That last contrast in itself—between the idiosyncratic and the popular—is basic to another dialectic: For certain purposes, it is important to emphasize that the shared patterns we call culture tend to be remarkably persistent. But in addressing other questions, we recognize and even stress that change is a characteristic of every culture, however archaic or tradition-bound it may appear in relative terms or at a given moment in history.

These kinds of dialectic—and probably others I have not articulated—lend a special vitality and variety to the anthropological literature. It does much to sustain interest in some, but we must recognize that it is suspect in the eyes of those who readily discredit what they see as inherent contradictions. If we have learned anything in the course of research on human behavior, we have learned that people have a remarkable capacity for holding and acting on what look like contradictory beliefs. With special reference to the study of alcohol, we have come to recognize that multidisciplinary perspectives are not merely a catchphrase to be ceremonially invoked as a remote ideal from time to time, but an indispensable part of the analytic tool kit that all must share if any is to learn.

In this chapter, I will try to highlight some of the commonalities and contrasts that have emerged from this broad sampling of American experiences with alcohol. In doing so, I will pay some attention to recurrent themes, especially striking in view of the fact that these chapters were prepared independently by authors who, in many instances, have never discussed the subject among themselves. It seems important to pay attention to a few distinctive features that occur in various chapters at the same time. In both of these connections, my frame of reference will be comparison not only among the chapters that comprise this book, but in a larger sense, I will try to set the discussion in a broader context of issues that have been dealt with elsewhere by investigators who have paid special attention to alcohol in sociocultural perspective.

RESEARCH METHODS

In a volume that emphasizes the work of anthropologists, it is striking that such varied approaches were applied to the understanding of American experiences with alcohol. Few individuals who are not actively engaged in research on some aspect of alcohol use and its outcomes are aware that there is a strong and rapidly growing community of scholars, truly international and multidisciplinary, who contribute systematically to "the field of alcohol studies." (Although the literature is scattered, diverse, and largely identified with more traditional fields of study, there is sufficient shared interest that some colleagues have even proposed "alcohology" as a new discipline, but that still seems remote.)

The chapters in this volume represent an important new development in terms of the anthropological contribution to alcohol studies. Until the early 1970s, most of the anthropological works on alcohol were serendipitous by-products of research that had other foci. Ethnographers who were studying a population with another theme in mind often discovered, quite unexpectedly, that alcohol was important to the people among whom they were working and so it assumed some importance in the analysis of the larger culture. By contrast, the case studies reported here are almost invariably the products of careful and systematic research by social scientists who set out with the intention of arriving at a detailed understanding of the meanings and uses of alcohol in a given population.

In some respects, it is a sign of maturity in the field that only a few of the authors specifically mention what used to be so often cited as "the traditional ethnographic methods of participant observation and non-directed interviewing." It does not in any significant measure represent a repudiation of those approaches, however; most of the authors appear to have combined them with directed interviewing, documentary analysis, surveys, and other methods.

Chapter 11 by Gaines briefly mentions some of the organizational and administrative problems that occur in connection with alcohol-related research. There is also a brief critique of the limitations that are inherent in reliance on a survey instrument, however helpful it may be in yielding data in an easily and consistently quantitative form. It is striking that so many of the authors of these chapters did use surveys, however, less often alone than, as both Room and I have recommended in other contexts in recent years, in combination with observational and other methods (e.g., Freund, Ch. 6; Page et al., Ch. 17; Simboli, Ch. 5; Sue et al., Ch. 19; Topper, Ch. 13; Westermeyer, Ch. 20).

The reservations expressed by various contributors with respect to heavy reliance on surveys do not all refer to the special values of emic data (formulations couched in terms that are meaningful to the people being studied), as contrasted with etic data (formulations couched in terms that fit the conceptions of the investigator). Anyone who recognized the pervasiveness of cultural differences must wonder about the validity of data collected cross-culturally with a uniform instrument, however reliable and statistically elegant the tabulations may be. In a more pedestrian and practical vein, it is coming to be recognized that sound observational studies can be conducted much more rapidly and inexpensively than can surveys that rely on rigorous sampling, especially in areas in which the relevant skills are in short supply.

By contrast, one of the most serious reservations that has been expressed by others with respect to heavy reliance on observational methods is that their oversampling of "normal" workaday contexts and their emphasis on positive social relations underestimate the problems associated with drinking. It is true that most of the authors here communicate some of the favorable views that people have toward drinking and the zest that it brings to at least some special events. However, almost every author also mentions one or more kinds of difficulty that members of a given population experience, or carefully try to avoid, in connection with alcohol. The kinds of problems with which these people are concerned include many of the problems with which social workers and policymakers are concerned: spouse abuse or child neglect in Irish households, *faux pas* for the Japanese, cirrhosis of the liver among old Italian men and increasingly among blacks, overexpenditure on drinks at a bar resulting in budgetary constraints at home for poor Hispanics, and traffic accidents among Indians of all ages and suicides among adolescent Indian males. Although the kinds of data reported here do not readily yield epidemiological rates, it seems pretty clear that sons of Irish mothers are at relatively high risk for incurring certain alcohol-related problems (Ablon, Ch. 21; Stivers, Ch. 8), and Indian males (Weibel-Orlando, Ch. 12), especially Navajos (Topper), are at risk for other kinds of alcohol-related problems. Other categories appear to be at relatively low risk, women in any of the groups (especially black and Hmong) and Jews of either sex, as well as Chinese and Japanese.

The question of sampling appears to have been systematically addressed in only a few of the studies reported here. Even in those instances in which it is mentioned, we are not given sufficient information on which to judge the appropriateness of such sampling. Page *et al.* are explicit about the selective quality of their sample, which includes only informants who were also known to be habitual users of illegal drugs. Rodin's account (Ch. 4), using information from and about affiliates of Alcoholics Anonymous, is selective in another way. Bennett's overview (Ch. 22) of an Episcopalian diocese apparently relies heavily on the impressions of a few high-status individuals in each parish, and Westermeyer also mentions dealing with a disproportionate number of community leaders among the Hmong. For most of the other case studies, we have little indication of the extent to which the respondents reflect the overall population, and none of the contributors reports having attempted any systematization in terms of sampling behaviors.

In a few instances, the patterns of belief and behavior reported in these case studies are compared to those reported by earlier observers (e.g., Ablon, Simboli, Stivers, Sue *et al.*, Ch. 19). In some other instances, the method of reporting does not lend itself to the resolution of some relevant questions: For example, how much of what Gaines characterizes as urban black patterns would be found similarly—as he himself points out—among Southern whites? Were the economic factors cited by Edwards (Ch. 9) in accounting for twentieth-century moonshining in Appalachia equally pertinent 100 years or more earlier? What is there in Kitano's chronicle of John's drinking career that distinctly relates to his Japanese background?

Ablon's work, and more recently Ames's (Ch. 23), have been recognized as unusually valuable in alcohol studies because they have been able to combine a concern for the family as a socially important unit with observations in a "naturalistic" setting, that is, the home. Ironic as it may seem, this kind of research is rare; the inherent difficulty of doing it effectively is one important reason. A "natural history" approach would seem an integral part of attempting to understand any kind of behavior, but the human animal sets many obstacles in the way of ethological study. The degree of acceptance and rapport that are involved in such work as that described by Ablon and Ames is exceptional. Beyond that, the fact that they have sustained such investigations adds a longitudinal dimension that is almost unique. The insights and information that people provide in their own homes over successive years give a very special meaning to the often perjorative term "qualitative data."

A contribution that is unique in many respects is Rodin's: She alone deals with what sociologists would call "achieved" (rather than "ascribed") status—a category in which individuals choose to be members rather than assigned to be members by others. More than most authors in this volume, she relies on a semiotic approach, emphasizing the interpretation of meanings. Some would find it paradoxical that such data, comprising little other than symbols derived from open-ended verbal improvisations by informants, are then juxtaposed with a more deliberate focus on physiology than occurs in most of the other chapters. But the resulting interpretations have that special quality of "making sense," even to some of the actors who could never quite articulate some of the anomalous meanings themselves; that is a valuable test of social science insights with respect to behavior that people often accept "on faith" because of its heuristic value, even without fully understanding why it works.

In several of the chapters, it is apparent that the smooth exposition has been carefully organized and integrated from vast amounts of data, both qualitative and quantitative, some garnered from questionnaires and tightly structured interviews and some from casual conversations and observation. A few of the authors have set the patterns they describe in historical context, and all reflect and refract what has too often been treated as a unitary American experience with alcohol.

HISTORICAL PERSPECTIVES

It is noteworthy that many of the authors here have avoided a pitfall that is commonplace in anthropology. There is some historical context for many of the case studies, and a few even focus on change as a central theme for understanding the distinctive patterns of belief and behavior with respect to alcohol as they occur and interact in a given population.

PERSISTENCE AND CHANGE IN DRINKING PATTERNS

Chapter 10 (Herd) on ambiguity in black drinking norms is explicitly "an ethnohistorical interpretation." The interplay of intellectual, political, economic, demographic, and other factors is succinctly demonstrated in what she characterizes as a shift from a predominantly temperance-oriented population in the nineteenth century to a more permissive and consumption-oriented population in the twentieth. The changing symbolic associations with slavery, ascetic religious fundamentalism, disfranchisement, high life, and so forth are striking, but it is never clear whether many individuals actually changed their views about drinking or whether different kinds of source materials available at different times emphasized the views of different constituencies. The ambiguity or ambivalence that she describes today probably refers to the range of responses that emerge from numerical analysis of large-scale surveys and not to any inconsistency in values or attitudes that are held by any significant number of individuals.

Stivers's discussion of Irish-American drinking has special value as a corrective to the widespread misconception that the stereotype represents a continuity from the old country. Edwards's contribution, by contrast, implies that the beginnings of moonshining in Appalachia had more to do with the customs brought by Scotch-Irish pioneers than with economic and ecological factors; ironically, those reasons that seem at least equally plausible for the eighteenth century are cited for the twentieth.

Significant change can, of course, take place in considerably shorter periods of time. Page *et al.* mention different patterns in the use of alcohol and other drugs among Cuban immigrants within a generation; the changes that Gordon (Ch. 16) found among three different Hispanic immigrant populations took place in less than a generation.

It is also a truism, important to anthropologists, but not always widely recognized by others, that various aspects of a culture can change at very different rates. Thus, Freund is struck not only by a remarkable degree of persistence in many areas of Polish culture, but also by the considerable restrictiveness in relation to alcohol use that has emerged in this country. Weibel-Orlando effectively reminds anyone who may not remember that American Indians are not a culturally uniform population, and never were throughout recorded history; both tribal variation and changing governmental policies are reflected in the diverse roles that alcohol has played in the Native American experience. Westermeyer is careful to make the point that the drinking patterns of the Hmong were undergoing rapid change in Laos even before their relatively recent immigration to the United States.

ACCULTURATION AND ADJUSTMENT

Just as cultural change in general has become a dominant theme in many realms of anthropological interest, acculturation has been recognized as an important force for understanding what happens to populations that are in close and sustained contact with each other. Those of us who have critically examined the concept and processes of acculturation with respect to large and varied realms of culture have tended, since the 1950s, to be preoccupied with how it works; by contrast, those who focus on acculturation with special reference to drinking have paid little attention to the process. The outcome has been a couple of unusually mechanistic models of acculturation in the alcohol field. One such view is as a sort of inevitable and irresistible force reshaping the customs of a relatively powerless population until they conform with the dominant norms of their more powerful neighbors. Kitano et al. (Ch. 18) refer specifically to "acculturation of drinking styles" to account for the way in which Japanese in this country appear to have shifted progressively from traditional patterns and toward some American patterns; although there is little discussion of how or why this occurs, it is quite apparent across successive generations. With reference to Italian immigrants, Simboli finds that such changes are even measurable among generational cohorts of Italian immigrants and their descendants. Freund notes that Polish informants take pains to mention that their attitudes toward drinking are less permissive (and hence "more American") than those of their ancestors, whereas Gilbert (Ch. 14) makes the point that

Mexican–American women, once employed, are less accepting of drunken rowdiness and other behaviors that they consider irresponsible on the part of their male companions. The idea that minority populations progressively—and almost automatically—adjust to what are presumed to be some sort of "mainstream American" norms (whether in drinking or in any other aspect of culture) appears to be a vestige of the "melting-pot" theory of assimilation, which has been generally discredited in recent decades. For that matter, Ames's characterization of "middle-class Protestant" attitudes and behaviors with respect to alcohol may surprise many who presumed that they were very different, especially in terms of permissiveness.

What is particularly striking about some of these cases is that they contradict another simplistic popular conception about the psychodynamics of acculturation. According to this view, acculturation is a sort of overwhelming pressure on members of an underdog population, creating great psychological stress. This occurs because they are subjected to different (and conflicting) sets of norms, and as a result, they are driven to drink to relieve tension. It is noteworthy that many of the authors explicitly rejected the relevance of such an interpretation for their case studies. With reference to urban blacks, Gaines makes the point that it is those who are most embedded in ancestral cultural patterns who have the highest rate of drinking problems. Similarly, despite the many and varied new stresses to which working women in the Mexican population are subject, Gilbert observes that they themselves drink only rarely, in keeping with traditional norms. Page *et al.* explicitly reject the thesis that acculturative stress is a major factor in drug use among Cuban immigrants. With reference to three other Hispanic populations, Gordon emphasizes variability, including a dominant pattern among Dominicans of drinking less here than they used to in the Caribbean. Another population presumably undergoing significant acculturative stress, but drinking less in the United States than in their home country, are the Hmong described by Westermeyer. Topper offers an unusual insight to the effect that drinking parties can be important contexts for continuity as well as for change. The recognition that Navajo transmission of traditional cultural norms and ritual information often occurs in informal drinking groups could only have emerged from the qualitative data that observational methods can yield.

One need not be a champion of cultural pluralism or ethnic pride to recognize the irony that many older Hmong men and women, who are presumably undergoing significant acculturative stress, are drinking much less, and less often, here than they did in Laos or Thailand, and that many of them have even become abstainers, or that Japanese men drink more, and more often get drunk, in Japan these days than in the United States.

NORMS AND SOCIAL CONTROLS

A major portion of the sociological and anthropological literature on alcohol is couched in terms of norms (in the sense of rules, both prescriptive about the ways in which things should be done and proscriptive as "Thou shalt nots") and the social controls by which such norms are enforced. The emphasis on norms has both advantages and disadvantages. A positive value is that such values and attitudes have much to do with individual motivation and the rewards and punishments that various behaviors elicit; thus they help in understanding how and why people learn to behave as they do. Eliciting normative statements from people of different cultures can reveal rules that account for different patterns of behavior. Most of the contributors to this book discuss norms at length, although few use the term. Such an emphasis on consensus can, however, be misleading if it conveys an unrealistic sense of uniformity, and the range of real variation is lost.

Among the authors who pay special attention to norms and social controls is Herd, whose attribution of ambivalence to contemporary blacks is interpreted in terms of historical change. Gilbert specifies that Mexican-Americans are often careful to manipulate drunks in ways that avoid social problems; Kitano *et al.* say the same of Japanese-Americans, and Sue *et al.* of Chinese in this country. Glassner and Berg (Ch. 7) point out that the Jewish view of drunkenness as a sign of moral weakness, and as non-Jewish behavior, helps account for the low rate of alcohol-related problems that has long been noted for Jews everywhere. Topper sees a high risk in the fact that young Navajos are in a culturally alien context, remote from normal social controls, when they drink in Anglo towns around the reservation. Weibel-Orlando goes a step further in pointing out that Indians in Los Angeles often seem to use public drinking as a political statement, asserting their ethnicity in "white" territory. Ames sketches the historical and theological roots of ascetic Protestantism more fully than do many who fail to make clear why norms of abstinence can be a risk factor for various alcohol-related problems.

Incorporation of quantitative data from surveys helps some authors avoid the fallacy of reporting only modal patterns and ignoring violations of or variations around the norms of a given population. Even qualitative data, when amply presented, provide clear evidence of the fact that real patterns of behavior often diverge to a significant degree from the ideal.

The concept—and the reality—of cultural integration is aptly illustrated in linkages between attitudes and actions with respect to alcohol and attitudes and actions with respect to other things. Illustrative of that point is the association, consistent across all of the Hispanic populations described in these case studies, of temperance and modesty on the part of women. The pervasive and powerful correlation of sobriety and sacredness among several American Indian groups appears to be salutary, not only in terms of ethnic identity and pride, but also in terms of providing therapeutic support to some individuals who might otherwise have greater difficulty in avoiding drinking problems.

The normative approach to describing drinking patterns has had another important impact in both anthropology and alcohol studies. Ethnographers have sometimes been taken to task for "problem deflation," dealing with alcohol in a way that fails to reflect the grave deleterious consequences that sometimes result from its use. It is striking how many more positive than negative aspects of drinking are cited by the authors and eulogized by their informants. This does not necessarily mean, however, that our contributors have systematically distorted the data, consciously or unconsciously. On balance it would appear that most of the people described in most of these populations view alcohol as a normal—sometimes important—part of their lives, and deplore "excessive" or "inappropriate" use of it as "abnormal," even "problematic" if it results in harm to the drinker or to others. Unfortunately, there is little detailed discussion in these papers about how norms are communicated and enforced—except for Rodin's insightful treatment of Alcoholics Anonymous. Further aspects of the normative approach are discussed below, as they relate to the important phenomenon of intracultural variation.

INTRACULTURAL VARIATION

In a sense this is a book as much about cultural pluralism as it is about alcohol. The case studies deal with a wide range of populations— what some would speak of as "ethnic groups," others as "subcultures," "minorities," "special populations," and so forth. In other contexts, it is often appropriate to belabor definitions of such terms and to pay close attention to specifying criteria for labeling the various populations. In this context, it may be sufficient to avoid the fallacious pitfall of comparing apples with oranges and to recognize that these chapters reflect several very different levels of generalization.

Bennett focuses on a single Episcopalian diocese, whereas Glassner and Berg generalize about Jewish-Americans. Edwards talks about various populations within a geographic region, whereas Rodin focuses on affiliates of a single self-help movement. The category *Hispanics* that is so often treated as if it were unitary is here portrayed in striking diversity by Gordon, Gilbert, Trotter, (Ch. 15), and Page *et al.* Herd's historical perspectives on black drinking complement Gaines's comments on contemporary attitudes. These and other variations illustrate that the strength of this collection is not its encyclopedic coverage, but rather in its rich sampling of the many kinds of diversity that interact to constitute the complex reality of contemporary America. They do not add up to a convenient sum, nor can the data be manipulated in any meaningful way to yield any kind of "average" or "typical" or "normal" or "representative" characterization of "American drinking culture." Some readers may find this disappointing, but they should recognize that it reflects the pluralism that is both a strength and a weakness of this country.

The national motto *"E pluribus unum"* originally referred to the federal unit that was composed of the several states, each of which retained considerable autonomy. In recent years, it can as well be interpreted as referring to a national population who share some key concerns, but whose workaday allegiance and relations are often focused within relatively small segments of the population that constitute reference groups in which membership is based on various criteria.

Not long ago, it was considered an appropriate advance when the spotlight of attention was focused on "minority populations" to the extent that someone would write a paper on "Hispanic drinking," with a generalized description of predominant patterns of belief and behavior and, from an exceptionally conscientious author, some caveats about a few pitfalls to be avoided or hints that might be helpful in dealing with Spanish-speaking populations; another on "Indians"; still another on "blacks"; and so forth. It is a measure of both maturity and realism that the editors and publisher of this volume are comfortable in having four essays, on more than twice that many Latino populations; separate chapters on Chinese and Japanese rather than a single discussion on "Orientals"; and even two very different treatments of Irish-Americans (although the patterns described by Ablon and by Stivers are remarkably similar).

Such striking variation among American experiences with alcohol strongly underscores the fact that acculturation to a "mainstream American" mode of drinking is questionable. Even if such a pattern could be delineated in broadly generalized terms, the realities of everyday social life remind us that such norms would not apply equally to the sexes, to individuals of different ages, to members of numerous occupational or socioeconomic classes, or across other categories that have varying relevance in various contexts.

SEX SIMILARITIES AND DIFFERENCES

Even within a given population, the norms for many kinds of behavior differ according to sex, and that appears to be generally true with respect to alcohol use. Although researchers in many other disciplines have only recently begun to pay systematic attention to drinking among women, ethnographers have done so fairly consistently.

A striking regularity in many of the populations described here is the sexual double standard, whereby men are allowed—even sometimes expected—to drink, even to excess, whereas women would be condemned for similar behavior. The idea that drinking is a shameful thing in women is sometimes implicit and sometimes explicit. Usually that view is linked with the presumption that a woman who drinks would also be wanton and sexually promiscuous (e.g., Ablon, Page et al.); the distinction between private and public realms of behavior is relevant here, with female drinking sometimes allowable at home, but not elsewhere (e.g., Gilbert, Topper). Herd and Gaines both remark on the unusually high rate of abstention among black women, although both predict this will diminish in the future. Chinese and Japanese populations are far more tolerant of men's drinking (Kitano et al., Sue et al.). The Hispanic macho image appears still to be important in some populations, although not in the crudely stereotypic sense that led to the adoption of that word into English: Gilbert emphasizes similarities to Anglo culture in this respect; Gordon notes its diminished importance for Dominicans and Puerto Ricans; Trotter stresses contextual variation. A few of the groups are described as having become more permissive in this respect as they become increasingly acculturated, although Westermeyer remarks on a significant increase in abstinence among female Hmong immigrants and diminished drinking among their male companions as well.

CLASS AND OTHER SOCIAL CATEGORIES

An unusual strength of this book is the unique emphasis on variation with respect to alcohol use, an aspect of culture that is often treated as if it were remarkably uniform throughout the United States. The results of surveys have long shown that such is not the case, but these case studies give a much more vivid and coherent picture of such variation than could be gleaned from out-of-context statistical comparisons of disparate elements. To be sure, there have been several previous efforts at describing and interpreting variations in experience with alcohol among diverse segments of the United States—sometimes in terms of geographical regions, other times in terms of biological "races" or of ethnicity, minorities, or a number of other categories, recently lumped under the convenient but vague label of "special populations." Among such efforts that have effectively portrayed some of the variation within the American experience at large, however, only a very few have had the additional strength of emphasizing variation *within* such segments. It is another of the pioneering features of this book that many of the authors do offer such characterizations, which take us far beyond stereotypes and avoid the misleading uniformitarian imagery that has often been associated with normative descriptions that are painted with a broad brush.

Even in dealing with the limited "official" viewpoint of a metropolitan Episcopalian diocese, Bennett discerns significant differences among socioeconomic classes and racial categories. In her discussion of California, Gilbert is careful to point out that some similarities hold for social classes in a way that cuts across "racial" boundaries; Gaines makes the same point about the Southeast. According to Gordon, aspiring Dominicans copy what they perceive to be "American" norms for drinking, as Simboli, Kitano *et al.*, and Sue *et al.* imply is being progressively done by successive generations from Italy, Japan, and China; in the last two cases, the hereditary biochemical "flushing" reaction is sometimes considered racially distinctive. Interestingly, when Indian drinking is discussed from the viewpoint of insiders, there is little mention of "the firewater myth" (Topper, Weibel-Orlando), although both authors point out that norms about drinking and drunkenness vary markedly along lines of age and sex and also very much depend on the situational context. Among Jews, it appears as if the risk of incurring some alcohol-related problems were in almost inverse relation to the degree of religious orthodoxy (Glassner & Berg). Herd's ethnohistorical interpretation of changing black views toward alcohol is not couched in terms of subgroup variation, but it seems probable that the ambiguity she notes may have at least as much to do with the relative dominance of different media at different times as it does with actual changes in the views held by significant numbers of black individuals. In taking pains to delineate variation within categories that have too often been portrayed as unitary, Trotter goes so far as to characterize six "life-style subdivisions" among Mexican-Americans, based on a combination of income, cultural–ecological, and linguistic variables; with reference to Navajo drinking, Topper identifies "five subcultures."

If this volume serves no other purpose, it will have been worthwhile for effectively demonstrating the reality and complexity of cultural pluralism within contemporary American society, on a subject that is often discussed as if the norms were uniform and supported by an overwhelmingly popular consensus. And never again will the work of a social scientist on patterns of behavior, attitudes, and outcomes of drinking enjoy an uncritical reading if it neglects to mention intracultural variation.

SETTINGS FOR DRINKING

In attempting to understand the interrelations between alcohol and human behavior, social scientists have paid increasing attention in recent years to the settings in which drinking takes place. As in many other studies of Jewish drinking, Glassner and Berg cite the fact that children are introduced to alcohol at an early age within the supportive context of the family at home, and specifically with religious ritual significance, as probably important in shaping the pattern of long-term sobriety that characterizes that population. The sacramental symbolism of wine is also integral to Irish Catholicism and Episcopalian Protestantism, however, where it provides little defense against the development of drinking problems among adults, as indicated by Ablon, Stivers, and Bennett.

The positive social functions of public drinking places have been cited by many authors over the years, and that theme recurs in many of these case studies. Ablon and Stivers both cite the importance of male companionship in bars frequented by the Irish; ethnic social clubs serve these and other functions for senior generations of Polish- and Italian-Americans (Freund, Simboli). Cubans in Miami were uncomfortable in new kinds of drinking establishments, according to Page *et al.*, whereas some Mexican men described by Trotter seek out occasional opportunities for their traditional rowdy *pachangas*. Guatemalans were the only ones, among three Hispanic migrant groups studied by Gordon, who retained home-country patterns of episodic round-drinking in a cantina-like atmosphere. Public drinking establishments are so important to some Mexican-Americans that Gilbert offers a typology and points out the very different clienteles and comportment that characterize different settings. Herd cites the rapid evolution of a cosmopolitan and permissive "nightclub culture" in the first quarter of this century as a major determinant of what she sees as ambiguity in black drinking norms today. When Weibel-Orlando characterizes urban Indian bars as "social service institutions," she presupposes that readers are familiar with a rich literature about how much job- and apartment-hunting, social credit, and other non-drinking activities that deal with how to survive in the city take place in such a context. Her insight about the ethnically assertive quality of Indian drinking in public places is important, as is her careful delineation of the careful sacred segregation practiced by various Indian tribes, which effectively insulates religious practice from any influence of alcohol. The Navajo manner of compartmentalizing activities is somewhat different; they are careful to make sure that drinking does not interfere with ritual, but those performances themselves often provide the occasion for groups of men to drink together at some distance from the sacred activity. As described by Topper, such parties, secular in themselves, are important opportunities for the transmission of sacred knowledge among the participants. It is little wonder that young Navajos are far more likely to incur problems with drinking among strangers in town than are their elders among friends in such a typically Navajo context.

Drinking in private settings is less often studied by social scientists, so that an important part of the American experience with alcohol tends to be grossly underreported. In this collection, however, some useful points of comparison emerge. Part of the double standard that obtains among several populations is that women may drink at home, but are suspected of immorality if they drink in public (Ablon, Gilbert, Topper, and others). The specific propriety of drinking as a gesture of hospitality is noted among Polish-Americans (Freund) and Dominicans (Gordon); the risks of solitary drinking are cited by Ablon (Irish males), Gilbert (Mexican–Americans), and Topper (Navajos). The degree to which white, middle-class Protestant wives and mothers are carefully given "time-out" for solitary drinking, even in the face of strong normative opposition, reflects the extent to which familial systems can adjust to aberrant alcohol use (Ames). Although these case studies do not consistently report on the specifics of contextual variations in drinking and drunken comportment, it is apparent that most of the populations do have norms that differ markedly from one setting to another.

ROLES OF THE FAMILY

There have been only a few societies in human history—many of them short-lived experiments—in which some form of the nuclear family of woman, man, and their offspring is not the keystone institution of manifold and enduring significance. Often it is the unit of primary importance in terms of subsistence, education, and social support as well as various other crucial concerns, so its relevance to alcohol should not be underestimated.

Long before many of her colleagues in other disciplines recognized the importance of the family in relation to alcohol, Ablon made it the focus of her painstaking longitudinal research. As a result, her studies provide uniquely rich insights into Irish-American life, from both a socio-cultural and psychological perspective; although Stivers relies on more conventional historical methods, his interpretations of how and why men in that population are at exceptionally high risk for drinking problems fit well with hers.

The value of learning to drink in a supportive family context is attested by the experience of Polish-Americans (Freund) and Jews (Glassner and Berg); ascetic Protestants sometimes suffer from the opposite view (Ames). Gilbert, Topper, and Simboli are among the many authors who point to the widespread belief that, although it may be all right to drink among kin, it is dangerous to do so among strangers. Page notes the irony that Cuban youths have easier access to "hard" drugs that to alcohol, but that their parents are often permissive, or even encourage drinking, without recognizing its potential danger as a drug.

Disruption of the nuclear family is cited as an important antecedent of problem drinking among both Cubans (Page *et al.*) and Guatemalans (Gordon). Trotter's account of Mexican migrant farm workers illustrates a pattern that is familiar to anthropologists, but that often seems paradoxical to others—an innovation that was intended to ameliorate some social problems, but that aggravated others. Adaptive role changes by white middle-class Protestants may hide a woman's drinking for a while, but result in a variety of problems for other members of the families (Ames). Regulations requiring the housing for families be improved sound beneficent enough, but the practical outcome—easily explained in terms of the interconnectedness of various aspects of culture—was the virtual elimination of such housing, disruption of familial work patterns, and increased drinking in all-male barracks.

The importance of the family as a social support system is a recurrent theme in the best of times—and a variety of deleterious consequences are suffered by most of the individuals who are close to a problem drinker. For these and other reasons, it is often helpful if "significant others"—a convenient term presumably devised by social workers in recognition of the diverse meanings and forms that "the family" assumes in various populations—be actively engaged in various forms of treatment when a member's alcohol-related problems affect them.

IMPLICATIONS FOR ACTION

There is one aspect of the American experience with alcohol that most researchers know first hand, but that is rarely mentioned in print. It is remarkable how many people in the United States immediately and unwittingly transform the word "alcohol" into "alcoholism" when they are told the subject matter that one is studying. This may result from a misguided sense that "everybody knows about normal drinking" and that "the only real mystery is why some people can't drink without suffering the consequences." An alternative interpretation might be that "social science is just a fancy name for social work," so one "must be wrestling with the 'problem' of alcoholism."

Such a commonplace view reflects some ironic ambiguities and ambivalence in the American experience as it relates to science. In a society that obstensibly adheres to scientific principles more generally than most, there is little understanding of or patience with "pure research" outside of a narrow circle of investigators. Related to this view is the attitude that "something has to be done," whether that something is to "close the saloons which are dens of iniquity" in the 1920s; to "break up the bootleggers" during Prohibition; to "give the states control" with repeal; to "help those who suffer from the disease of alcoholism" in the 1970s; or to "get the drunks off the road" in the 1980s. We must not forget that the American experience with alcohol includes not only drinking, but also abstaining, not only what drunks do, but what is done to them. It also includes a vast body of laws and regulations, front-page news almost every week, whether in terms of fires, accidents, drownings, and legislative controversies over the minimum drinking age.

The fact that Herd uses the term "nightclub culture" to characterize the ambience of uptown Manhattan in the early part of this century is a telling comment on the pervasiveness of alcohol in some realms of the American experience. Advertising in the mass media is not the uncontrollable demonic force that some seem to believe, as demonstrated in the leveling off of per-capita consumption of most alcoholic beverages during the past couple of years, after one-quarter of a century of steady and rapid growth. As we have seen in these case studies, most of the members of most of those populations have more that is good to say about alcohol than they have that is bad.

Nevertheless, there is no escaping the lamentable fact that a small portion of those who drink suffer deleterious consequences.

It is in connection with such problem drinking or alcohol-related problems that many social scientists feel a practical and moral imperative to spell out some of the implications for action that they feel are inherent in or can be derived from the data they have collected. In fact the chapters in this volume are unusual in the extent to which they mention particular aspects of drinking that are viewed as problematic by the members of the populations that are being described (e.g., drunkenness among Jews, cirrhosis among Italians and increasingly among blacks, spouse-abuse among the Irish, economic deprivation among Guatemalans, family disruption among Cubans, and accidents and suicide among American Indians).

Of special interest for clinical applications are some of the emic approaches to treatment—that is, the "folk wisdom" or "common knowledge" about how to deal with those persons who do have troubles as a result of drinking. The Jewish pattern of denial can hardly be considered therapeutic, but it has apparently been remarkably consistent over time (as noted by Glassner & Berg); a striking contrast is the rapidly growing use of confrontation techniques and referral to Alcoholics Anonymous by Episcopalians (described by Bennett). Although Alcoholics Anonymous has often been described as relatively ineffectual among various Hispanic populations, it obviously offers valuable support to some of the Guatemalans in New England (Gordon). Weibel-Orlando devotes considerable attention to a number of reasons why treatment programs have had little success with Indians, even those programs that were specifically designed and staffed to be "culturally sensitive"; at the same time, her focus on sacred segration as an ethnic principle seems to hold significant promise for helping Indians who need it. Topper's clinical experience lends special weight to his forceful reminder that cultural relativism can be overdone—specifically, he insists we recognize that certain kinds of Western treatment *are* appropriate for certain Navajos.

A few of the authors go so far as to list some specific recommendations that have come out of their research. Ablon's call for the fostering of communication between Irish spouses, involvement of the entire family in treatment for problem drinking by any member, and use of existing organizations for education about alcohol is such a practical conclusion. Similarly Topper drew on his ample experience as a practitioner in the Public Health Service to outline some of the dimensions that must be weighed in order to evaluate Navajo patients and to tailor aspects of treatment for an individual; his thoughtful suggestions for a prevention program similarly reflect a realistic awareness of distinctive structural problems that would have to be taken into consideration. Weibel-Orlando shows a similar combination of awareness, concern, and practicality in her approach to successes and shortcomings among alcohol programs that have been developed for various American Indian populations.

In a context like this, it would be presumptuous of me to try to point out further practical implications with respect to the specific populations, which are far better known to and understood by the contributing authors. Similarly, it would be inappropriate to contradict the pluralistic sense that gives special unity and meaning to this book by offering some vaguely universalistic admonitions to greater cultural awareness and willingness to adjust programs to accommodate to different values and attitudes. Those points are of immense importance, but they have been forcefully made in other contexts, and are probably accepted by most readers of this volume. Perhaps the greatest value that will come out of this examination of *The American Experience with Alcohol* is the forceful realization that there are a great many very different American experiences with alcohol. Old stereotypes can be misleading, especially when we are dealing with the richly varied responses of people from immensely diverse backgrounds as they relate to alcohol, which is at the same time a food, a poison, and a drug, a natural substance and a highly elaborated artifact, deeply embedded in the intricate web of complexly integrated and ever-changing cultures.

Index